LEAN ENTERPRISE SYSTEMS

LEAN ENTERPRISE SYSTEMS

Using IT for Continuous Improvement

STEVE BELL

WILEY-INTERSCIENCE

A JOHN WILEY & SONS, INC., PUBLICATION

Published by John Wiley & Sons, Inc., Hoboken, New Jersey.
Published simultaneously in Canada.

For general information on our other products and services or for technical support, please contact our Customer Care Department within the United States at (800) 762-2974, outside the United States at (317) 572-3993 or fax (317) 572-4002.

Wiley also publishes its books in a variety of electronic formats. Some content that appears in print may not be available in electronic formats. For more information about Wiley products, visit our web site at www.wiley.com.

Library of Congress Cataloging-in-Publication Data:

Bell, Steve, 1960 Sept. 30-
    Lean enterprise systems : using IT for continuous improvement / by Steve Bell.
      p. cm.
    Includes bibliographical references and index.
    ISBN-13: 978-0-471-67784-0
    ISBN-10: 0-471-67784-1
    1. Manufacturing processes. 2. Process control. 3. Management information systems.
I. Title.

TS183.B444 2006
658.5–dc22   2005048987

Printed in the United States of America.

10 9 8 7 6 5 4 3 2 1

Society has reached the point where one can push a button and be immediately deluged with technical and managerial information. This is all very convenient, of course, but if one is not careful there is a danger of losing the ability to think. We must remember that in the end it is the individual human being who must solve the problems.

Eiji Toyoda, 1983

Contents

Foreword, by Carol Ptak

Lean Enterprise Systems: Using IT for Continuous Improvement describes the application of Lean principles, with the aid of information technology, to improve the performance of *any* business in *any* industry. Lean methods first emerged in manufacturing with a laser-like focus on waste reduction. Not only was waste reduced, overall productivity and quality improved. The greatest gains were realized in those companies where the primary focus was holistic demand flow rather than simply cost reduction. These companies looked beyond the islands of shop floor productivity, creating real value for the customer by enabling the smooth reliable flow of material and information across the entire enterprise.

Recent events cause us to examine the continued rise in productivity worldwide and question how a competitive advantage can be won and maintained. USA employment in manufacturing peaked at 19 million in 1979 and has been on a downward trend since. The most common cause cited is outsourcing and offshoring to countries like China and the Far East. However the real situation is more overwhelming than that. Between 1995 and 2002 over 31 million factory jobs disappeared from the top 20 global economies. During those same years global productivity increased by 30%, while American productivity increased 20%. This pattern has been seen before in the agriculture industry. In 1810 the population in the US was 11 million with 85% of people in agriculture—it took 9 million people to feed 11 million plus providing substantial exports. In 2001 only 4.8 million US agricultural workers fed 290 million while continuing to provide substantial exports. In addition to this continued rise in manufacturing productivity, significant capacity has been added in China, Korea, Malaysia, Thailand, Vietnam, and the eastern European countries. Not surprisingly the world of scarce capacity in the mid 1990s has been turned upside down, and now manufacturing capacity is plentiful around the world. (Data source: US Census Bureau and Rochester Center for Economic Research).

The technological world has also evolved dramatically in the last five decades. A tight relationship exists between computing power and the availability of new technological tools. Rudimentary MRP (Material Requirements Planning) systems emerged in the 1950's, and evolved to closed loop MRP as computer systems increased in power to include capacity planning. When financial capabilities were integrated in the 1980's, comprehensive systems developed called MRPII (Manufacturing Resource Planning). Soon after, computers continued to increase in power, making it possible to manage and track all the resources across an enterprise using ERP (Enterprise Resource Planning) systems. By the mid 1990s, the software industry recognized that if scarce capacity could be kept working on the most profitable parts, the manufacturing enterprise should realize dramatic bottom line results. Due to the memory resident calculation capability that was now possible, sophisticated APS (Advanced Planning and Scheduling) systems were developed.

On a parallel path, new ways of doing business have developed. With lessons learned from the early Henry Ford manufacturing days combined with the quality lessons of W. Edward Deming, post-war Japan began to redefine the manufacturing industry. Taiichi Ohno and Shigeo Shingo launched what would later become known as the Toyota Production System. By the late 1970's in the USA the emergence of Just-In-Time was seen with great successes at early adopters such as Hewlett Packard. Manufacturing costs began to shift from labor to materials as manufacturers focused on improving productivity and reducing cost. A few lone voices in the wilderness advocated this different vision of manufacturing. John Costanza began to evangelize Demand Flow™ manufacturing and openly criticized the MRP systems of the day with his "No MRP" buttons. Dick Ling developed and advocated the idea of sales and operations planning to truly exploit capacity for profits—an idea only now seeing support from commercial software. Dr. W. Edwards Deming came back to the USA to begin his quality crusade work after his amazing success transforming the meaning of "Made in Japan" from cheap, poor quality goods to a "Lexus quality" standard. His work was the foundation behind the popular Six Sigma improvement concept today. In 1984, Dr. Eliyahu Goldratt shocked the world with his business book that was a novel, *The Goal*, introducing the Theory of Constraints (or was that a novel that was also a business book?). In either case, it taught the lesson that a focused goal and the constraints to achieve that goal must be identified and managed. How many forget this common sense and suffer for it?

Early adopters of these emerging ideas from the past few decades leveraged pilot projects to learn how to embrace these new business rules. The early results were nothing short of amazing. However, as quickly as the champion for that specific approach left to pursue new opportunities, or as companies were merged and acquired, these successful pilot projects fell by the wayside and the early improvements quickly deteriorated. Technology was often viewed as part of the non-value added baggage to be eliminated, rather than as simply a tool to help achieve and sustain the positive change. Unfortunately, these innovative approaches often failed to become common practice and the company suffered as a result.

Today, labor is less than 10% of manufacturing cost—down from 60–70% just 50 years ago. The focus on improving labor productivity now often yields insignificant marginal benefit. Companies must learn to compete on their ability to identify profitable opportunities in the marketplace and respond more quickly than their competition. Lead time is now a great challenge; expectations of months are now weeks, weeks are now days, and days are now hours. We are witnessing a startling convergence today of fundamental issues into a perfect storm. Around the globe many companies in many industries are struggling with the very same competitive factors:

- **Customer Power**—access to real-time information through the Internet has irrevocably shifted the global balance of power to the customer. Customers can now demand what they want and the price they are willing to pay.
- **Worldwide Overcapacity**—This is due to productivity gains of established companies from operational improvement and incorporation of automation combined with the addition of significant new capacity in Latin America, China, Asia and Eastern Europe.
- **Market Volatility**—due to the significant reduction of transactional friction from the advances in technology, the world has become a buyer's market. Now there are constantly emerging new demand patterns for sourcing and outsourcing which extend the physical supply chain while simultaneously compressing overall product lifecycles.

There is only one way to establish lasting competitive advantage in this new reality. Each company must exploit their unique capability to develop a win-win relationship with the customer that solves customer problems while providing profit for themselves. The relentless compression of product and transaction lifecycles means that the complex and iterative forecasting, planning, and push scheduling approach must be replaced with a more strategic planning process supported by quick response, demand-driven, Lean manufacturing throughout the supply chain. To survive and thrive in this new world, a company must combine this vision of how their unique capabilities can be profitably exploited to provide value for their customers with clearly aligned business practices and supporting technology.

Lean Enterprise Systems: Using IT for Continuous Improvement describes the synergistic impact of technology and Lean business practices. This book provides in-depth discussion of Lean as well as the requisite technology necessary to sustain the improvement momentum. No longer is it possible to exclude technology from the Lean approach. However, a different kind of technology is needed. This book describes in depth what that technology should be.

Although the Lean improvement process has its roots in manufacturing, *Lean Enterprise Systems* expands the application of these techniques to all industries. The pragmatic approach taken in this book incorporates best practices and ideas from other management disciplines like Six Sigma, Theory of Constraints, and Sales and Operations Planning into a blended approach. The overall implementation process is fully described with expectations and pitfalls clearly outlined.

Lean Enterprise Systems: Using IT for Continuous Improvement provides a very complete summary of current Lean improvement techniques as well as providing innovative thought leadership. This is a book that should be on the desk of every manager thinking about a Lean project or in the process of implementing Lean. It is a reference that you will consult often. The author has a genuine passion for the subject and it clearly comes through in this work. Read and enjoy!

Carol Ptak, CFPIM, CIRM, Jonah

Carol Ptak is a past president and CEO of APICS and former Vice President of Manufacturing Strategy for Peoplesoft Corporation. She is the author of MRP and Beyond, *and* ERP, Tools, Techniques and Applications for Integrating the Supply Chain *(Second Edition).* Necessary but not Sufficient *was co-authored by Dr. Eli Goldratt, Eli Schragenheim and Carol Ptak. Most recently she was integral in the update of John Constanza's book* Quantum Leap. *She is the 2005–2006 Executive in Residence at Pacific Lutheran University in Tacoma, Washington.*

Preface: The Goal of This Book

During the two years spent researching and writing this book, I was often asked: "Who is your audience, and what will they take away from this book?" After all, this topic is so vast, bridging the disciplines of operations, information systems, and business management, that without a clear focus it could easily consume hundreds of pages without delivering specific value to an individual, team, or enterprise.

In this book I will demonstrate how the techniques learned from the evolution of Lean Manufacturing, combined with *Lean IT* practices, will continuously improve Lean Enterprise performance in any industry. The goal of this book is to help *all* enterprises, not just those in manufacturing, leverage Information Technology (IT) to improve business performance in ways that add significant value to the customer. IT alone will not solve a company's problems; in fact, if not judiciously applied, IT can introduce more problems than it solves. For an enterprise seeking to achieve *sustainable* competitive advantage, the foundation of all solutions may be found in the continuous improvement of *people, processes, and technology–in that order*.

This book serves as a practical guide not only for large enterprises, but for small and medium-sized companies that nurture entrepreneurial spirit and innovation. These smaller companies face the same complexity as their larger counterparts, yet they lack the resources to afford dedicated change management teams and expensive enterprise information systems. They cannot absorb the impact of a significant project failure, or even second-rate results.

Lean is no longer just for repetitive manufacturers. Lean techniques and supporting software capabilities have matured, and many enterprises are now extending the benefits first realized in Lean Manufacturing into all industries, including low-volume and high-mix job shop manufacturers, distributors, retailers, service providers, and others. Necessary for this Lean evolution is the effective and flexible management of information. Information technology tools and techniques have matured, and an enterprise can now achieve agility and return on investment without the frequent and traumatic software replacement cycles of the past. New approaches to Lean IT, many derived from the lessons of Lean Manufacturing, allow us to build long-lived and adaptable information systems that stimulate continuous improvement.

How do we build and then continuously improve IT, so it is capable of enhancing Lean performance without introducing unnecessary complexity and waste? How do we design an information system that enables the enterprise to adapt quickly to sudden threats and market opportunities? How can *Lean IT* help companies deliver excellent customer service and value, and create competitive advantage? How do we develop and nurture an integrated environment of people, processes, and technology that enables us to continuously improve?

Follow me, I'll show you.

HOW THIS BOOK IS ORGANIZED

This book is divided into three parts, exploring how people, processes, and technology combine forces to enable continuous improvement:

In Part 1: Building Blocks of the Lean Enterprise we'll examine how to *improve processes* throughout the value streams of the Lean Enterprise. We'll look at the essentials of Lean, explore continuous improvement techniques, and the advancement of Lean techniques from the shop floor to the global supply chain. We'll discover where, when, and how Lean IT can add substantial value to the Lean Enterprise through integrated processes of planning, scheduling, execution, control, and decision-making, across the full spectrum of operations.

After reading Part 1 you should be able to:

- Develop teams and begin mapping your own value streams, illustrating and quantifying the complementary flows of material and information throughout the enterprise.
- Understand how Lean principles may be applied to reduce supply chain waste and improve performance.
- In Chapters 4 and 5 (which focus on Lean *Manufacturing* techniques) learn to deploy a variety of scheduling, flow, demand pull, and kanban techniques (with appropriate application of software tools) across the entire product/process continuum from repetitive manufacturing to job shops.
- Simplify—and improve—any process, using the power of Lean IT to reduce waste.

In Part 2: Building Blocks of Information Systems we'll examine the many ways that *information technology* can support Lean performance. We'll explore the primary components of an enterprise information system and explain how these components may be integrated to improve the flow of information supporting value streams. We'll also examine how information systems can help to organize and deliver knowledge when and where needed.

After reading Part 2 you should be able to:

- Understand the general structure of business information systems, developing insights that will help you realize substantial business benefits and ROI from your IT investments.
- Recognize the vital components of an enterprise information system, and interpret the alphabet soup of information technology tools and techniques.
- Consider the fundamental challenges when integrating fragmented systems and processes to build effective value streams.
- Capture, manage, and deliver structured and unstructured information to the right people, at the right place, at the right time, and in the right format, enabling Lean performance through effective knowledge management.
- Understand how the future of the Internet will enable the small or medium-sized enterprise to compete effectively in the global economy.

In Part 3: Managing Change with IT we'll explore how the skillful combination of process and information technology improvements can *empower people* to continuously improve the Lean Enterprise, delivering value to the customer, while enabling the development of competitive advantage.

We'll explore a comprehensive framework for performance measurement and management that aligns strategy with the initiatives of continuous improvement teams, focusing the energy of the enterprise where it matters most—enabling breakthrough performance.

We'll learn how to build real value into IT systems, capitalizing on emerging information technology tools and change management methods, to build a platform upon which components can be added or removed to meet changing needs, continuously improving IT and enterprise agility. We'll explore how to apply continuous improvement techniques to our now-adaptable IT systems to create *Lean IT*.

After reading Part 3 you should be able to:

- Develop an integrated performance management system to guide continuous improvement team initiatives in alignment with strategic goals and objectives.
- Use a new approach to measuring ROI on investments in Lean information systems. This includes balanced measures of operational effectiveness, customer service, and innovation, supplementing the traditional financial measures.

- Demonstrate how decision-support and event-driven exception management tools and techniques can help find and eliminate wasteful practices.
- Design and build agile and continuously improving Lean IT operations.
- Energize individuals and teams through continuous improvement efforts, joining people, process, and technology into a holistic environment for sustainable Lean performance.

WHO SHOULD READ THIS BOOK?

- **Executives and Managers of Lean Enterprises and Their Supply Chain Partners** seeking a bootstrap education on the application of Lean operations and Lean IT principles to improve performance, apply knowledge, add value, and create competitive advantage.
- **Lean Practitioners** seeking to enrich their knowledge of value-adding IT tools and techniques. By reading this book Lean practitioners will learn to work in partnership with IT teams to enhance performance.
- **Information Systems Practitioners** desiring a richer understanding of Lean tools and techniques so they will more effectively support and sustain continuous improvement initiatives. By reading this book, IT practitioners will benefit by learning to craft value-adding IT initiatives to support Lean operations, while at the same time learning how to develop Lean IT practices.
- **Consultants, Project Managers, and Software Designers** seeking to enhance the value they offer clients by applying these Lean operations and IT techniques.
- **Educators and Students** desiring a comprehensive and practical guide to this rich subject.

The spirit of continuous improvement urges us to keep an open mind and a curious nature, asking questions and exploring new avenues for improvement across the entire Lean Enterprise and the global supply chain. If you are just beginning the journey to Lean, in this book you will learn that IT offers many tools and techniques for improving Lean performance. And if you are an experienced traveler on this path, you will learn that careful consideration of IT may introduce new ideas to advance your improvement efforts, adding substantial value while enhancing competitive advantage.

Let's get started.

STEVE BELL, CFPIM
www.steadyimprovement.com

Preface: The Goal of this Book

1. Lean and IT: The Human Factor

I Building Blocks of the
Lean Enterprise

II Building Blocks of
Information Systems

2. Realizing the
Value of Lean

6. Charting the Enterprise
Software Universe:
ERP
CRM
PLM

3. Three Stages of
Lean Evolution

4. Fundamentals of
Production and Inventory
Management

7. Integrating
Value Streams

5. Lean Planning
and Execution

8. Managing Knowledge for
Competitive Advantage

III
Managing Change with IT

9. The Event-Driven Lean Enterprise

10. Linking Strategy with Action:
Performance Management

11. Lean IT: Applying Continuous Improvement
to Information Systems

Postscript: Zen and the Art of Lean

Part One

Building Blocks of the Lean Enterprise

Chapter 1

Shouldn't we wait to see what the other team does?[1]

Lean and IT: The Human Factor

We sat at opposite ends of a chipped Formica table. Through the window I could sense the Chicago wind gusting; inside the thin walls of the small break room vibrated with the muffled din of heavy equipment. Coffee steamed in Styrofoam cups, and scattered about the table lay newspapers, year-old magazines, and a half-empty box of stale donuts. The plant manager swirled the coffee in his cup, then he looked me straight in the eye. "Just keep that @#$% ERP system away from my Lean shop floor!"

Well then, I thought, where do we go from here?

Two months earlier our consulting firm had been hired by a multinational manufacturing enterprise to facilitate the selection and installation of a new

Lean Enterprise Systems: Using IT for Continuous Improvement, by Steve Bell
Copyright © 2006 by John Wiley & Sons, Inc.

Enterprise Resource Planning (ERP) system for their North American operations, with several plants spanning the continent from Canada to Mexico. Although many talented individuals representing years of specialized industry experience worked at these plants, each individual rarely communicated with his or her peers at the other sites. This was partly due to their geographic separation, but the most significant cause was more subtle and difficult to overcome. Each of these plants had been an independent business acquired by the parent company. Each location enjoyed a proud heritage, where local managers and employees maintained their own customs and business practices. The parent company sought to blend these entities, but although every location offered many ideas for collective improvement, each was unwilling to surrender its ways to standardized business processes.

Executives hoped that a new ERP system would be the catalyst to bring these disparate sites together—sharing ideas to develop enterprise-wide best practices. By marshalling their considerable design and engineering talent in collaboration with their customers, they would develop a coordinated supply chain enabling them to better service national accounts, thus establishing a competitive advantage in what was a relatively unsophisticated and localized niche industry.

During the initial interviews, our firm met with the management team at each site and learned of their relative strengths and weaknesses. The Chicago plant was a particularly interesting story. Years before they were anything but Lean, with no standard work or visual management, staggering lead times, poor quality, and mountains of inventory as far as the eye could see. And to top it off, they spent considerable resources on the care and feeding of an old MRP system that scheduled their incapable processes in great detail, causing perpetual turmoil on the shop floor.

Then they were exposed to Lean. They invested in training and launched several kaizen projects involving batch size reduction, cells, and a simple kanban system. Low and behold inventory suddenly dropped, while productivity and quality improved. They hired a Lean consultant, carried out a few more projects, and the results kept coming. Ready to perform without a net, they switched off their old MRP system, using instead several visual control and pull mechanisms, while 'empowering' the workers with large spreadsheets for planning and scheduling.

The plant had much to show for their efforts. The shop floor was clean and orderly, the staff seemed well trained and supportive, quality was up and rework was down. The sales team frequently led tours of their spotless plant to impress potential customers. However, not everything was rosy. Even though the shop was using a kanban system at various stages of material handling, they were far from reaching their inventory reduction target. Yet paradoxically, even with too much inventory they continued to experience frequent stock-outs. This led to late deliveries, busy expeditors, and frequent spasms of disorder on the shop floor. Inventory variances yo-yoed every month and no one could explain why.

Although the Chicago plant was in far better condition than before the Lean initiatives began, improvements were still needed, but efforts had apparently stalled. When we asked the plant manager what value he felt an ERP system could offer, he would not even consider the possibility. To me this was a puzzling attitude within an organization seemingly motivated by the inexhaustible possibilities of continuous improvement. He acknowledged they were not perfect, but he was confident they would sort these issues out in time through ongoing Lean transformation. Other than a simple order entry system, no computerized planning, scheduling, or execution software was necessary to support the Lean operations, period.

As we probed, we discovered that the Chicago plant manager was particularly concerned that a new ERP system would push more work onto the shop floor than it could handle, while at the same time requiring unnecessary data capture activity—muda. He had successfully eliminated many wasteful behaviors over the past two years, and he wasn't about to let an outsider introduce new ones!

The interview progressed to production scheduling. Customers wanted delivery commitments for their orders, and salespeople were frequently interrupting his busy staff with questions on availability. Occasionally they were forced to juggle the production schedule to respond to an important and unexpected customer situation. "This is just another source of waste," lamented the plant manager. "We need an Available to Promise (ATP) function in our order entry system," he insisted, "to help the salespeople manage customer expectations and delivery schedules so that we can maintain level production and avoid stockouts."

I saw an opening. To me this request was not surprising: ATP is a common capability of an order processing system, especially in a make to order job shop environment like this one. But wait—to calculate a valid promise date for an order you typically need some form of a production planning and scheduling system. And in an operation of his scale and product/process variability, that most likely required software support spanning the entire value stream from design to delivery. Didn't the Chicago plant manager say just moments before that software wasn't welcome in his Lean shop?

I became convinced that the tiger was chasing his own tail.

A HEALTHY PROCESS

The Chicago case will seem familiar to many, since it exposes the seemingly great divide between the "Lean camp" and the "IT camp". ERP initiatives are often sponsored by business management, while Lean transformation and continuous improvement initiatives are usually driven by operations management. These two constituencies must somehow learn to cooperate towards the continuous improvement of the entire enterprise—no more opposing camps.

I suspect that any change agent hired to bring a new approach to an organization believes they are there to do good. They are bringing a better way, perhaps in their mind "the" better way. To many change agents, it seems the only way to deal with resistance is to overcome it—through sheer force of will if necessary. But there is another perspective. Sometimes those living within the system recognize some things the outside change agents do not: their history, their culture, and recollection of many top management interventions and their unintended consequences.

Disagreement and negotiation can be messy and uncomfortable, but they are also a healthy means of constructive change. Every department or functional unit within an enterprise may have its own point of view, with objectives that, while not directly in opposition with the others, create subtle conflict. For example, the objectives of a traditional finance department may include the utilization of assets and maintenance of shareholder value, causing a focus on short-term financial results, often driving sales and production to sub-optimal decisions. Similarly, the sales department may have incentives to stimulate short-term sales revenue causing lumpy demand especially at month, quarter, and year end, despite the fact that this behavior results in un-level demand. In both cases, these departments may have internal goals that they believe are in everyone's interest but in fact do not advance the overall success of the enterprise. The interplay of conflicting and misconceived priorities within an organization is often guided by human and political influences, and the balance of power can shift like the ebb and flow of tides. IT, Six Sigma, Theory of Constraints, and Lean initiatives are no different in this sense—each appeals to a particular point of view and will elicit positive support from some and be rejected by others.

Ideally, continuous improvement is part of a culture in which individuals and departments are aligned toward common goals—first and foremost is customer satisfaction. When effectively managed, and with the right people engaged in meaningful ways, continuous improvement can stimulate reconciliation among organizational silos, leading to holistic change. When the cycles of continuous improvement trigger disagreement then we know the process is stimulating sensitive nerve endings; this is healthy as long as the final outcome is consensual and constructive.

So what is the reliable and objective mechanism to mediate disagreements, to overcome localized motivations, to develop fact-based consensus? Continuous improvement encourages analysis of a problem using all *relevant* facts and experience. Clearly this process begins with analysis, which requires good information, and thus an effective information system. When an information system provides multiple perspectives on the same situation based on a single source of fact-data, the problem-solving team begins to feel solid ground beneath their feet. Whether that information system is simple and visual, or complex and electronic, is another matter.

Continuous improvement encourages placing relevant information and decision-making authority into the hands of the people doing the work. At

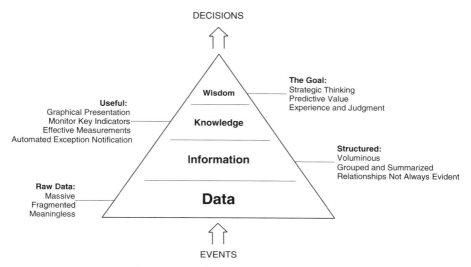

Figure 1-01. The Pyramid of Knowledge

Toyota, this begins with *genchi genbutsu*, or *gemba*, which means literally "go see it for yourself." Taiichi Ohno, a founding father of Lean, once said, "Data is of course important in manufacturing, but I place the greatest emphasis on facts."[2] A direct and intuitive understanding of a situation is far more useful than mountains of data.

The raw data stored in a database adds value for decision-making only if the *right* information is presented in the *right* format, to the *right* people, at the *right* time. A tall stack of printout may contain the right data, but it's certainly not in an accessible format. Massive weekly batch printouts do not enable timely and proactive decisions. Raw data must be summarized, structured, and presented as digestible information. Once information is combined with *direct experience*, then the incredible human mind can extract and develop useful knowledge. Over time, as knowledge is accumulated and combined with direct experience and judgment, wisdom develops. This evolution is described by the classic pyramid of knowledge shown in Figure 1-01.

BACK TO CHICAGO

So what happened in Chicago? We can speculate upon several possible perspectives for why the team and its change leader were far from a true Lean system, yet they refused any help from IT providers:

1. They feared wasteful IT systems and procedures would be foisted on them.

2. They were focused on Lean implementation and did not want to be distracted by new technology at that time.
3. They had bad experiences with the corporate office trying to provide "help," and viewed it as an invasion of their autonomy.
4. They had embraced a narrow interpretation of Lean and viewed all IT intervention as synonymous with waste.

There may be a degree of truth to all these perspectives. In fact the shop floor was functioning much better than at had in the past; it was designed around value streams with several simple visual systems that seemed to be working. On the other hand, during our interviews we found non-production managers at the Chicago plant were less satisfied with the lack of good information systems to help them make informed decisions. The data presented by their current system was inappropriately formatted, its validity was often questionable, and it was usually obtained too late for preventative or even corrective action. And beyond the walls of the enterprise, customers and supply chain partners were constantly requesting availability, production, and delivery information; valuable human resources were consumed responding to these requests, and the information delivered was not always accurate or timely.

Many information needs beyond the shop floor were left wanting, however the plant manager wielded the power. Having demonstrated substantial Lean performance improvements on his shop floor, executive management didn't want to hinder his continued efforts. In fact, they were hoping that other sites could learn from his Lean deployment experience. But arguably this single-minded focus on Lean shop floor techniques was leaving on the table much opportunity for the broader enterprise.

The Chicago plant had made a sincere and enthusiastic initial effort to transform their operations. However as is often the case, especially with smaller companies who don't have the benefit of a Sensei (master) on loan from a key customer, or a full time value stream manager with prior Lean experience, they were just dabbling with tactical improvement initiatives (process kaizen) without looking at the overall value stream effect (flow kaizen). They were not "seeing the whole" within the context of the enterprise; the plant was in its infancy of holistic Lean transformation.

The plant manager was justified in eliminating the former IT system: it caused too much push while inhibiting pull, it was not user friendly, did not support visual management, nor did it support any type of production leveling. As a result however, they were too eager to banish IT entirely, since they had been warned of the evils of "push systems" and the waste that traditionally rigid software systems can cause. If they looked deeper they would have found that Toyota, the creator of Lean production, used quite sophisticated IT systems—for planning demand and supply, for leveling the schedule, for ordering long lead time items, and even more recently for scheduling regularly used parts to the line, supported by a backup manual kanban system.

But the Chicago team was not open to seeing any possible synergies between Lean and IT, how the flow of information and materials along a value stream are simply two sides of the same coin.

There were two common disconnects at work here. First, there was uncertainty regarding the boundaries between Lean flow methods that controlled production activity within the short-term time horizon, and the traditional Master Production Scheduling (MPS) and Material Requirements Planning (MRP) systems (components of an ERP system) that plan resource requirements over the intermediate and long-range horizons. This led to an assumption that all IT systems involve scheduling and push. Secondly, there was a mistrust of the complexity and mystery of IT, leading to avoidance and underutilization of its capabilities.

We'll explore the first disconnect, related to how IT systems can be effectively used to plan and schedule Lean Manufacturing operations, in Chapters 4 and 5. But for now let's consider the second disconnect, which may occur in any industry, caused by general mistrust of IT. We are, after all, human. Each of us acts in our unique way, based on our disposition, attitude, background, education, experience, karma, baggage . . . call it what you will, we each carry it with us, and it subtly and often unconsciously influences our every action and decision. It's perfectly valid, in fact necessary, to attempt to understand these natural tendencies as we work with diverse groups of individuals toward common goals. With that in mind, let's take a look at motivation and individual points-of-view.

Point of View #1: Executive Management

Lean performance of operations (whether the shop floor of a manufacturer, or the operational function of any industry) has become very important to executives as they have seen productivity and quality go up and delivery schedules improve. But it is well known that an effective enterprise is more than operational excellence, and the best run manufacturing plant is worthless unless the right products are being produced at the right time, and for a profit. Consequently there is more to a Lean-friendly enterprise information system than those tools directly responsible for running the shop floor. Making a profit is important. Keeping customers happy is important. And creating a learning and innovative organization is important, if not critical, to long-term viability. Operational excellence, financial performance, customer satisfaction, and innovation should all be kept in balance; business strategy should drive appropriately balanced goals, objectives, and measures. Executives must think about the enterprise from a top down value stream perspective (flow kaizen) while problem solving and performance improvement initiatives in alignment with these measures should percolate up from teams and individuals (process kaizen).

Holistic enterprise performance management is a team sport, requiring a clearly articulated strategy, balanced and focused measures, with alignment and regular communication throughout the entire organization. An information system (comprised of the appropriate balance of visual, manual, and electronic data capture and presentation tools) is an essential enabler of this process. When the information system is fragmented, unreliable, overly complex, and burdensome to operate, then it is counterproductive and wasteful. Such a system may even cause feelings that it is merely a tool of authoritarian control. When properly designed, on the other hand, an enterprise performance management system supports the free flow of knowledge and continuous improvement, orchestrating the efforts of teams and individuals toward a shared purpose.

Point of View #2: Production Manager

After years of pushing the schedule, expediting orders, chasing demand, troubleshooting unexpected problems, managing out of control situations, and living in a state of nearly perpetual turmoil, seasoned manufacturing professionals have learned to appreciate the simplicity and effectiveness of Lean methods. As we learned from the Chicago story, for many production managers Lean is a response (perhaps rejection is a better word) to the unnecessary complexities introduced by industrial engineering theorists and software designers intent on planning and controlling every last detail.

Although we may argue the benefits of integrating and orchestrating the information flow across enterprise value streams, it's clear that overly-ambitious systems can add complexity and waste without delivering sufficient benefit. Flash back to the 1960s and the emergence of powerful business computers and MRP programming logic. According to George Plossl, a founding father of MRP:

> Operations researchers were intrigued by the problem of determining "optimum" work sequence in complex manufacturing environments. MRP program designers, obsessed with the potential power of computers and software, attempted to build into MRP capabilities to cope with every eventuality in manufacturing, and to include every known technique, however little use these would be.[3]

Chaos theory and ordinary experience have since taught us that a complex system is virtually impossible to fully understand and predict, and that each attempt to control its outcome creates additional complexity and unintended consequences. In other words, meddling with a complex environment can trigger a nasty cycle of spiraling instability. On the shop floor, unexpected events arise constantly—every day, every minute. As Lean practitioners know from experience, complex scheduling and execution software (known as *push systems*) tend to propagate pools of inventory and other waste around every unplanned event and interruption of flow on the shop floor.

Although *properly* designed and installed information systems can deliver numerous benefits to a Lean operation, the unfortunate fact is that information systems are often not well designed and installed. Legendary tales of costly ERP project failures have become business folklore, causing fear, mistrust, and avoidance. Although the mere existence of an information system encroaching upon production operations may not produce complexity and waste, this may be the natural assumption of many Lean practitioners, causing information systems to be targeted for elimination rather than improvement. For example, a memorable 1999 Industry Week article titled *Lean vs. ERP* declared:

> Who would have imagined that enterprise resource planning (ERP) and its forerunner, manufacturing resource planning (MRP II)—the core of manufacturing information systems for the last three decades—would one day be viewed as the enemy of streamlined production?[4]

The salient point is that there can be a natural state of conflict between the paradigms of IT and Lean practitioners: complexity versus simplicity, planning versus acting, one side pushing while the other is pulling—a curious tug of war.

Point of View #3: IT Manager

Now let's sit on the other side of the table: IT professionals suffer their own challenges. They are responsible for an amalgamation of hardware, communications, databases, software, design, maintenance, support, upgrades, budgets, and the occasionally unsympathetic manager or uncooperative user.

Most companies have multiple software systems in operation, managing various aspects of the organization such as marketing, sales, engineering, planning, production, quality, compliance, finance, and human resources; a value stream generally flows across these boundaries. For example, a customer order may originate in estimating, moving through engineering, scheduling, production, and shipping, concluding in customer service. At the same time, data related to the transaction life cycle moves across system boundaries, passing through separate software applications, databases, and user communities. At each handoff the information flow can break down, causing invalid, redundant or missing data, delayed activities, lost steps, dead-end processes, and security threats.

Integrated information systems *must* be designed to overcome these challenges. But while tight integration among systems and processes results in ease of operation, well-integrated and fault-tolerant software design adds a layer of hidden complexity and cost.

Furthermore, within an already complex environment there may also be a harmful tendency for overdesign. This can be motivated by an enthusiastic user community seeking to reduce the apparent complexity of a process. Overdesign may result from an unconscious longing to mimic the behavior of the old system because of its familiarity or the fear of change. Overdesign may also result from the passionate efforts of misguided software designers, striving to deliver the most aesthetically pleasing, intellectually stimulating, or career-enhancing software creation. Whatever the motivation, overdesign masks process problems that should be simplified or eliminated, while creating an inflexible system that is costly to maintain, leading to neglect, misuse, lost benefits, and premature replacement.

To complicate matters further, information technology evolution is rapid, relentless, and unpredictable. New and unproven technologies are constantly emerging, some that may solve existing problems while instantly creating new ones. It is the responsibility of the information systems department to evaluate the merit of each new technology, weighing the potential benefit (and the likelihood that the benefit will actually be realized) against the cost and risk of integrating the new technology within an already complex environment. The unintended consequences of a seemingly innocent and minor change can be costly. For this reason, information systems managers may be accused of being conservative, cautious, or even anti-progressive. This apparent attitude stems from an acute awareness of the dangers and unpredictability of emerging technologies. In his classic book on information technology adoption, *Inside the Tornado*, author Geoffrey Moore comments on the behavior of IT professionals:

> In a classic human response, they form support groups. IT professionals are experts at networking with each other, even across company and industry boundaries if need be, to discuss the ramifications of the latest technology. These groups are united by a need to answer a single question: Is it time to move yet? [. . .] being pragmatists they will operate like herd animals, and now they have gotten nervous because some unknown scent is in the air. Should they ignore it or should they stampede? If the IT community moves too soon, they incur all the trials of an early—which is to say premature—adoption. If they move too late, they expose their company to competitive disadvantages as others in the industry operate at a lower cost and greater speed by virtue of their more efficient infrastructures. Worst of all, if they move way too late, they run the risk of getting trapped in end-of-life systems that, with alarming rapidity, become almost impossible to maintain.[5]

WHERE POINTS OF VIEW INTERSECT

Now let's join these perspectives and see the whole—but melding these perspectives to come up with value-adding IT solutions requires a change in mindset.

Within the complex and sometimes fragile world of information systems, one small glitch can potentially destabilize an entire system. For this reason the IT department is charged with maintaining *stability*. They must keep the value streams, the lifeblood of the enterprise, running smoothly. To satisfy this objective, systems are planned in painstaking detail over long time horizons, and are supported for as long as economically and practically possible. Rig-

Attribute	LEAN	Traditional IT
Change Management	Organic, incremental and continuous	Engineered and planned large events
Organization	Cross-functional teams	Central command and control
Measures	Top-down and bottom-up performance measures linking improvement initiatives to strategic goals	Cost containment and uptime
Knowledge Management	Generalization	Specialization
Education	Process focus	Task focus
Definition of Success	Speed and Agility	Stability

Figure 1-02. Contrasting Attributes of Lean and traditional IT

orous planning, control, and change management are traditionally strict requirements for survival in the world of IT.

In stark contrast, Lean practitioners place their faith in fluid and organic systems, adaptability, and continuous improvement. They reject the notion that a production system should be designed, deployed, and then left alone. In fact, the emphasis of Lean is for each team member to take personal responsibility to search for new ideas, incrementally and continuously improving the product and process every minute of every day.

It is clear that Lean culture can be in conflict with the traditional rigidity of IT; this conflict is illustrated in Figure 1-02.

On the other hand, the convergence of Lean practices with the Internet and other maturing information technologies has created many opportunities for rapid innovation. Value streams are becoming faster and more variable, global supply chains more capable and complex, and customers more demanding, ruthlessly driving out every trace of waste. Enterprises that are unprepared to leverage information technology to exploit these market challenges and opportunities may find themselves at a significant competitive disadvantage.

The ultimate message of this book is that *Lean IT* can be a powerful tool to aid the continuous improvement of any enterprise in any industry. However IT is simply a tool that enables people to improve processes. IT may be used skillfully to simplify processes and add value, but if it is used poorly IT may obscure or institutionalize the very waste that must be eliminated to achieve breakthrough performance.

Potentially isolated IT professionals need exposure to the realities of the social and physical world of making things and providing services. And the use of IT in a Lean enterprise extends far beyond the operations center, where many who are first exposed to Lean principles tend to concentrate.

For lasting Lean transformation, we must focus on the whole enterprise, understanding the synergistic flows of value and information across the entire value stream. Despite their differences, it is clear that business management, Lean and IT practitioners must work together, learning to speak each other's language. Education is the starting point for developing effective cross-functional teams, with minds that are open to understanding alternative points of view, defining shared goals, and nurturing an environment where the spirit of innovation and continuous improvement can thrive. That is where the journey of this book begins, with the turn of this page.

Chapter 2

Realizing the Value of Lean

Question: How do you carve an elephant from a block of marble?

Answer: Using a hammer and chisel, simply remove anything that doesn't look like an elephant.

As you can see by this riddle, it is difficult to define a concept by what it *isn't*. Lean is about eliminating waste, but that's essentially a definition about what Lean isn't. So what *is* Lean?

A statesman once said, "I can't define it, but I know it when I see it." Even that may not be true about Lean. Witnessing a Lean process, an outsider may remark, "What's so special about that, it looks so simple?" Of course it is, nothing could be simpler than Lean. But then Olympic athletes make their actions look simple only after years of practice, focused on a single goal: the elimination of all wasteful motion. The fact that something appears to be simple does not make it easy.

Anyone who has stood at the counter of a fast food restaurant has witnessed Lean Manufacturing. Demand pull begins at the cash register. Immediately upon entry of your order, an electronic signal is given to the shop floor (the food assembly line, with a computer display of orders and assembly sequence overhead) and production begins. There you see a simple and flexible cell design, where all materials are stored at point of use in the appropriate assembly sequence. Trained workers assemble each item using standardized work procedures, and production flows according to a standard drumbeat—a

Lean Enterprise Systems: Using IT for Continuous Improvement, by Steve Bell
Copyright © 2006 by John Wiley & Sons, Inc.

chicken sandwich and a burger require the same time to assemble. Each item is assembled to order (one-piece flow), so special requirements (configurable items) are satisfied without disrupting the flow of production ("♪ hold the pickle, hold the lettuce, special orders don't upset us . . . . ♪"[6]). Certain finished good items are made ahead in batches (coffee and fries) and meet up at picking, where the server hands the completed order to the customer with a smile. What could be simpler than that?

ELIMINATION OF WASTE

What is waste? My favorite definition is *anything your customer would not be willing to pay for.* Another way to state this: Waste is any activity that you would rather not tell your customer about. If you're inclined to conceal it, then it's probably waste.

Here's a simple exercise in what Zen masters call *beginner's mind.* Pretend that you've never been inside your workplace before. Now imagine that you're a customer asking the question: "Do we want to do business with this company?" Then take a slow walk with a notepad and a camera, carefully observing *everything* throughout the life cycle of a single order through the curious eyes of your customer. Do you see activities that surprise or displease you? Those may be signs of waste.

Why is waste important? Rather than eliminating waste, why don't we just sell more and make up the difference? After all, if we sell a large quantity we can afford to waste a little, can't we? First of all, if your contribution margin* is 20%, then you must sell $5 to offset $1 of waste. Furthermore, by eliminating waste, you're removing the root cause of problems that may cascade throughout your company, generating costs seen and unseen, measured and unmeasured. By eliminating waste you improve performance in ways that are meaningful to your customers, like quality, cost, speed, and variety. As a result you naturally improve your potential for increasing revenues, because you enhance your productive capacity as you simultaneously become a more desirable supplier to your customers.

An interesting relationship between revenue and waste thus appears. By increasing revenues while ignoring waste, waste naturally increases because you create more inefficiency by forcing an already inefficient process to work harder. With a focus on eliminating waste, however, you naturally create the conditions that lead to additional capacity and increased revenue potential.

Muda is the Japanese word for waste; another common term is *Non-Value-Added* activity (NVA), defined as:

---

* Contribution margin is the profit after deducting variable costs, including direct labor and materials.

Any activity which clearly creates no value, which can be removed immediately with minimum or no capital investment, and with no detrimental effect on end value. This is classified as "Type Two Muda" by Womack and Jones in *Lean Thinking*.

There is an even more curious term that arises in many Lean discussions: *Necessary but Non-Value-Added* activity (NNVA), which is defined as:

Any activity which again creates no value but is unavoidable, given the current operating constraints of technology, production assets, and operating procedures. This is "Type One Muda."[7]

There are many tasks that are required for a company to function but that do not directly add value to the product. The activities of the human resources department may be a good example—do they directly add value to production? Generally no, at least not directly, but can they be eliminated? Some activities may be improved or eliminated through changes in policy; others may be outsourced, but the activity still exists and creates cost. Another major source of NNVA is the effort required for compliance with various regulatory agencies; while these regulations may contribute to the health and welfare of society, they do not add value to the product.

In their book *Reengineering the Corporation*, Michael Hammer and James Champy describe a visit by Ford Motor Company executives to Mazda, a slightly smaller competitor. In Mazda's accounts payable department Ford executives were stunned to witness a staff of five, where in the same role Ford employed a staff of five hundred. Ford executives were compelled to rethink the entire process, so that the receiving dock verified receipt of the goods against the purchase order, which automatically signaled release of payment to the supplier. No invoice was ever printed, mailed, matched, or entered by the accounts payable department.[8]

This story shows that NNVA may be eliminated (made *non*essential) through creative process redesign, enabling human resources to be reassigned to more value-adding and potentially rewarding activities. In any process there is always more waste to uncover, and you can be sure there are similarly disguised NVAs lurking within your own enterprise.

The Seven Forms of Waste

During the development of the Toyota Production System, Taiichi Ohno and Dr. Shigeo Shingo identified seven distinct forms of waste.

Inventory. When Taiichi Ohno spoke to a group, his message was always simple, and he would often start with his famous river example: Inventory is like a river of water, and as it flows through the plant it hides problems and waste. Ohno recommended that we slowly reduce inventory, lowering the level of water in the river, exposing the rocks and hazards previously concealed. "As you lower the level of inventory, problems rise to the surface of your awareness," he used to say.[9]

Although traditional accountants define inventory as an asset, Lean accountants consider it a liability. Excess inventory ties up cash and creates waste in many forms including storage facilities, tracking, transactions, movement, damage, obsolescence, and the physical counting and adjustment of records.

Delay. According to Ohno and Shingo, after inventory the next waste to focus upon is delay, the unnecessary wait time that occurs throughout the production process. In particular, they emphasized setup time reduction. Setup time represents a primary fixed cost component of the manufacturing equation, and it is the basis for the economy of scale assumptions that have justified mass production since the start of the Industrial Revolution.

Unnecessary wait time may be caused by improper scheduling, causing people, tools, and materials not to appear in the right place at the right time. Poor material planning and procurement, late deliveries, quality problems, unnecessary inspection, and searching for information or work instructions can also cause wait time. Attempting to reduce the number of setups leads to large production batches and mass production; large batch sizes may cause some workcenters to be overburdened while others are simultaneously starved and waiting for work. Because the traditional work ethic requires each worker to focus on productivity, these idle workcenters continue working, producing excess inventory.

Motion. Unnecessary human motion—the ergonomics of walking, bending, reaching, twisting, lifting, handling, requiring two hands instead of one—not only wastes effort, but may cause health and safety issues. It may be helpful to film an operation, and view it in fast forward with the operators, so they may discuss their own activities objectively. In addition to human motion, this form of waste also includes unnecessary machine motion, which causes additional maintenance, energy cost, and machine wear, leading to quality problems.

Transportation. Unnecessary movement of materials, supplies, and resources is waste. Material may move from receiving to a storage location, from one storage location to another, or from a storage location to the point of use, before it is finally consumed in production. Whenever practical, materials should be received in the appropriate quantities directly to the point of use.

Overproduction. Overproduction waste is caused by making more and/or sooner than the customer demands. This may be caused by improper demand planning, long setup times, large batch sizes, inappropriate kanban sizing, or quality rejects. Overproduction leads to consumption of too many resources, people, machines, inventory, storage space, energy, and cash tied up in these assets. Producing earlier than necessary creates excess inventory and consumes material and capacity that may be required by higher-priority work. In addition to these direct costs, overproduction creates congestion on the shop floor and can mask inefficiencies in other processes.

Overprocessing. While overproduction involves producing too much or too soon, overprocessing waste is caused by performing unnecessary work. Processing waste may be caused by using wrong or poorly maintained tools, improper work instructions, and inadequate training. Processing waste may also be due to inappropriate product design caused by a lack of communication between design and production engineering. It may also be caused by failure to understand what the customer wants, doing more work than the customer requires or is willing to pay for.

Improperly designed information systems can cause overprocessing in the form of unnecessary *transactions.* When a process is simplified, it requires less data to monitor and control. If information systems aren't simplified in conjunction with the processes they model, "transaction" waste results, a form of overprocessing. Effort to capture production data that does not add value is wasted processing and motion. When this information is recorded by hand and then later entered into a computer, the waste is multiplied. The time spent storing, managing, printing, distributing, and analyzing unnecessary information only adds to this waste.

Defects. This is the cost of poor quality that may result from faulty product design, insufficient training, lack of standardized work methods and instructions, improper tooling or workcenter preparation, unnecessary inspection and quality countermeasures, and excess processing caused by repair and rework. Defect waste includes interrupted schedules, missed due dates, uneven production flow, inspection to catch defects, and unnecessary setup and runtime caused by unscheduled repair and rework. Quality problems are often concealed by excess inventory and large batch sizes.

When you scrutinize any value stream, you may discover that less than 10% of the activity adds value to the finished output; the rest is NVA or NNVA. Consider the fact that in most factories you can single out a job and expedite it, and the job will finish in a fraction of the time indicated by the standard routings and lead times stored in the planning system. This proves that the throughput time of any single job is considerably less than the standard, but when we schedule we must plan for the average rather than the ideal time for the schedule to be "reliable." This highlights a key argument that Lean practitioners make against traditional manufacturing software—push scheduling systems factor inherent waste into the planning and scheduling of work, thus institutionalizing rather than eliminating waste. We'll explore this issue carefully in Chapters 4 and 5.

The Elusive Eighth Form of Waste

During my research I encountered many efforts to identify an eighth form of waste. For instance, I have heard the eighth form of waste defined as *complexity*, the harm caused by processes and information systems that are more complicated than necessary. Is complexity the eighth form of waste? Let's consider cause and effect. Complexity can cause waste to occur in the other seven forms. For example, an overly complex shop floor layout may cause unnecessary movement, and an overly complex product design may cause processing waste. But complexity itself is not waste.

Similarly, I have seen the suggestion that *knowledge disconnect*, the improper availability and flow of information to support a process, is another form of waste. Examples of this include: the wrong tool or part showing up at the wrong place or time due to improper scheduling or routing of work, rework caused by an obsolete drawing issued to the floor, over-processing due to faulty work instructions, or making products that customers don't wish to purchase due to faulty demand management, forecasting, replenishment, planning, or scheduling practices. Knowledge disconnect may be caused by an information technology problem, poorly defined processes, or simply disorganization and lack of communication. But is it another form of waste? Once again, I conclude that it is not, but its existence causes the other seven forms to occur.

In my opinion, the eighth form of waste should be the *Loss of Human Potential*.[10] During a visit to a supplier, Ohno stopped to observe a process. An operator stood watching his machine. After watching several cycles, Ohno asked him, "How often does this machine break down?"

> "Never," the worker replied.
> "Well, what do you do all day?"
> "I watch this machine, Ohno-san."
> "All day long, you watch this machine, which never breaks down?"
> "Yes," said the worker, "that is my job."
> "What a terrible waste of humanity," thought Ohno.[11]

In this example, the waste of overprocessing (the individual's unnecessary activity) could have been redirected to a value-adding task. But while this individual's productive effort was being misused, Mr. Ohno clearly understood there was also the waste of his human potential. Inside each of us there is a spark, a desire to do something of value, to know that what we do has meaning. We are able to overcome many hardships in order to achieve valuable results, as the heroic efforts of countless individuals have shown throughout history, around the world, and within our own communities. In his book *The Art of Happiness*, psychiatrist Howard Cutler writes:

> Victor Frankl, a Jewish psychiatrist imprisoned by the Nazis in World War II, once said, "Man is ready and willing to shoulder any suffering as soon and as long as he can see a meaning in it." Frankl used his brutal and inhumane experience in the concentration camps to gain insight into how people survived the atrocities. Closely observing who survived and who didn't, he determined that survival

wasn't based on youth or physical condition, but rather on the strength derived from purpose, and the discovery of meaning in one's life and experience.[12]

We all perform necessary tasks we don't enjoy. Whether it's the grimy machine overhaul, counting inventory in dusty back rooms, or facing up to the quarterly budget review meeting . . . it's always something. We are able to endure discomfort and adversity as long as we have purpose.

But what if there is no purpose? What if the task we do each and every day adds no value? What if we know there's a better way, but no one is listening? Perhaps we need the job—to house, clothe, and feed our family, and rocking the boat might risk too much. Yet work without satisfaction leads to apathy and resentment, which in turn lead to poor productivity and quality. Over time they may also lead to health, emotional, and family problems.

Each time an individual is required to knowingly engage in a wasteful action (one of the first seven types) a by-product is the eighth form of waste. The individual is reduced to a machine, and an expendable one at that, for his action serves no purpose.

DR. DEMING AND CONTINUOUS IMPROVEMENT

Dr. W. Edwards Deming is considered by many as the father of the quality revolution. He taught a systematic and team-based quality management technique that became the foundation for continuous improvement. In the late 1940s he traveled to Japan, a country ravaged by war and starved for resources, to assist in reconstruction. In his book *Dr. Deming, The American Who Taught the Japanese About Quality*, author Rafael Aguayo describes Deming's introduction of these important concepts to Japan:

> A good place to start is Toyota's headquarters in Tokyo. The striking thing one first notices in the main lobby is larger than life pictures of three individuals. One is of Toyota's founder, another of the same size is of Toyota's current chairman, and a third, much larger picture, is of W. Edwards Deming. The picture is there out of respect for the man they acknowledge as having started it all.
>
> When Deming joined the U.S. Census Bureau in 1939, he was already the acknowledged world expert in sampling. After World War II, Deming visited Japan and at the request of the Japanese Union of Scientists and Engineers gave a series of lectures on quality control to Japanese engineers and to top management on management's tasks and responsibilities.
>
> As Deming says, "Where is quality made? Quality is made in the boardroom!"[13]

Dr. Deming taught that most quality problems are caused by process, policy, and procedure issues, rather than by people. His principles are summarized in Figure 2-01.[14] Note their similarity to the principles recommended by Lean Manufacturing texts today—some things never change.

Traditional Company	Deming Company
Quality is expensive	Quality leads to lower costs
Inspection is the key to quality	Inspection is too late. If workers can produce defect-free goods, eliminate inspections
Quality control experts and inspectors can assure quality	Quality is made in the boardroom
Defects are caused by workers	Most defects are caused by the system
The manufacturing process can be optimized by outside experts. No change in system afterward. No input from workers.	Process never optimized; it can always be improved
Use of work standards, quotas, and goals can help productivity	Elimination of all work standards and quotas is necessary
Fear and reward are proper ways to motivate	Fear leads to disaster
People can be treated like commmodities - buying more when needed, laying off when needing less	People should be made to feel secure in their jobs
Rewarding the best performers and punishing the worst will lead to greater productivity and creativity	Most variation is caused by the system. Review systems that judge, punish, and reward above or below-average performance destroy teamwork and the company.
Buy on lowest cost	Buy from vendors committed to quality
Play one supplier off against another	Work with suppliers
Switch suppliers frequently based on price only	Invest time and knowledge to help suppliers improve quality and costs. Develop long-term relationships with suppliers
Profits are made by keeping revenue high and costs down	Profits are generated by loyal customers
Profit is the most important indicator of a company	Running a company by profit alone is like driving a car by looking in the rearview mirror. It tells you where you've been, not where you are going.

Figure 2-01. Deming's principles for quality improvement

To guide change efforts, Dr. Deming introduced a technique he learned from his friend Dr. Walter A. Shewhart at the Bell Telephone Laboratories of AT&T. The Deming–Shewhart Cycle of Continuous Improvement contains four steps:

> The first is to plan a change of whatever you're trying to improve. The second is to carry out the change on a small scale. The third step is to observe the results. The fourth step is to study the results and decide what you've learned from the change.
>
> The cycle is then repeated again and again. One doesn't make a change in one cycle and then undo it in the next cycle—that's just a waste of time. When you plan a change, you are saying, "I believe this change will make things better." If it doesn't, you've learned a great deal.
>
> As you improve your process, you improve your knowledge of the process at the same time. Improvement of the product and process goes hand in hand with greater understanding and better theory. Maybe this is nothing more than the application of the scientific method to business.[15]

It is challenging and often threatening for a team to face a complex situation they do not understand. That's why the scientific method exists—to *reliably* test hypotheses and determine what is really happening, compared to what we think might be happening. By using the scientific method we are virtually assured of discovering the underlying cause and effect relationships within a puzzling situation, although it may require time and patience.

The Deming–Shewhart Cycle of Continuous Improvement is such a method of inquiry and discovery. It is *iterative*, meaning that it repeats, building upon each new fact and insight gained from the last experiment. And it is a *continuous* discovery process, leading to incremental *improvement*. The Deming–Shewhart cycle has become popularly known as the Deming *Plan–Do–Check–Act* (PDCA) cycle illustrated in Figure 2-02.

The PDCA cycle seems simple, but it is not easy. Why? Because the problems we face usually aren't simple; they can be clever and deceptive. And

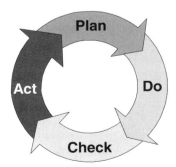

Figure 2-02. The Deming PDCA Cycle of Continuous Improvement

once you introduce people, you also introduce perceptions, assumptions, and emotions. Group dynamics make the situation even more complicated when people affiliate with departments and disciplines and begin to consciously or unconsciously defend their positions. A team may quickly become a disorganized rabble unless a leader with clear goals and a disciplined problem-solving method guides them. That's what PDCA is for.

Implementing the PDCA Cycle

Plan. Once the team has identified a problem they may begin to formulate ideas about how to fix it. This is the time when root cause identification techniques are useful. In scientific terms, these ideas become hypotheses, which may then be tested empirically. Process mapping, value stream mapping, and narratives are vital during this stage, because the effort to describe a process visually and verbally helps to clarify the situation, separating facts from assumptions. Any assumption or rationalization that resembles "because we've always done it that way" should be carefully scrutinized.

It is a common mistake to address too many variables at once. If the team limits the variables for each PDCA cycle, they can be assured of identifying true cause and effect relationships without becoming confused and wasting cycles. Once the team has determined the parameters of the test cycle, they should take baseline measurements of the current state and identify measurement targets representing the anticipated future state. When designing measurements, it is important to use both *result* and *process* measures. Result measures describe the end result (for example, a reduction in lead time from order to delivery), and process measures describe the various inputs to the process (for example, the average time required to input an order or the setup time for a bottleneck operation) that contribute to the result. If the result is unsatisfactory, the process measures help determine what went wrong and where. For example, a result measure could be the percentage of on-time shipments. Process measures then focus on specific variables that may cause due dates to be missed, such as rework, late receipt of materials, machine downtime, and so on. To truly solve a problem, root causes, not symptoms, must be identified and corrected.

Finally, before beginning the first test cycle, it is important to assess the capabilities of the process and the people involved. Does everyone know what they're doing and why? Has everyone been trained properly? Is the test environment in place?

Do. A common misunderstanding of PDCA is that we execute the change at this point. The Do phase is a pilot (test, prototype) phase where the proposed change is tested under carefully controlled yet realistic conditions. Creating an appropriate test environment can be challenging. For example, testing the conversion of a production line to a cellular layout may require setting aside the appropriate equipment and resources; this may restrict plant capacity and revenue. Creative simulation techniques using software or physical devices such as dice or block games may be used instead of physical reorganization to test ideas. The pilot phase should be quick and focused—by limiting the variables in each rapid cycle, the team is able to test each variable independently, measuring the results at each step to prove or disprove a hypothesis.

Check. Now the team compares the results against their hypothetical target measurements. The extent of the difference may be attributed to a number of factors, including a failed test, poor measurements, or an invalid hypothesis. If the team has been careful with their methods, the latter should be the only cause.

The Check phase occurs when the team evaluates the success of the pilot, and it is nearly always successful. Why? There is only one form of true failure: the failure of the testing and measurement process itself. If the hypothesis is proven *valid*, the test is a success. If the hypothesis is proven *invalid*, the test is also a success, although the team may be disappointed because they now must start another cycle with a new hypothesis. In either case, the team has learned something useful to build upon. During the iterative PDCA cycle, if a single hypothesis is proven correct, then it is time to test the next variable, and the one after that, moving patiently and incrementally toward the anticipated end result developed during the planning phase. It is important to note that variables may not always be singled out for test purposes, and statistical process control (SPC) and Six Sigma techniques may be helpful to address the inherent complexities of a process with interdependent variables.

Act. At the conclusion of the test process, when the team has clearly defined, tested, and validated the desired future state, it is time to implement (standardize) the new current state. The change may be quick and simple, or lengthy and complex, requiring diligent project management. The change may involve documenting new procedures, moving equipment into place, training employees, and doing whatever else is necessary to enact the improvements and make them stick. It may be helpful to leave certain process and result measures in place temporarily, to make certain that the change is implemented properly, and to ensure that the test process didn't fail to identify an important issue.

Once a PDCA cycle is completed, the next follows immediately, with ideas captured in one cycle flowing smoothly into the next. In this way, the PDCA cycle can be viewed more as a never-ending corkscrew as shown in Figure 2-03.

The sophistication of the PDCA process should match the magnitude, complexity, cost, and risk of the problem to be solved. We're not looking for overkill, just a *reliable* method for solving our problems. Regardless of the simplicity or sophistication of a particular PDCA cycle, problem-solving should be *patient and disciplined.* In our practice we encounter change efforts (even at the executive level) that have spun in circles for months or even years without achieving results. Individuals and organizations can get tangled up in

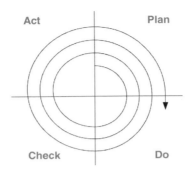

Figure 2-03. The PDCA corkscrew

confusion, frustration, mistaken assumptions, bad data, poor measurements, misleading goals, politics, and bad habits. These scenes are frustrating to watch, and they are even more frustrating to be a part of. When we encounter such a situation, we go back to the basics of problem-solving methodology, assembling the team at square one to assess the root causes of the problem. Does this require patience and discipline? Of course!

Where PDCA May Lead You

The problems we tackle with continuous improvement are often issues our workers confront every day, which is why they're usually in the best position to develop and test ideas for their improvement. This is the basis of the *visual workplace*: If the team can see, touch, feel, and sense how the plant is operating without complex analyses, they can quickly identify improvements. Orderly shop floors with minimal inventory levels will make a problem stick out like what it is, a sore thumb.

Management plays an important role in continuous improvement—guiding, educating, and encouraging, while removing barriers and ensuring that the initiatives are in alignment with corporate strategy. Often the improvements that result from cross-functional teamwork require little capital investment, just changes in company policy and behavior. In fact, it's a hard pill for management to swallow when a team suggests the *elimination* of an expensive machine or long-held policy.

There are countless stories of Shingo (and other less famous shop floor champions) reducing a machine setup from hours to minutes, and then finally eliminating the setup task altogether. This sort of breakthrough improvement requires relentless change efforts, a never-ending curiosity, and the questioning of assumptions. For example, how long will it take you to change four tires, fill your automobile with gasoline, and clean the windshield? If all goes well, maybe 45 minutes. How long does it take at the Indianapolis Speedway? In 2003 it took 12 seconds.[16] Of course, this requires a dedicated setup team, training, practice, special equipment, and redesign of the automobile and its fixtures. Expensive? Yes. Justified? Only if you want a chance to win the race.

Do you feel you can't justify dedicated floating changeover crews or a significant investment in retooling to achieve breakthrough setup time reduction in your own plant? Do you feel you can't start reducing your batch sizes, just a little at a time? Maybe not now, perhaps first you have to go through several cycles of improvement over weeks or months. But there may come a time when the low-hanging-fruit improvements are behind you, momentum for continuous improvement now exists, your customers are responding, your competition is lagging, and you see the benefit to the next breakthrough through a significant investment or possibly the removal of one. Not possible in your plant? How can you really know until you get there?

Consider the story of Pratt and Whitney's 'Billion Dollar Room' described by Womack and Jones in *Lean Thinking*. This is the room where turbine blades for jet engines are manufactured. Often jet engines are sold at or below cost, and the profit comes through the sale of the replacement blades. In this room, an $80 million, fully automated, multistage grinding machine was a bottleneck, requiring extensive preprocessing operations and causing excessive throughput time and accumulation of WIP inventory. In the Lean vernacular this is called a *monument*; the entire shop floor and all material flow are subject to its performance. After investing in many other Lean improvements, what did Pratt and Whitney finally do? They replaced this monument with nine sequential workcenters staffed by skilled technicians, raising labor costs and significantly increasing the processing time for each unit. These nine new machines were similar to the workcenters the $80 million monument had replaced just five years earlier. Womack and Jones describe the result:

> By increasing processing time from three minutes to seventy-five minutes, the total time through the process could be reduced from ten days to seventy-five minutes. Downtime for changeovers could be reduced by more than 99 percent, as each of the nine machines was changed over just-in-time for the new part coming through. The number of parts in the process would fall from about 1640 to 15—one in each machine plus one waiting to start and one blade just completed. The amount of space needed could be reduced by 60 percent. Total manufacturing cost could be cut by more than half for a capital investment of less than $1.7 million for each new cell.[17]

Are we courageous (or foolish) enough to launch into change on a relatively massive scale such as this? You shouldn't have to, until you can prove that *not* doing it might be foolish. Continuous improvement is not a leap of faith, but slow and steady progress where you question assumptions, test hypotheses, measure results, and embrace change. And when it's time to take the big step you should have a satisfactory degree of confidence to move forward.

Continuous improvement never ends. There are always more opportunities just around the corner, and you won't know what's around that far corner until you and your team members are peeking around it yourselves. Most importantly, be ready for the long haul, far beyond the low-hanging-fruit successes you may discover at the beginning. According to Dr. Richard Schonberger:

If your company claims to have Lean well in hand but has not maintained improving inventory numbers for at least five—better yet, ten—years, you probably don't. Above all, install discipline to make sure that Lean is not here and there, not a flash in the pan. It is not very Lean if it does not stick.[18]

THEORY OF THE MONTH CLUB

The principles of Lean are not new; many have been around in one form or another for a very long time. Just like the latest diet fad, however, it is human nature for us to hope for a quick fix. Although any approach may be legitimate, if implemented carelessly it will have about as much lasting effect as the latest grapefruit and yogurt diet. When the Vice-President-of-something-or-another strides in Monday morning and announces that he's just read *the book that will change everything*, the seasoned folks mutter under their breath "*Here we go again . . .*"

Often improvement techniques and theories develop followers and factions, all competing for attention. Arguments over nomenclature (what is and isn't "Lean" for example) can sometimes be silly, and they're usually a waste of time. Nevertheless, these rivalries are easy to understand—when we invest energy in learning a new skill, our ego invests in this new knowledge and we begin to identify with it. As we earn a CPIM, CIRM, CPM, Jonah, Six Sigma Black Belt, or similar certification in a rigorous operations or engineering discipline, we tend to defend these labels as somehow unique or special. Steven Thompson of Boeing Military Programs offers these insights into the apparent conflict:

> The reality is that each particular technology has provided tremendous gains for the organizations that have successfully applied it. This results in tremendous passion and bias towards that approach. I believe Dr. Edward De Bono's work on thinking and learning might help explain the conflict. People learn by developing patterns of thought from prior experiences.[19] The profound consequence is that given a successful implementation your mental pattern will favor the principle you were taught first. If you learned Theory of Constraints, Six Sigma, or Lean, first you most likely will favor that approach and try to fit the others into your existing mental framework. Unless given a serious situational conflict, you will likely succeed in this fit-up and create a rational logic to support your mental framework.[20]

In the pursuit of the simple yet comprehensive goal of Lean, which is the elimination of all forms of waste, we should make a sincere effort to set egos and labels aside. In this book you will find my bias is toward a very open-minded definition of Lean, tied to philosophy and principle rather than tool and technique. If any particular method can help us to improve, we should be willing to embrace it, no matter what its name. And if it proves ineffective, then we should drop it like a piece of deadwood, rather than continuing to identify with and defend it. There is an old saying that "to a man with a hammer everything looks like a nail." The antidote to this bias is simple: Analyze the problem first, then objectively determine what tool is needed.

To avoid getting bogged down in these semantic arguments, in our consulting practice we usually lead with the all-embracing term "continuous improvement." Why? It's possible to take a position and argue for, or against, a particular technique like Lean, Theory of Constraints, or Six Sigma. But who can argue against continuous improvement? And from a broad perspective, continuous improvement encompasses a wide variety of activities, including the following.

Kaizen Strike

A Kaizen Strike is a spontaneous exercise begun the moment a problem is detected. For example, a machine shuts down or begins performing out of specifications, and a team quickly gathers to troubleshoot the problem. Although a Kaizen Strike may start and finish in just minutes or hours, it must be a well-organized event, and it requires a leader and a disciplined problem-solving methodology. A Kaizen Strike follows the same PDCA rules as a preplanned event, and all participants must be trained in problem solving and continuous improvement techniques. In a critical situation and under time pressure, departure from a disciplined approach may quickly degrade into a goat rodeo.

Kaizen Blitz

A Kaizen Blitz is a planned event usually completed in less than a week. It is important to have a specific problem for an effective Kaizen Blitz, for example, the reorganization of a troublesome stockroom. Quick changeover or quality improvement on a single workcenter may also qualify as a Kaizen Blitz, and may be used to generate momentum and ideas before launching a longer-term improvement initiative.

Kaizen Event

A Kaizen Event is a carefully planned event that may require several weeks or months to complete, often focused on an area where problems are not clearly understood. Selection of the project focus is important: If you ask the wrong question, you may receive a correct but useless answer. The team should invest sufficient time in grasping the situation and analyzing the problem before doing, checking, or acting. It's compelling but usually counterproductive to pursue the first cause that comes to mind, as that cause may be only a symptom of an underlying root cause. Careful design of measurements and collection of baseline data may flush out illogical causal relationships.

Theory of Constraints

Every system contains a bottleneck (primary constraint) that limits the throughput of the entire system. Theory of Constraints (TOC) advises you identify the constraint, understand it, then minimize or eliminate it. When you have done this, another primary constraint moves up to take its place. You should eliminate each new constraint in turn, increasing throughput and revenue potential of the entire system with each step.

TOC reduces the apparent complexity of a system by focusing effort on optimizing and/or eliminating the constraint that restricts the entire process. This is especially useful for problem solving in a complex environment. We'll examine TOC applied to a Lean job shop in Chapter 5.

Six Sigma

Complex situations often require sophisticated empirical and statistical techniques to eliminate the sources of *process variation*. Six Sigma is a rigorous application of the scientific problem-solving method to a large and complex problem domain, emphasizing disciplined project management, measurement, and analysis. Six Sigma uses the Define–Measure–Analyze–Improve–Control (DMAIC) methodology, which is similar to but more rigorous than the Deming PDCA cycle. The foundation for Six Sigma is information, according to the authors of *What is Lean Six Sigma?*:

> There are a lot of good reasons why data and facts form the true foundation of Lean Six Sigma. Want to know who your customers are and what they want? You need to collect data. Want to improve processes? You'll need to collect data on variation, defects, and process flow. Want to avoid the kind of needless arguments and squabbling that destroy teamwork? Have a rule that people must support their opinions with facts.
>
> At the end of its project, one team working on a purchasing problem realized that it had spent 75% of its project struggling to get good, reliable data. When some people hear a number like that, their first reaction is "we can't afford to spend that kind of time just gathering data!" That kind of reaction is short sighted. It was because of the time they invested (that the team) could solve a problem that had been around for years.[21]

With the strength of Six Sigma comes a cost, so it is important to strive for a proper balance between intuitive Lean methods and the rigorous scientific approach of Six Sigma. According to Pete Pande, co-author of *What is Six Sigma?*:

> Lean [is] an experiential approach to finding the causes of problems. In other words, you get the people in a room who know the process, who know the operation, and you try to generate ideas as to what the real issues are in the process. The Six Sigma approach starts with the assumption that even our experience may not be adequate to really understand what's going on. Therefore, there's much more emphasis on gathering hard data about the causes and the background of the problem, and to try to base conclusions and solutions on a little harder data-edged analysis.
>
> People who get trained in [Six Sigma] problem-solving methodology often try to solve every problem with data when experience may be enough. And people who

have a Lean capability in place oftentimes are able to put quick, meaningful solutions in place more effectively than people who are just doing Six Sigma. So there's really a balance between the two approaches.[22]

How Do These Methods Work Together?

Note the relationship among these and many other continuous improvement methods suggested in Figure 2-04. The Lean goal is waste elimination through the continuous improvement of all processes, and the methods of attaining this goal fall along a spectrum from the simplest to the most sophisticated initiatives. Regardless of the method used, the team must first comprehend and describe the situation, using a variety of analytical, problem-solving, and mapping skills, then select the appropriate approach to improvement.

Note that the last item in each category in Figure 2-04 is *Gemba*—going where the action is—an essential aspect of Lean problem solving. Sheltered executives cannot make holistic decisions without interaction with the moving parts of the organization. Overemphasis on disembodied facts and statistics can lead to impractical solutions, whereas direct experience and clear awareness lead to insight. In the same regard, workers must understand the strategic goals and objectives of the organization to identify and prioritize improvement efforts. This requires clear and articulate strategy, communication, and performance measurement linkages from top to bottom and back again.

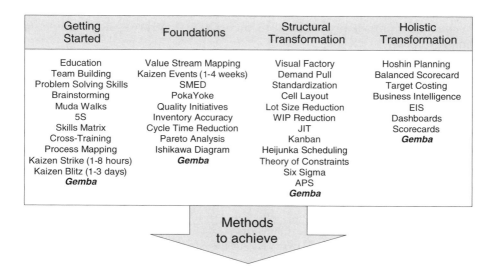

Getting Started	Foundations	Structural Transformation	Holistic Transformation
Education	Value Stream Mapping	Visual Factory	Hoshin Planning
Team Building	Kaizen Events (1-4 weeks)	Demand Pull	Balanced Scorecard
Problem Solving Skills	SMED	Standardization	Target Costing
Brainstorming	PokaYoke	Cell Layout	Business Intelligence
Muda Walks	Quality Initiatives	Lot Size Reduction	EIS
5S	Inventory Accuracy	WIP Reduction	Dashboards
Skills Matrix	Cycle Time Reduction	JIT	Scorecards
Cross-Training	Pareto Analysis	Kanban	*Gemba*
Process Mapping	Ishikawa Diagram	Heijunka Scheduling	
Kaizen Strike (1-8 hours)	*Gemba*	Theory of Constraints	
Kaizen Blitz (1-3 days)		Six Sigma	
Gemba		APS	
		Gemba	

Methods to achieve

Waste Elimination and Continuous Improvement

Figure 2-04. Spectrum of continuous improvement methods

Don't Start with IT!

A final word should be said on where to start with a continuous improvement initiative, or rather how *not* to start one. There are several reasons why information technology solutions are *not a good place to start* a Lean initiative:

- Automated solutions tend to mask cause and effect relationships from the people closest to the process, which diminishes their problem-solving ability and limits individual involvement—the very spirit of continuous improvement.
- Automation before analysis and simplification will only serve to entrench faulty processes.
- Automated solutions usually take time to design and implement. An incremental approach to continuous improvement encourages frequent experimentation and adaptation, whereas IT projects *traditionally** lock in scope early and do not allow for correction during implementation without costly scope creep.

An enterprise should initially focus on process improvement and simplification, which may lead to the elimination of legacy-technology monuments. However, continuous improvement efforts may eventually reveal more challenging issues that require increasingly sophisticated tools. That is not to say that information technology is inevitable, and the goal should be continuous simplification and flow. But success leads to growth, which often leads to complexity; whenever this is the case, then perhaps information technology *is* inevitably required to orchestrate enterprise-wide continuous improvement.

THE ECONOMIC IMPACT OF LEAN

Before we close this chapter, we should briefly explore the significance of Lean in a broader context than our own enterprise, employees, suppliers, and customers. Lean may have a lasting impact on the economy, health, and welfare of our local communities, our nations, and the world. Does this seem like a bold or a naive assertion?

It is impossible to accurately estimate the overall magnitude of investment in Lean, or its global impact. Indeed, entire enterprises, industries, and economies have shifted to these new practices. During the 1970s the Japanese industrial phenomenon took everyone in the United States by surprise. Many asked: "What happened?" and "How did we fall so far so fast?" In Dr. Deming's opinion, the answer was quite simple:

---

* We'll revisit this traditional limitation in Chapter 11.

In the decade after the [Second World] War, the rest of the world was devastated. North America was the only source of manufactured products the rest of the world needed. Almost any system of management will do well in a seller's market. Success in America was confused with the ability to manage.[23]

When Taiichi Ohno published *Toyota Production System, Beyond Large-Scale Production* in 1978 it was met with great disapproval by the management at Toyota.[24] They did not want to share this competitive information with other automobile companies. Some suspect that the publication of this book is one of the reasons that Mr. Ohno was exiled to a relatively minor Toyota company for the remainder of his days, rather than rising to the helm of the enterprise he led to greatness. Fortunately for the rest of the world we now understand and appreciate the breakthrough efforts of Ohno, Shingo, and many others, and we are learning to apply these insightful techniques within our own enterprises. With results spanning over fifty years, the macroeconomic effects of Lean are now becoming evident.

Robert Parry, President of the United States Federal Reserve Bank of San Francisco, representing the largest district in terms of geography and economy, said in November 2003, "It is true that the United States has lost manufacturing jobs for the past fifty years. The boom following World War II was the peak manufacturing employment, and we have never returned to that. But it's important to consider also that the share of GDP represented by manufacturing output during this same period has remained stable. The answer is clearly productivity."[25]

BEYOND MANUFACTURING

Lean is experiencing a renaissance in new sectors of the global economy. The eight wastes are alive and well within most businesses and institutions. Waste elimination goes beyond the shop floor to the administrative "paperwork factory"[26] of any traditional office. One industry that is particularly burdened with waste, offering significant societal impact as a result of its improvement, is healthcare. In *Lean Thinking*, Womack and Jones lay it on the line:

> When you visit your doctor you enter a world of queues and disjointed processes. Why? Because your doctor and health care planner think about health care from the standpoint of organizational charts, functional expertise, and "efficiency". Each of the centers of expertise in the health care system—the specialist physician, the single-purpose diagnostic tool, the centralized laboratory—is extremely expensive. Therefore, efficiency demands that it be completely utilized. Obvious, isn't it? To get full utilization, it's necessary to route you around from specialist to machine to laboratory and to over-schedule the specialists, machines, and labs to make sure they are always fully occupied.[27]

The *Industry Week* article "Lean Health Care? It Works!" describes a Lean initiative in the recovery room of a Montana hospital:

> After one Lean event, the doctor, anesthesiologist and hospital are more productive. The patient's bill (which adds up by the minute in the intensive care recovery room) goes down. The nurses feel better about their work because they're not so frustrated by the work-arounds. Access to care for the patient—who now waits six weeks to get in—improves 20%. And the quality of care is improved.[28]

This is an idea whose time has come. Lean principles evolved gradually in the manufacturing industry, but perhaps these techniques can be quickly applied to health care. We can only hope that government and education are not far behind, because their massive waste is only equaled by the potential for positive societal impact should they become more Lean and effective.

And what about the environment? Should we expect there to be a positive environmental impact of Lean? Absolutely. With the overwhelming evidence of global warming, vanishing rainforests, and countless other assaults upon our one and only planet, sooner or later something must be done. When *will* something be done? The answer seems to be only when it is in our immediate and individual economic interests to do something. Perhaps Lean offers valuable insights here.

Let's choose an example of an environmental industry suffering egregious waste: commercial fishing. Having worked on a commercial fishing vessel during my college days, I know that it is messy, physical, cold, and dangerous work. It's also very wasteful. When the nets come in there is a deluge of fish, so the workers pay little attention when a few wash back overboard and later when more are lost in various stages of processing. There seems to be such abundance, so why does it matter to pay attention to a few fish lost here and there?

It matters because the magnitude of waste in this industry is staggering. Seventy percent of the world's fish populations are overfished, according to a study sponsored by the United Nations and the World Bank. For every pound of shrimp caught, seven pounds of other sea life are killed. A biological observer living and working aboard Russian, Korean, and Japanese trawlers estimated that a fleet of six trawlers threw out *100,000 tons* of sea life in one 6-week season. The oceans are being emptied up to 150 times faster than forests are being cleared. Even species that are not wanted for food are at risk, endangering biodiversity and whole ocean communities.[29]

Bob Kerr, partner with High Performance Solutions, a Lean consulting firm in Ontario, Canada, shared the story of a commercial clam-harvesting vessel that applied Lean techniques to improve their performance. The staff of this particular vessel value stream mapped each operation, from when the harvesting gear was pulled on board until the inventory was stowed in the freezer hold. Not only did they identify considerable movement and processing waste, they also discovered that at each stage as much as 5% of the yield was lost.

Before	After Lean Improvements
Concentrate on fishing	Concentrate on recovering more protein
32 crew members working hard	32 crew members slowing down
Trip duration 44 days	Trip duration 33 days - 25% shorter
Crew wage $8,500	Crew wage $15,000 - 75% more earnings
	50% scrap savings
	> $2M additional annual revenue for fleet

Figure 2-05. Commercial fishing yields before and after lean improvements

By the end of the process, yield losses mounted to between 50% and 60% of the total catch. As a result of this value stream analysis, equipment was reorganized and workers were retrained. Not surprisingly, product yield improved considerably, as shown in Figure 2-05. The fleet made more money, workers went home a week earlier with more money in their pockets, and there was less environmental waste for the same yield. Everyone wins, especially the environment.

Jim Womack offers a strong endorsement for Lean as an enabler to environmental *and* economic sustainability in his article "Is Lean Green?":

> We've learned that consumers are resistant to paying higher prices for the same product just because it's "green" and that they are equally resistant to giving up products like big cars and large homes that are central to their enjoyment of life. But it's also apparent that many emerging product technologies and Lean Manufacturing and distribution concepts can dramatically improve our environment, if only they can be widely incorporated into products and production systems without needing to increase product prices.

> For example, hybrid motor vehicles are available right away, and highly compressed value streams with right-sized process technologies to locate production closer to the user can be introduced within a few years. If these technologies and methods are fully adopted, there is reason to think that the burden on the environment can go down even as consumption goes up.

> This means that Lean's role is to be green's critical enabler, as the massive waste in our current industrial practices is reduced to free up resources for improving product technologies and production locations for free (that is, at no apparent cost to the consumer). You'll remember how strange it first sounded when people began to realize that "quality is free." To say that "green is free" if we turn production waste into environmental value sounds equally strange today. But not, I think, for long.[30]

WHY IS IT IMPORTANT FOR LEAN?

In this chapter we've explored the value of Lean, and the significance of continuous improvement, but where does IT fit in?

Simply put, the right information, delivered to the right place, at the right time, and in the right format, is a powerful tool for continuous improvement. Information is the "C" in PDCA, the feedback that informs us whether our actions have achieved the desired results. The accumulation and treatment of the appropriate information leads to knowledge, and when harnessed properly this develops into individual and collective wisdom—our most effective competitive advantage.

Information is empowering. When an organization, whether a business or an authoritarian government, wishes to control its people, it controls information—this is a form of bondage. On the other hand, when an organization wishes to empower its people, to develop and harness their collective potential to improve their situation, it freely distributes information. A well-designed information system, comprised of an appropriate mixture of visual, manual, and electronic tools, is a key to continuous improvement.

Each step in the PDCA cycle requires information for guidance on what to test, and feedback on what has and hasn't worked. In a simple environment information may be visual and intuitive. But as situations become more complex because of geography, collaboration, volume, velocity, and variation, we need the help of computerized information systems. As we continuously improve our performance, as our enterprise grows and diversifies, and as the global economy continues to simultaneously expand and contract, our challenges become more complex and there is a greater need for skillful use of information technology.

Chapter 3

Three Stages of Lean Evolution

There are three stages most organizations will encounter on their journey to Lean:

Lean Operations—the elimination of waste and continuous improvement of production and service operations. I choose to call this *Lean Operations* instead of *Lean Manufacturing* since these principles can be applied to all industries.

Lean Enterprise—the elimination of waste and continuous improvement throughout the internal value stream of transactions and activities encompassing engineering, marketing, purchasing, planning, production, quality, distribution, service, finance, human resource, and administration.*

Lean Network—the elimination of waste and continuous improvement throughout the dynamic, global, electronic, demand-driven "supply chain."

Each stage must be considered both independently and interdependently with the other stages; focusing on a particular stage or technique to the exclusion of the others may yield only localized and temporary gains. When holistic thinking across these three stages is infused within the culture of the

---

* Whereas Womack and Jones' definition of Lean Enterprise presented in *Lean Thinking* encompasses the extended value stream from raw material suppliers to the end customer, for my purpose here the Lean Enterprise describes a single business entity, bounded by the information and transaction interfaces to external suppliers and customers.

Lean Enterprise Systems: Using IT for Continuous Improvement, by Steve Bell
Copyright © 2006 by John Wiley & Sons, Inc.

organization, however, enduring breakthrough performance and distinct competitive advantage may be achieved.

STAGE 1: LEAN OPERATIONS

This is not a book on the specific tools and techniques of Lean Manufacturing*, for there are already many available on the subject. However, it will be helpful for you to understand the essential concepts to get the most from this book. So if you would like an introductory or refresher course, I suggest you put this book aside for a short time and read one of the following:

- The quintessential book for anyone beginning the journey to Lean, in any industry, is Womack and Jones' *Lean Thinking*.
- And if you want to fully appreciate the inspiration that guided Toyota to develop these concepts and to sustain a culture of continuous improvement for the past fifty years, I encourage you to read Jeffrey Liker's *The Toyota Way*.

For an exploration of specific Lean Manufacturing methods and implementation approaches, I suggest one or more of the following:

- *Lean Manufacturing Implementation: A Complete Execution Manual for Any Size Manufacturer*, by Dennis Hobbs
- *Lean Manufacturing: Tools, Techniques, and How to Use Them*, by William Feld
- *Lean Production Simplified, A Plain-Language Guide to the World's Most Powerful Production System*, by Pascal Dennis

Go ahead, I'll be waiting for you . . . back already? Good, let's continue.

Lean Manufacturing practitioners often begin their journey with a focus on production activity. You can see, feel, touch, and sometimes trip over Lean Manufacturing—it's tangible work. Although the traditional Lean Manufacturing focus on waste reduction and value creation on the shop floor has led to significant benefits, a common pitfall is the development of a single-minded focus on the tools themselves, rather than the development of systems thinking and a learning culture. A company may proudly declare itself "Lean" because of enthusiastic initiatives in key areas, such as a burst of 5S† house-

---

* With the exception of Chapters 4 and 5, which provide a comprehensive treatment of detailed planning, scheduling, and execution techniques, illustrating the essential interface between traditional and Lean manufacturing software.

† 5S is the systematic practice of workplace organization that leads to waste elimination, derived from the Japanese words seiri (organization), seiton (tidiness), seiso (purity), seiketsu (cleanliness), and shitsuke (discipline). In English 5S is loosely translated as sort, straighten, shine, standardize, and sustain.

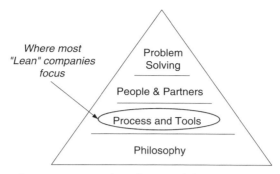

Figure 3-01. Four foundations of the Toyota Way

keeping, inventory reduction in a highly visible stockroom, or setup time reduction on a perceived bottleneck. As the sincerest form of flattery, they may emulate Toyota's historical methods regardless of the applicability to their own requirements.

Jeffrey Liker is Director of the Japan Technology Management Program at the University of Michigan and four-time recipient of the Shingo Prize.* For over twenty years he has worked intimately with the automotive industry both in the United States and Japan. In his book *The Toyota Way*, Liker identifies fourteen management principles of the Toyota Production System, grouped into four foundations that exemplify this broad systems thinking (Fig. 3-01):

Long-Term Philosophy

Principle 1—Base your management decisions on a long-term philosophy, even at the expense of short-term financial goals.

The Right Process Will Produce the Right Results

Principle 2—Create continuous process flow to bring problems to the surface.

Principle 3—Use pull systems to avoid overproduction.

Principle 4—Level out the workload (*heijunka*—work like the tortoise, not the hare).

Principle 5—Build a culture of stopping to fix problems, to get quality right the first time.

---

* In 1988 Utah State University recognized Dr. Shigeo Shingo for his lifetime accomplishments with an Honorary Doctorate in Business, and announced the creation of the Shingo Prize. Dr. Shingo was so honored that upon his death he was interred in his cap and gown. *Business Week* called the Shingo Prize the *Nobel Prize of manufacturing* while *Industry Week* described it as the *Malcolm Baldridge Award equivalent to Lean*. For more information visit www.shingoprize.org.

Principle 6—Standardized tasks are the foundation for continuous improvement and employee empowerment.

Principle 7—Use visual control so no problems are hidden.

Principle 8—Use only reliable, thoroughly tested technology that serves your people and processes.

Add Value to the Organization by Developing Your People and Partners

Principle 9—Grow leaders who thoroughly understand the work, live the philosophy, and teach it to others.

Principle 10—Develop exceptional people and teams who follow your company's philosophy.

Principle 11—Respect your extended network of partners and suppliers by challenging them and helping them improve.

Continuously Solving Root Problems Drives Organizational Learning

Principle 12—Go and see for yourself to thoroughly understand the situation (Gemba).

Principle 13—Make decisions slowly by consensus, thoroughly considering all options; implement decisions rapidly.

Principle 14—Become a learning organization through relentless reflection and continuous improvement.[31]

According to Liker, many companies make the mistake of thinking that Lean performance may be accomplished solely by application of the tools and techniques:

> Most attempts to implement Lean have been fairly superficial. The problem is that companies have mistaken a particular set of Lean tools for deep "Lean thinking." [They] have embraced Lean tools but do not understand what makes them work together in a system.[32]

This emphasis on holistic Lean thinking, and especially the role of IT in its realization, is a recurring theme throughout this book.

The Lean Development Life Cycle

There is a natural life cycle to every product, process, and system, and the evolutionary development of Lean Manufacturing within an organization is no exception. Speaking at the 2003 Shingo Prize Conference, former award-winner Tim Costello offered his thoughts in a presentation entitled "Lean Software Does Not Mean No Software."[33] In this presentation Costello described four phases that a company may experience if it does not properly integrate information systems into its Lean program: Beginning the Journey, Widespread Adoption, Plateau, and War of Attrition.

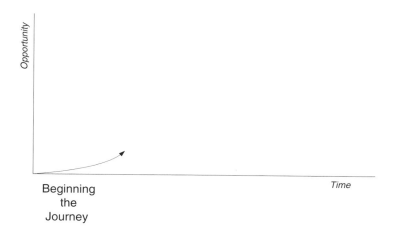

Under the guidance of an enthusiastic champion, the organization invests in limited education on Lean Manufacturing tools and techniques. Lean Manufacturing activities begin as pilots and isolated conversions of existing business processes, usually focused on shop floor activities. Little attention is given to broad systemic implications that impact the overall value streams within the organization or supply chain; the focus remains on localized techniques. Measurements are often limited and localized in scope, emphasizing isolated processes rather than results of the overall value stream.

During this first phase, limited information technology workarounds are developed, because the existing IT framework is firmly rooted in the bedrock of legacy business practices and software. Workcenters use offline systems such as spreadsheets for planning, scheduling, and analysis. Redundant and overlapping data proliferate, requiring considerable effort to manage, validate, and reconcile. Data does not flow smoothly to support the processes but moves in an erratic fashion. Nevertheless, because of the limited nature of the implementation, little systems pain is felt, and significant but localized Lean Manufacturing accomplishments are achieved. There is well-earned celebration, and the early success seems repeatable.

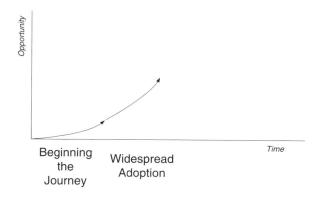

In the aftermath of the initial pilots there is great enthusiasm. Newsletter articles are written, rewards are distributed, new goals are set, and expectations are high. After years of disappointment with failed change efforts there is hope that success may be sustained over the long term. Lean Manufacturing activities expand as each small success is publicized. Other groups within the organization copy the early adopters, spreading the methods and offline systems throughout the plant and across the enterprise.

Because of the lack of integrated software tools and the understandable reluctance to invest in the costly and risky modification of existing software, offline systems become sanctioned. The IT department is enlisted in the development, proliferation, and support of these disintegrated islands of information. Daring feats of design wizardry create massive spreadsheets with countless tabs, complex formulas, lookup tables, macros, and connections to ERP and other databases. These tools become so complex that they seem to develop a life of their own, while on the shop floor efforts to achieve material flow are bombarded by the effects of chaotic information flows.

Disintegrated data becomes the standard. Management visibility of the information is reduced. The savings of Lean Manufacturing efforts, and the focus of key individuals, is partly offset by the costly administration and distraction of these fragmented systems. The ongoing effort necessary to reconcile the offline systems with the backend order processing and inventory control systems begins to increase exponentially. As the complexity and volume of these offline systems swell, the validity of their data suffers. Frustrations mount. Disputes arise over which version of data is correct. Errors are made in purchasing and production. Concern develops for the sustainability of the information infrastructure. And all the while, shop-floor driven Lean Manufacturing initiatives continue to spread and gather momentum.

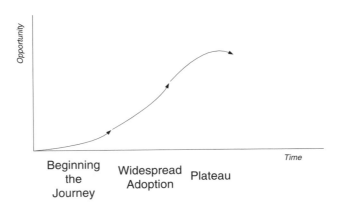

Despite productivity gains, localized pockets of information disintegration, communication disconnects, counterproductive top-down measurements, and waste beyond the shop floor hinder holistic Lean transformation of the enterprise. Periods of rapid gains give way to long dry spells and occasional conflict among factions within the enterprise. At the same time project budgets may be stretched thin. Concern and disillusionment begin to arise that overall benefits have not appeared as quickly as expected, and it begins to feel like the Lean teams are swimming against a strong current. Enduring Lean transformation appears threatened.

It is possible that Lean consultants, who have been hired to provide training, software tools, and guidance through the initial improvement efforts, may reach the limits of their skills and experience. The isolated and rapid operational improvements they consistently achieve with each new client may give way to the more difficult challenges of fusing these isolated gains together into a smooth flow across the entire organization. Although an outside consultant can provide teams with a valuable boost early on, if the teams don't take responsibility and begin to lead the improvement initiatives throughout the internal communication framework of the organization, then ultimately the Lean initiatives are not self-sustaining. Suddenly the teams may find themselves without an effective leader, while feeling themselves not yet ready to lead, and with the organizational groundwork not yet laid.

On the other hand, it may be tempting but not justified for the company to blame the consultants for this malaise. Most consultants will agree that the most important elements of sustained transformation are cultural and organizational, rather than inherent within any particular tool or technique. However, consultants are often introduced during a crisis, when the organization is focused on fixing a specific problem and not receptive to homilies about leadership and communication. As a result, consultants must often lead with a particular problem-solving tool or technique to achieve an early success in a symptomatic area (nonroot cause), gaining a foothold of trust and credibility with their new client. According to Leon McGinnis, Professor of Manufacturing at the Georgia Institute of Technology:

> Often the client will dictate their approach to Lean, and what they expect to get out of it—immediate returns or long term improvements, and these require different approaches. You can't fault consultants for doing the things that have the least risk and pay off the quickest. A savvy consultant can ease their client's pain immediately, and over time help them to recognize the deeper issues that require long term solutions.[34]

For holistic Lean transformation to occur, experienced consultants understand that this isolated focus must eventually expand. But if the initial problems are deeply entrenched and the crisis not easily solved, then the consultants may be tossed out as the company searches for the next quick fix.

At this fragile plateau the fragmented software tools that have blossomed around the expanding Lean initiatives may also begin to stretch beyond their limits, as spreadsheets, desktop databases, and the users that run them cannot scale in transaction volume and complexity to meet the mounting challenges. Furthermore, because they were developed as independent point-solutions, these applications are unable to extend across the organization to address the entire value stream. This perpetuates the isolated view of Lean improvements, and what little interaction there may be with other parts of the enterprise may be clumsy and error prone rather than supportive.

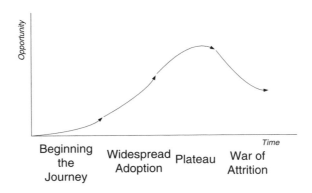

At some point teams and management may begin to feel that they have reached the limits of Lean capability. This is a dangerous time, because the enterprise has just reached a significant developmental milestone, and the entire continuous improvement culture is at risk of losing momentum. Management must now make a choice: to lose interest in the journey, or push ahead, promoting Lean thinking beyond the shop floor and throughout the enterprise. All of the components of the value streams must join with shop floor improvements to enable a truly *Lean Enterprise*. And what is the thread that weaves throughout the enterprise, orchestrating events and enabling decisions? The flow of information.

Tim Costello offered this insight at the conclusion of his presentation:

> Companies get these early successes caused by kaizen activities, and everybody's thrilled. But suddenly it looks like they run into a limit, they've done all they can do. In many cases backsliding begins to occur, inventory levels start to creep up again, productivity actually goes down, and a whole variety of symptoms appear. It's just the limit of the current system—they don't actually have standardized work, they don't have the software and systems tools to support what they're trying to do. These aren't limits to Lean, they're limits to the work architecture and systems support they have deployed to support Lean.[35]

Of course, everyone needs to start somewhere, and the shop floor may begin with some 5S housekeeping, a Kaizen Blitz, or reduction of setup time on a critical operation; these are great places to start because they encourage individual and team awareness of the immediate environment. But to achieve *sustained* Lean Manufacturing performance over the long run, it is important to emphasize a holistic systems perspective from the boardroom to the shop floor. This takes us to the Lean Enterprise.

STAGE 2: LEAN ENTERPRISE

The shop floor is the engine room of Lean Operations, but who steers the course? Ultimately, the customer does. In their book *Lean Thinking*, James Womack and Dan Jones emphasize that you must first understand clearly *what does and does not create value from the customers' perspective*. They describe the *Lean Enterprise* as an organization that reduces waste across *all activities*, which they collectively call the *value stream*:

> The set of all specific actions required to bring a specific product through the three critical management tasks of any business: the *problem-solving task* running from concept through detailed design and engineering to product launch, the *information management task* running from order-taking through detailed scheduling to delivery, and the *physical transformation task* proceeding from raw materials to finished product in the hands of the customer.[36]

Similar to the enthusiastic adoption of Lean Manufacturing in the West, during the 1990s there was an emphasis on Business Process Reengineering (BPR) popularized* by Michael Hammer and James Champy in their 1993 book, *Reengineering the Corporation*. Although BPR was predominantly focused upon administrative rather than manufacturing processes, its message is very similar to the tenets of Lean Manufacturing. BPR strives for new levels of productivity, breaking the rigid and hierarchical management and work methods of the past, by emphasizing the following points:

- Process orientation
- Rule breaking
- Hybrid centralized-decentralized operations
- Workers making decisions
- Several jobs combined into one
- Steps in the process are performed in a natural order.
- Work is performed where it makes the most sense.
- Checks and controls are reduced.
- Jobs change from simple tasks to multidimensional work.
- Individual roles change from controlled to empowered.
- Job preparation changes from training to education.
- Focus of performance measures and compensation shifts from activity to results.
- Values change from protective to productive.

---

* The concepts of work simplification and process improvement predated Hammer and Champy. For a brief history of these early endeavors see Allan Mogensen, Improving Productivity by Involving Your Work Force: The Work Simplification Story, Institute of Industrial Engineers Fall 1981 Conference Proceedings.

- Organizational structures change from hierarchical to flat.
- Executives change from scorekeepers to leaders.
- Managers change from supervisors to coaches.
- Creative use of information technology[37]

Reengineering was partly responsible for the surge of ERP popularity during the 1990s, and is also considered partly responsible for the general growth in productivity in many sectors of the economy that continue to this day. Whereas BPR primarily focused on process improvement in the office, Lean Manufacturing was at the same time improving productivity in the factory.

Mapping Information Flow Within the Value Stream

When we speak of production flow, the movement of material generally comes to mind. But there is another necessary flow, of information that instructs each individual and operation what to do next. In a simple kanban environment the appearance of the container is all the information required to instruct the process, while in other situations the need for detailed information can be far greater. According to Rother and Shook in *Learning to See*, "Material and information flow are two sides of the same coin. You must map both of them."[38]

Lean and BPR similarly focus on the elimination of waste by improving the overall value stream. To improve we must understand the existing business processes, which are often guided by the functional and departmental silos and local optima institutionalized within the culture, compensation policies, procedures, and information systems of an organization.

Then there are a multitude of informal and inflexible *glue* software applications that can lock cumbersome or counterproductive policies and procedures into place. These are usually spreadsheets but may also include small desktop databases and handwritten lists pinned on noteboards. These fragmented islands of information are usually inserted into the gaps and cracks

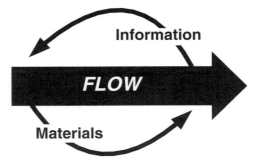

Figure 3-02. Complementary flow of information and materials

between larger systems and information flows. It's not just small companies that resort to these methods; this is a common phenomenon in even the largest. In fact, the problem is often more acute in larger organizations, especially if they have achieved their size through acquisition of many companies in widespread geographies and cultures, each with their own legacy systems and business practices that must be glued together (integrated) into smooth business flows.

By helping systems and people communicate more effectively, *Enterprise Integration* smoothes the flow of information as the value stream crosses system boundaries. We'll explore Enterprise Integration tools and techniques in Chapter 7, but first we must learn how to understand and document the current processes and information flows so they may be improved.

Understanding the flow of information requires that we first document and quantify corresponding physical processes. This process of unraveling and documenting an organization's processes has been called *peeling the onion*, because cross-functional teams methodically examine successively deeper layers of the organization's structure, value streams, and intricate processes. It is generally impossible for a single individual to understand all the essential processes and interdependencies within even a medium-sized organization, but through mapping teams learn to holistically and visually describe the flow of material and information along the entire value stream. This places the teams on a strong footing for effective enterprise-wide continuous improvement.

Value stream mapping was introduced by Mike Rother and John Shook in *Learning to See* after studying Toyota's improvement methods. In this book they recommend the development of current-state value stream maps to illustrate the current flow of materials, work, and information, followed by future-state maps describing the desired state:

> The first step is drawing the current state, which is done by gathering information on the shop floor. This provides the information you need to develop a future state. Future state ideas will come to you as you are mapping the current state. Likewise, drawing your future state will often point out important current-state information you have overlooked. The final step is to prepare and begin actively using an implementation plan that describes how you plan to achieve the future state. Then, as your future state becomes reality, a new future state map should be drawn. That's continuous improvement at the value stream level. *There must always be a future-state map.*[39]

Value stream mapping is essential to continuous improvement, and efforts at process improvement before mapping may be misguided by a narrow focus on tasks, resulting in localized and nonstrategic waste reduction that does not improve the value stream. According to Rother and Shook, "Too many Lean implementation efforts have been *seven-waste scavenger hunts.*"[40]

In our practice, one of our first objectives is to help the cross-functional teams understand the organizational and transactional relationships of their

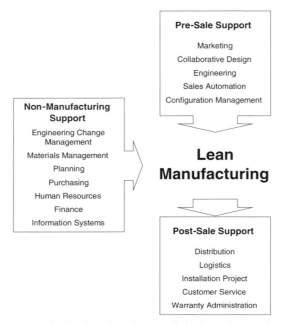

Figure 3-03. Functional map of the Lean Enterprise

enterprise. To do this we build a high-level map similar to Figure 3-03. Note that the core Lean Manufacturing operations are supported by presale, non-manufacturing operations, and postsale activities, all of which are essential components of the overall value stream.

Once the team understands the relationships at this highest level, they can drill down to another layer, developing process maps that illustrate the flow of materials and information at the *entity* level. These may be presented as an Entity Relationship Diagram (ERD) such as the example shown in Figure 3-04. Note that in an ERD the boxes represent organizational entities or functions and the arrows represent relationships rather than distinct processes and information flows. Each individual can usually identify with one or several functions within the entity diagram, helping them to visualize their role within the overall enterprise and its value streams.

We developed the flowchart in Figure 3-04 when our firm was asked to design a stand-alone (nonintegrated) central purchasing system for a multisite manufacturer with an obsolete and incomplete MRP II system. Although the client planned to implement a new ERP system, they were hoping to delay this expensive investment for a year or two by installing several smaller applications at their key pain points. This diagram was useful in communicating the numerous cross-functional relationships and interdependencies necessary for

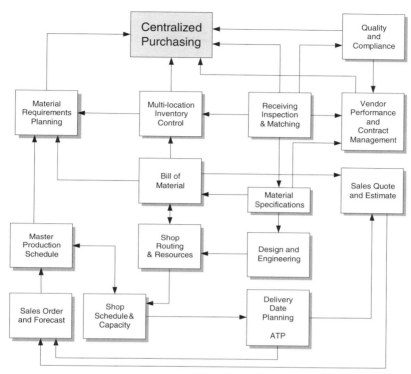

Figure 3-04. Entity Relationship Diagram from a purchasing perspective

the central purchasing project to be successful. In other words, a stand-alone central purchasing software investment was not going to provide the desired quick fix. For a company that is used to operating with a nonintegrated system, or as traditional functional silos, the entity mapping process can illuminate the root causes of nagging problems that are hidden from plain view.

As the teams drill to the next level of detail beyond entity relationships, they begin to focus on distinct processes and subprocesses; within a medium-sized organization there may be hundreds of small diagrams created at this time. Figure 3-05 shows an example of a simple sales order entry process using standard flowchart symbols.

As the teams scrutinize the details of each process, they may begin to classify specific tasks in terms of Value-Added (VA), Non-Value-Added (NVA), and Necessary Non-Value-Added (NNVA) activity. These tasks may be quantified in terms of time, motion, cost, inventory consumption, information flow, and other relevant measures of value and waste. At this next detailed stage, more articulate value stream mapping tools and techniques become useful; several commonly used icons are illustrated in Figure 3-06.

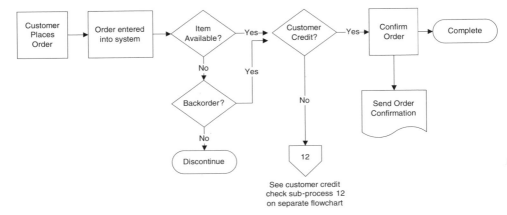

Figure 3-05. Sales order process flowchart

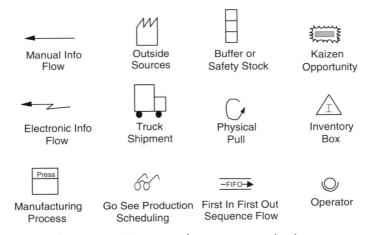

Figure 3-06. Common value stream mapping icons

A portion of a value stream map is illustrated in Figure 3-07.* Note that the Manufacturing Process Box contains values such as C/T = cycle time, C/O = changeover (setup) time, and EPE. These terms will be explained in Chapter 5. Also note the proportion of VA and NVA time described on the elapsed time bar.

Rother and Shook recommend that when developing detailed maps it is important to go to the source (Gemba) to see the work being done. The team

---

* For a full exploration of value stream mapping tools and techniques, see Rother and Shook, *Learning To See*. To purchase a value stream mapping template for Microsoft Visio diagramming software visit the Lean Enterprise Institute at www.lean.org.

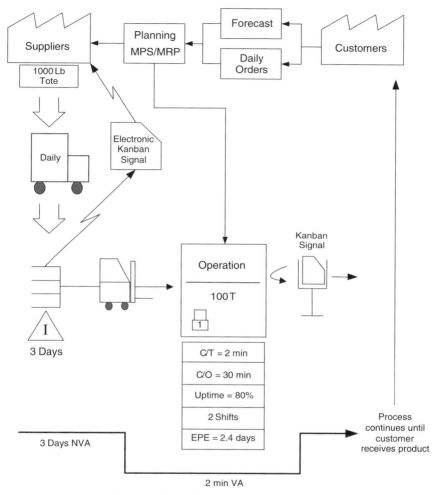

Figure 3-07. Value stream map fragment

should walk the process from end to end, asking questions while taking notes, photographs, and videos. It is often helpful to start at the end of the process, working backward to visualize demand pull and material flow from the customer's perspective. At this level of detail it may also be useful to quantify the VA, NVA, and NNVA components of each particular step, resulting in the tabular worksheet shown in Figure 3-08.

As this worksheet illustrates, it can be enlightening to measure the relative amount of VA contained within any process, which is commonly estimated at less than 5–10% of the total activity. And it is particularly interesting to measure the effectiveness of the *information flows* in support of those

Order Receipt and Processing	Distance (feet)	Time (min)	Operation	Transport	Wait	Inspection
Delay until order is picked up from fax machine		30			NVA	
Carry from fax machine to desk	20	1		NVA		
Wait until convenient to process order in batch		30			NVA	
Enter order, verify pricing and credit		5	VA			
Send copy to legal department for review		5		NNVA		
Legal department review (4 hours)		240				NNVA
Order release to scheduler		5	VA			
Total	20	316	10	6	60	240

	Total	%
VA	10	3%
NVA	61	19%
NNVA	245	78%

Figure 3-08. VA, NVA, and NNVA analysis

processes. One study, published by the International Journal of Logistics, entitled "Lean Information and Supply Chain Effectiveness"[41] reported:

> The findings suggest a VA—NVA—NNVA profile of 1—49—50 percent in the information dimension. This differs markedly from the 5—60—35 percent profile established for world-class supply chains. Such a metric has major implications for investment decisions. Based on the argument that physical and information capabilities are mutually dependent and that physical capabilities are more developed than their information counterparts, an organization is likely to obtain a better Return On Investment (ROI) for a targeted waste elimination program in the information domain because there is a wider scope for improvement.[42]

This study suggests that in many organizations 99% of the information processing activity is wasteful, caused by unnecessary data capture and entry, fragmentation, redundancy, errors, missing data, timing problems, reconciliations, security precautions, and the countless hours spent analyzing data that are incomplete, inaccurate, or irrelevant. One conclusion you may draw from this is that IT itself is not the enemy of Lean, but the *improper use of IT is*. Although your results may differ from this study, you may find it worthwhile to measure the waste within your physical processes, and separately measure the waste of the information flows supporting them.

The Missing Link in Value Stream Mapping

When the time comes to begin mapping and improving processes, theoretical definitions of value streams meet with a more complex reality. In *Lean Thinking*, Womack and Jones define the value stream as being comprised of three elements in a single company-wide value stream:

- **Problem-Solving** that leads from concept through design and engineering to product launch.
- **Physical Transformation** from raw materials through production and delivery to the customer.
- **Information Management** of the entire process.

This definition of a value stream may be so general that it can be difficult for a kaizen team to get their arms around it, especially if an enterprise has multiple locations, products, services, and markets. And when you see a value stream map by Rother and Shook in *Learning to See*, or examples in many other Lean texts, it usually describes a relatively simple and linear process flow from customer order through production and delivery. Beginning at the upper right with the customer order, the value stream flow moves counterclockwise through order management, production control, supplier delivery of materials, stages of production, and inventory buffers, finally coming full circle to customer delivery. High-level value stream mapping generally disregards the "supporting" functions of the organization such as marketing, sales, engineering, quality, customer service, finance, and so on.

When mapping a value stream of a complex production operation in such a simplistic fashion you reduce operations and routings down to a simplified set of icons at critical buffer and movement points. Here lies the magic of value stream mapping, reducing complexity to simplicity, while developing the team's ability to prioritize and incrementally improve the entire process. Womack and Jones assert that this general mapping is an important first step, because the initial objective is to develop a "breakthrough in shared consciousness of waste and to identify systematic opportunities for eliminating the waste." They further suggest that "mapping the value stream of every component going into the product is time consuming and costly and we have found that it overwhelms managers with too much data."[43]

Although this high-level map is useful to illustrate the flow of the process in general, it is terribly abstract when compared to the specific transactions and tasks that flow across the enterprise. To lead improvement teams to the next level of detail and expand into the nonmanufacturing support functions of the enterprise, another perspective is often needed, on the specific flows of business processes that represent transactions or events.

Deconstructing an enterprise into its component processes and subprocesses helps the teams to clearly identify and quantify all the moving parts and interrelationships. Many in the cross-functional team *learn to see* for the first time their interrelationships to the other activities of the company. The teams can now begin to visualize and formulate their desired future state in sufficient detail to guide their improvement initiatives. From this current- and future-state mapping process a requirements list can also be developed to aid in the selection and implementation of appropriate software tools. At some point the teams are ready to reassemble the many distinct process and subprocess maps into a comprehensive map of each value stream. Whether teams

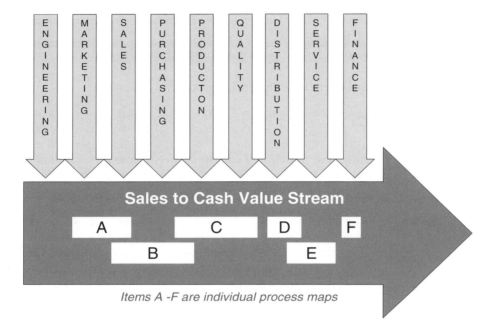

Figure 3-09. Value stream with individual processes identified (A–F)

are able to fit all of the company processes and subprocesses within a *single* comprehensive value stream schema or they choose to divide the company activities into several value streams (i.e., sales to cash, design to build, purchase to pay) is less important than clearly establishing the relationships and flow among the individual processes and subprocesses as shown in Figure 3-09.

It is often necessary to map all of the processes and subprocesses so the individual teams can identify *key control points* that simplify the control of each process and subprocess. It is only at these critical points that *performance measures* should be assigned, developing a limited number of Key Performance Indicators (KPIs) that effectively measure and manage the performance of each process and contribute to the results of the overall value stream. These KPIs become the prioritized process improvement targets that are used to guide the continuous improvement teams from the bottom up.

It is important to note that these detailed process and subprocess diagrams are necessary for measurement and control at a *tactical* level, for establishing KPIs, improvement activities and targets, documenting specific and standardized work instructions, and for documenting and supporting the end user software tasks that automate these processes. However, this level of granular detail would distract from top-down management focus on the few strategic priorities guiding the overall enterprise, and any results at the granular process and subprocess level should be aggregated, if presented at all, for executive-level performance management. We will revisit this critical issue of strategic focus and team-based improvement linkage in Chapters 10 and 11.

STAGE 3: LEAN NETWORK

The Lean Network is commonly called the *Supply Chain*; however, there are two connotations of this term that I find inappropriate:

1. The Lean Network isn't about "supply," it's *demand driven*. Although the term *Supply Chain* has apparently found a permanent place in our vocabulary, it does not convey the absolute importance of demand, which triggers all value-adding activities.
2. This new business model isn't a *chain*, it's a *network*. A chain is linear, sequential, and slow. Lean Network relationships are fluid, organic, and multinodal. There are infinite potential interrelationships, some permanent and others temporary, which are all spontaneously reconfiguring. Relationships, product life cycles, and transactions appear and dissolve in a flash. Like the neurological patterns of the brain, the ultimate adaptive network, the pathways of the Lean Network develop and reformulate according to changing circumstance.

How is the Lean Network different from the Supply Chain Management (SCM) practices that have been evolving for the past several decades? The Lean Network is a natural extension of traditional SCM, *but with a primary focus on eliminating waste throughout the intercompany value streams*. And there is plenty of waste to be eliminated. Authorities estimate that many billions of dollars are lost each year. To eliminate this massive waste, in their book *The Lean Extended Enterprise* authors Terrence Burton and Steven Boeder suggest that an enterprise should:

> Think about developing a profit-and-loss and a value stream map for your own Lean Extended Enterprise. Based on our benchmarking and client experiences, you would find that as much as 70% to 95% of product cost is generated outside of your company. You would find that as much as 75% to 95% of lead time is consumed outside of your company. You would learn that 95%+ of the key activities of design, supply chain planning, and manufacturing (and the associated employees) are outside of your company. And finally, you would wonder why your organization is not in hot pursuit of this gold mine of collective opportunity.[44]

Many would argue that SCM has focused on waste elimination from the beginning. Similar to the evolution of EDI (Electronic Data Interchange), however, many SCM initiatives have resulted in benefit primarily for the largest trading partner. For example, many smaller companies have had no choice but to comply with mandatory EDI requirements. Although it is true that some of these smaller companies took advantage of this disguised opportunity to improve their internal value streams and automate their processes, many others complied by simply adding another layer of wasteful activity and software to their already inefficient business processes. Just imagine how many small and medium-sized manufacturers still remove an EDI transmittal from their fax machine, then manually enter the data into their order processing system.

Although many early EDI initiatives may have been focused on waste elimination from a segment of the supply chain, they often resulted in waste *transference* to the smaller trading partners. The same is true in many cases with SCM today, where smaller suppliers build ahead inventory for Just-In-Time (JIT) customers, holding large stocks just waiting for a kanban signal. The same may also be true of many Vendor-Managed Inventory (VMI) relationships.

In contrast, the Lean Network looks beyond local improvement, aspiring to eliminate waste and enable smooth flow of material and information across the entire supply chain for everyone's benefit. After all, a customer usually pays for waste in its supply chain, directly or indirectly. If a supplier is forced to carry the burden of excess inventory, this makes the relationship less profitable, resulting in small compromises that add up to diminished innovation, quality, service, and longevity.

Toyota learned very early that strong suppliers make stronger manufacturers. It becomes a great impediment to Lean (and particularly JIT) when suppliers are unable to reduce their inventory and improve their agility. Toyota therefore invested heavily in the development of their suppliers over a long period of time. According to Ohno, establishing the Toyota Production System took over twenty years. Once Toyota had made significant strides, JIT was slowly introduced to their subsidiaries. To spread JIT to the major subcontractors, Ohno picked ten key people, one from each of the top tier suppliers, to lead the effort. These ten key people, known as the *Toyota Autonomous Study Group*, acted as consultants to all of the subsidiaries and worked closely together.[45]

This practice continues today within many supply chains and will continue to trickle down the Lean Network through layers of relationships and across many industries. In the *Information Week* article "Never Too Lean," Delphi Corporation describes their early success in supplier development:

> Delphi Corp. is taking Lean beyond its four walls and out to its suppliers. The $28 billion-a-year auto-parts supplier has 47 employees on loan at nearly 70 suppliers to help transform those factories' production facilities into Lean operations.

> Carlisle Engineered Products Inc., a supplier of rubber and plastic products primarily to automakers, became one of Delphi's first suppliers to reengineer its communication, production, and material-flow processes. "We're not finished

analyzing our processes, but the amount of waste we have identified so far is ungodly," says Bruce Wandyez, VP of Manufacturing. "When you look at a process in great detail, there's an upsetting amount of inefficiencies."

The company will migrate more than ten other facilities to Lean methodologies, which Wandyez says could take ten years. But this will give Carlisle Engineered Products time to establish the significant cultural changes that Lean Manufacturing requires, he says. Then the company can extend the principles beyond its four walls. "We would like to extend this methodology to our suppliers," Wandyez says, "but we're in the early stages and need to develop a higher level of expertise."[46]

This bears repeating—the Lean Network emphasizes the elimination of waste through the entire supply chain to the benefit of *all* trading partners. Like any other initiative involving information technology, we must focus on results rather than becoming seduced by the tools themselves. The phrase "improved visibility into the supply chain" has been used to justify enormous IT expenditures; the phrase sounds seductive, but to what end? The ultimate focus should be on eliminating inventory and other wastes, rather than simply visualizing them.

In *Seeing the Whole—Mapping the Extended Value Stream* (the sequel to Rother and Shook's *Learning to See*) Womack and Jones explain that the intercompany value streams must be mapped with the same diligence as internal value streams of the Lean Enterprise. To accomplish this they suggest that an individual from each firm should be responsible for the intercompany mapping process of each product line. They call this role the *product line manager* and suggest that the product line manager in the furthest downstream enterprise (closest to the customer) should drive the mapping process upstream through the suppliers, and the suppliers' suppliers, through an expanding web of relationships back to the multitude of raw material sources. This suggests the collaboration of numerous improvement teams sharing considerable knowledge within and across company boundaries, and the development of extended value stream goals, objectives, and measurements. This implies a new shape of collaboration involving rigorous program and project management, leaving one to wonder whether this structure of collaborative relationships will someday become the standard. And if so, how are we to keep all this knowledge and the countless moving parts orchestrated, if not through the use of skillfully designed information systems?

IT and the Lean Network

Many enterprises have been building the foundation for the Lean Network for years. They have invested massive resources in improving their information technology infrastructure, often in conjunction with BPR, to replace fragmented transactional systems with a centralized *system of record*.[47] It is the ERP system that automates, records, and integrates the core financial and operational events within a single transaction database model. Customer Relationship Management (CRM) and Product Lifecycle Management (PLM) systems have also received considerable investment, integrating marketing, sales, design, engineering, and customer service processes with the ERP core.

Thoughtfully implemented, integrated enterprise software advances the organization toward a basic prerequisite for the Lean Network—standardization, automation, and continuous improvement of the internal value streams. This prepares them for the greater challenge of extending those value streams, and perhaps outsourcing portions of them, across trading partner boundaries. According to Dave Caruso of AMR Research:

> Business processes good enough for a domestic market look feeble in the context of managing a global operation. Like innovation, globalization must become a core capability for manufacturers of all sizes. Globalization in this respect means strong, singular processes supported by effective technologies. IT delivers the fabric to speed innovation and allow differentiation.[48]

And according to Robert Kennedy, Professor of Corporate Strategy and International Business at the University of Michigan, commenting on the trend to global outsourcing:

> The unit of analysis is much smaller. In manufacturing, we move a whole factory or outsource a large part of the value chain. Now some firms are moving specific jobs, perhaps only three or four, which is blowing apart the value chain. How do we disperse all these activities around the world and then reintegrate them?[49]

The ERP system is the bedrock, the core, the inner framework for automating and controlling these increasingly fluid relationships and complex transactions. ERP enables the internal value streams to adapt and extend beyond the four walls of the organization, beyond the rigid traditional EDI agreements among legacy trading partners, toward the great unknowns of the Lean Network.

The very fact that the Lean Network encompasses the entire global supply chain means that it must include smaller enterprises. But how can a small or medium-sized manufacturer afford to play in the Lean Network? Rather, if a small or medium-sized manufacturer has no choice but to participate, how can they approach this arena where larger companies have spent countless sums just to lay the foundation for electronic commerce? The recoil from massive ERP implementations during the Y2K era has fostered a new approach to Lean Network investment, favoring smaller point-solutions that are often product-, market-, task-, or even business partner-specific. These smaller projects pose less risk, offer greater agility in their design and execution, and produce faster and more easily measured results. These projects are often accomplished with the aid of their larger trading partners.

Does the trend of these smaller investments and point-solutions suggest a backslide toward the fragmentation and disarray of information flows? Not if they support the value streams and carefully integrate with the core ERP framework, sharing its centrally managed data and logic. In fact, if the enterprise develops the agility to rapidly create and implement these agile point-solutions, responding to sudden demand and tactical shifts while maintaining stability and continuous improvement of the overall information system architecture, this could produce substantial competitive advantage. We'll explore this fundamental shift in Lean IT thinking in the last chapter of this book.

Internal business process improvement and systems integration are basic requirements for effective Lean Network execution: ERP, CRM, and PLM systems lay the foundation by integrating and automating the core value streams. This enterprise software foundation may be extended to external trading partners through systems that *Plan*, *Execute*, and *Inform*.

Planning Lean Network Events

China has emerged as a powerful force in the global manufacturing supply chain; however, for western countries China is distant and their operations are often not very Lean. Although developing countries like China offer inexpensive labor and low-cost products, the trade-off is usually for large lot sizes and long lead times. If developed countries extend their dependence on global suppliers as key partners in their supply chain, they must develop skillful forecasting and planning systems to compensate for these shortcomings.[50]

Collaborative Planning, Forecasting, and Replenishment (CPFR) is a software-enabled approach that aids trading partners in managing demand, supply, and delivery service levels.* But for these sophisticated collaboration tools to succeed, there must already exist the basic capabilities for demand management and planning. According to the *CFO* magazine article "Working on the Chain":

> "Best practice collaborative forecasting can be done in the here and now," says Kevin O'Marah of AMR Research. Successful collaborative forecasting rests on a rigorous Sales and Operations Planning process. That amounts to convening formal, regular meetings between sales managers, who know what customers will pay for, and operations managers, who can match that demand with sourcing, production, and logistics requirements. "The specific technology used to collaborate with suppliers doesn't matter," says O'Marah, "as long as it's simple and fast; EDI or E-Mail may serve as well as collaborative supply chain applications."[51]

A prerequisite to CPFR is effective Sales and Operational Planning (S&OP), a topic we'll explore in Chapters 4 and 5. For now it is important to note that you cannot electronically collaborate with your suppliers if you can't balance your own demand and supply, if you don't know what you need and when you need it.

* For a comprehensive look at CPFR tools and methods visit www.cpfr.org.

Along with the fundamental value stream management and S&OP capabilities, to develop a flexible supply chain an organization must develop trusting and collaborative relationships with its key trading partners. Many companies have managed EDI relationships for decades, but these are often rigidly defined and repetitive transactions. CPFR develops adaptive relationships, sharing vital market intelligence while attempting to guide rather than react to the future. This is a special relationship based on trust and cooperation, according to Michael Tanner of the Chasm Group:

> The value these systems bring to your organization is directly proportional to the amount of information your trading partners are willing to put into them. Similarly, the ability to gain ROI is directly related to the degree to which other organizations agree to use your systems. This means the people who manage these systems must be more than technology experts. They also must be marketing geniuses, able to convince outside parties to use their systems. The initial focus should be on the suppliers you could help the most—suppliers that in turn will recommend the system(s) to others. Sell the vision hard, but deliver short term results.[52]

Executing Lean Network Events

Demand pull signals should initiate the flow of products and services throughout the Lean Network. These signals usually require a preestablished contractual agreement between trading partners, which come in many forms. Among these, EDI continues to show strong growth and create new opportunities for smaller trading partners. In the *Industry Week* article entitled "EDI is Dead! Long Live EDI!", editor David Drickhamer notes:

> When it comes to information technology, it's rare to hear reports of 20 or even 25 years of experience. But that's how it is with EDI. Today, the total annual value of EDI transactions ranges from $1.8 trillion to $3.2 trillion worldwide, depending on how the data being exchanged between companies is tallied. According to technology research firm IDC, traditional EDI commerce will continue to have a compound annual growth rate (CAGR) of 8.4% through 2006, while Internet EDI commerce will have a CAGR of 52.1% through the same period. Traditionally EDI has involved the 25% of suppliers that account for 75% of a large manufacturers' raw material and part volume. Much of today's EDI activity has been aimed at increasing connections between large enterprises and their small and medium size suppliers that account for much lower volumes, but where manual paper processing leads to disproportionately high administrative expenses.[53]

In addition to EDI, other electronic procurement and inventory management methods are available:

- **Electronic Kanban Signals**—Automated release signals instruct a supplier to send inventory. These signals may be delivered to the supplier actively through e-mail, fax, EDI, or XML messages, or passively through Internet portals.*

- **Vendor-Managed Inventory (VMI)**—Customers provide their suppliers with the responsibility and authority to manage inventory levels and replenishment actions. This generally requires providing the supplier with near real-time visibility to the customer's forecasting, sales and returns transactions, inventory balances and policies, and current production schedule.

- **Exchanges**—an Internet site that allows buyers and sellers to spontaneously communicate, collaborate, quote, bid, and transact in a variety of ways. After a false start during the dot.com boom, exchanges and auctions have gathered momentum, according to the 2004 *Economist* article "A Perfect Market":

Before the dot-com bubble popped, the really big money in e-commerce was expected to be in business-to-business (B2B) websites, especially in online auctions. It did not work out like that. For one thing, companies were not particularly willing to sift through tenders from lots of suppliers they had never dealt with before. Most of them prefer to build stable longer-term relationships with a limited number of suppliers. But not all the early B2B exchanges floundered. In some larger industries, such as metals, chemicals and cars, they continue in various forms. The exchange is offering increasingly sophisticated services, such as auctions that factor in transportation costs, different currencies and even the notional cost of having to build a new relationship after switching suppliers. The process is also faster and more transparent than before it moved online.[54]

Finally, in the realm of execution there are the issues of logistics, warehousing, and transportation management. How will the material travel from its origin to its ultimate destination? How much waste must occur?

Developing countries may offer low-cost production but are disadvantaged by the lead time required to transport the product overseas. At the same time, however, many developing countries are building state-of-the-art communications and logistics infrastructures to support their burgeoning commerce. In a post-9/11 world, security and regulatory issues have greatly complicated global logistics cost and lead time planning factors. And in developed countries, outsourcing final assembly and warehousing operations offer numerous alternatives for configuring and locating products closer to the customer. As third-party logistics (3PL) service providers offer an increasing variety of sophisticated physical and electronic services to their clients, the power of choice adds more complexity to supply chain execution decisions.

Informing Lean Network Partners

The Internet will continue to introduce new possibilities for the Lean Network. In less than a decade, e-mail and instant messaging have changed

---

* We will discuss XML (eXtensible Markup Language) and portals in Chapters 7 and 8.

the patterns of business and personal communications for a significant percentage of the world's population. The Internet offers *information at our fingertips*, to borrow a 1990s phrase from Microsoft's Bill Gates. We now have the ability to provide our trading partners with secure channels of communication, collaboration, product information, design and engineering data, account history, transaction, and inventory status. Many logistics providers can instantly track a parcel around the world, and with satellite Global Positioning Systems (GPS) and Radio Frequency Identification (RFID) we may soon have real-time visibility down to the package and item virtually anywhere on the planet. Are we prepared to deal with this glut of information? How can we turn it to our advantage *without adding waste*? Throughout this book we'll explore these questions, using the following tools and techniques that inform and empower the Lean Network:

- Customer and vendor portals that help our trading partners to answer their own questions
- Product data and life cycle management systems that help manage the massive volumes of data needed by suppliers, customers, and regulatory entities
- Data warehousing, mining, and analysis tools that store and analyze massive amounts of transactional data to aid in decision-making
- Content management systems that store, manage, search, and retrieve massive distributed archives of unstructured and document-oriented information
- Performance dashboards that instantly communicate vital statistics, offering to drill down to identify root causes
- Automated event and exception notification tools that alert empowered individuals to a situation in time to take preventative or corrective action
- Automated data capture, item identification, and tracking systems to manage the massive flow of materials and information that grows exponentially with each passing year
- Enterprise Integration, eXtensible Markup Language (XML), Web Services, and Service-Oriented Architectures (SOA) that allow systems to communicate with each other

A Word About Wal-Mart

When it comes to supply chains, retail giant Wal-Mart Stores Inc. is clearly in a class by itself. In their 2003 and 2004 Ten Best Supply Chains report,[55] *Logistics Today* selected Wal-Mart as their retail category winner. With annual sales over $250 billion, Wal-Mart sells more products than the next five biggest retailers combined. That kind of market command has led to some unusual activities in the retail space. When retailer Toys "R" Us let it be known it might stop selling toys, which would basically cede the market to Wal-Mart, a group of top toy manufacturers agreed to produce and promote toys that would be exclusively sold at Toys "R" Us stores.

Wal-Mart might very well be unstoppable, says *Logistics World*. With revenues quickly approaching an unfathomable one billion dollars per day, Wal-Mart's strategy is deceptively simple—every move its supply chain makes has the ultimate goal of maintaining the lowest prices possible for the end consumer.

Logistics Today declares Wal-Mart the leader in the development of applied supply chain technologies and suggests they have single-handedly done more to drive these practices down to smaller manufacturers than any other trading partner. In 2001 Wal-Mart drove the adoption of the UCCnet data synchronization standard. In 2002 they directed 10,000 mid-sized suppliers to adopt the EDI-INT AS2 standard. And Wal-Mart is now tirelessly pursuing RFID tagging on all pallets and cases. Wal-Mart is even helping the U.S. Department of Defense develop its own RFID implementation strategy and standards.[56] It is evident that Wal-Mart and other massive supply chain partners are aggressively leading the design and development of the Lean Network infrastructure. Like the dissemination of EDI during the past three decades, the impact of their efforts will inevitably trickle down the supply chain to their smaller trading partners.

However, as with the emergence of EDI when many small suppliers were coerced into implementing expensive electronic order management systems when their internal processes and systems were not ready, the explosion of the Lean Network could have terrible consequences for those who are not prepared. But unlike the gradual adoption of EDI, with the current pace of global change and technology innovation, we should expect the Lean Network transition to occur much faster and with greater disruption and opportunity, reaching further inside the secure enterprise value streams and information systems.

Emerging enterprise software, integration, and electronic commerce tools may now be practical and economical for even a small company, as long as they are applied thoughtfully and are properly integrated with the core ERP system. If you are a small or medium-sized manufacturer, and have not yet extended your ERP framework with the planning, execution, and information-sharing interfaces required to participate in the Lean Network, you may expect that a significant trading partner will soon require this capability. This requirement may come in the form of EDI orders and shipment notifications, communicating transactions through a collaborative Web portal, or the RFID tagging of outbound materials. Even the smallest companies must now plan ahead, improving their internal business processes and automating key trading partner interfaces, developing the *Lean maturity* that will be required to survive and thrive. In the new global economy there may be few second chances to gain a foothold.

STAGE 4: IT AND LEAN MATURITY

Ma-tu-ri-ty, noun: The quality or state of being mature

Ma-ture, adjective: (1) Based on slow careful consideration, (2) Having completed natural growth and development[57]

Merriam-Webster Online Dictionary

What does it mean to say that an enterprise has attained *Lean maturity*? On what scale, and who is measuring? Perhaps the best indication of Lean maturity is how well the shop floor, enterprise value stream, and demand-driven supply chain activities are balanced and continuously improved according to an integrated business strategy.

Enterprises compete for market share, and so do nations. For developed countries, much repetitive production has been sent overseas and may never return. Developing countries have learned to satisfy the demand for low-cost and high-volume products by leveraging inexpensive labor, and through extraordinary economies of scale often coupled with aggressive government support. As a result of this shift in the global productivity balance, nonrepetitive variable production with high knowledge content, often located near the locus of customer delivery, may be the stronghold of competitive advantage for developed countries.

Managing this sort of low-volume, high-mix complexity must become a core competency for many manufacturing enterprises in developed countries. Although most Lean Manufacturing literature and education has focused on developing stable and repetitive production, this emphasis must shift to less repetitive methods for many practitioners. This is why we will invest substantial time in Chapter 5 looking at the complex requirements of job and project shops, mixed model, and postponed final assembly operations.

Managing Complexity

Coping with accelerating complexity may represent the greatest business challenge of our lifetime. Maturity is a measure of our ability to exploit this growing complexity, both inside and outside the walls of our plant. Technology has created the environment for complexity to flourish, and it's only through simplification of our own environment, supported by the sensible application of IT, that we can hope to be successful. This is where IT may play a *leading* role in Lean maturity.

In 1998 Tim Costello and Richard Lebovitz were launching FactoryLogic, a software company intent upon addressing the specific requirements of Lean Manufacturers.* While sitting in a restaurant, on the back of a napkin they

---

* In 2001 FactoryLogic was the first software company to be awarded the Shingo Prize for Research and Applied Programs in Manufacturing.

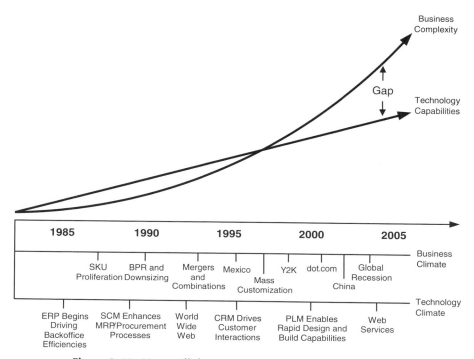

Figure 3-10. How will this image appear in another five years?

sketched a diagram that illustrated the widening gap between Lean Manufacturing requirements and enterprise software capabilities. This diagram identified growing complexity as the factor driving the need for change. With their blessing I updated this diagram (shown in Fig. 3-10) to reflect the dramatic changes that have transpired during the past five years.

Looking at the information technology climate timeline, you see the swift progression of *disruptive technologies*,[58] the introduction of new systems that supplement or displace existing methods: ERP, SCM, CRM, PLM, and the WWW . . . an acronym soup of rapid change for the enterprise. On the business climate timeline you see much the same, with economic swings, reengineering, down- and rightsizing, mergers and consolidations, agile business models and alliances, increasing global competition, intermixed with the new Lean standards of performance.

The world does not stand still, and complexity is accelerating. Many will argue that the sure path to lasting competitive advantage is through the skillful use of information technology. Then there are those who will argue otherwise. In his May 2003 *Harvard Business Review* article "Does IT Matter?", Nicholas Carr ignited a firestorm of debate by questioning the real value of IT, placing it in an historical context against the development of other infrastructure technologies. In his book of the same name that followed shortly after, Carr suggests many of us naturally make a basic assumption:

> . . . that as IT's power and ubiquity have increased, so too has its strategic importance. It's a reasonable assumption, even an intuitive one. But it's mistaken. What makes a business resource truly strategic—what gives it the capacity to be the basis for a sustained competitive advantage—is not ubiquity but scarcity. Information technology's very power and presence have begun to transform it from a potentially strategic resource into what economists call a commodity input, a cost of doing business that must be paid by all but provides distinction to none. Distinctiveness is what in the end determines a company's profitability and assures its survival.
>
> Information technology, in fact, is perhaps best understood as the latest in a series of broadly adopted technologies that have reshaped industry over the past two centuries—from the steam engine and the railroad to the telegraph and the telephone to the electric grid and the highway system. For brief periods, as they were being built into the infrastructure of commerce, all these technologies opened opportunities for smart, forward-looking companies to gain real advantages over their competitors. Early in the twentieth century, many large companies created the new management post of "vice president of electricity," an acknowledgement of electrification's transformative role in companies and industries. But as their availability increased and their cost decreased—as they became ubiquitous, they all became commodity inputs. They would often continue for many years to spur broad enhancements in business practices and to lift the productivity of entire industries. But from a strategic standpoint they began to become invisible; they mattered less and less to the competitive fortunes of individual companies.[59]

We are now in an era where IT regularly reinvents itself, always searching for what the software industry calls *the next killer app*—the next big thing. This rapid change will supply countless opportunities for companies to distinguish themselves, at great cost and risk, but only briefly. However, an unbalanced emphasis on an emerging information technology may sap the enterprise's focus and resources, creating an opportunity for a competitor to leapfrog when the lower-cost and more effective solution inevitably appears—the natural pattern of a disruptive technology. What Carr calls the *Technology Replication Cycle*, the cycle time for one disruptive technology to displace another, has shrunk to the point where technologies may come and go like fruit flies, without sufficient life span to achieve lasting advantage or payback. He advocates a fast-follower strategy for many companies, taking advantage of information technology as it matures while not paying the price of too-early adoption. Attempting to base lasting competitive advantage on advancing information technology is like building a structure upon a foundation of quicksand.

It is important to understand that the thoughtful application of any information technology, old or new, may add significant value by standardizing the core value streams and underlying business processes, just as standardized work on the Lean shop floor enables consistent performance. Standardization creates a stable environment that encourages continuous improvement, developing an agile environment capable of rapid change—so, paradoxically, standardization creates flexibility. Standardization of work through IT "best practices" establishes a baseline of performance that can be reliably measured. Standardization eliminates random noise in the environment, creating a manageable environment pervaded with a sense of discipline and confidence. This can free workers from anxieties and heighten their awareness, encouraging them to experiment and take calculated risks, leading to effective continuous improvement. Although standardized work, consistency, and efficiency may be obtained through the prepackaged best practices contained within enterprise software, *best practices themselves may be antithetical to the very idea of competitive advantage.*

While you may or may not agree with Carr's assertions, herein lies what I consider to be his key point. Bold new IT ventures rarely create *lasting* competitive advantage of themselves, because each new information technology has such a short half-life. And the acquisition of IT capability does not automatically confer advantage, because misguided IT investments often do more harm than good. *However, the thoughtful focus of IT on the improvement of people and processes can channel the competencies and creative energies of an organization, eliminating waste while continuously creating new value in the eyes of the customer.* Information technology can enable a company to do what it does, better, faster, cheaper, and to the recurring delight of the customer. Carr concludes, "Distinctive processes lie at the heart of competitive advantage. Success in the future will be less a matter of using information technology creatively than of simply *using it well.*"[60]

This message is especially clear when you look at the actions of Wal-Mart, which has apparently written the book on value creation in the Lean Network. In Michael Schrage's *Technology Review* article "Wal-Mart Trumps Moore's Law," he cites a recent McKinsey Global Institute report, in which MIT Nobel Prize-winning economist Robert Solow analyzes U.S. productivity growth from 1995 to 2000:

> By far the most important factor . . . is Wal-Mart. That was not expected. The technology that went into what Wal-Mart did was not brand new and not especially at the technological frontiers, but when it was combined with the firm's managerial and organizational innovations, the impact was huge. Productivity growth accelerated after 1995 because Wal-Mart's success forced competitors to improve their operations. In 1987, Wal-Mart had just nine percent market share but was 40 percent more productive than its competitors. By the mid-1990s, its share had grown to 27 percent while its productivity advantage widened to 48 percent. Competitors reacted by adopting many of Wal-Mart's innovations, including [. . .] warehouse logistics and purchasing, electronic data interchange and wireless bar code scanning. Consider Wal-Mart's $4 billion-plus investment in its "Retail Link" supply chain system. What's intriguing is not the multibillion-

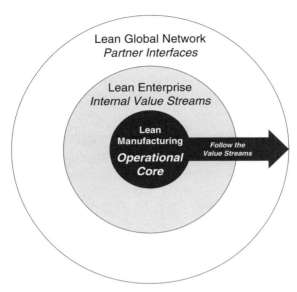

Figure 3-11. The core of Lean Manufacturing

dollar nature of the company's IT infrastructure initiative, but the fact that it has had at least an order-of-magnitude impact on its suppliers' own supply chain innovations. That is, Wal-Mart's own $4 billion expenditure has likely influenced at least $40 billion worth of supplier investments in systems and software. In terms of sheer economic impact, the single most important, dynamic, defining technological innovation in America hasn't been the silicon cliché of Moore's law; it's the relentless promotional promise of "everyday low prices." Microsoft and Cisco may set technical standards; Wal-Mart sets business process standards. Corporate IT departments may "care" about the latest Windows upgrade or faster microprocessor from Intel. But Wal-Mart's ongoing infrastructure innovation is what inspires their investments, actions, and fears. The result has been a genuine revolution in economic productivity. This revolution also reinforces a profound truth about the economics of innovation: *implementation matters far more than invention.*[61]

Focusing IT Toward Lean *Manufacturing**

The three stages of Lean evolution are interdependent. Although ultimately they should improve in unison, a Lean Manufacturer simply cannot become an effective Lean Enterprise, or a partner within a successful Lean Network, without first cleaning up its own nest—or else these extended efforts may amplify the internal waste. So a logical place to begin is by improving the core Lean Manufacturing operations, improving the value streams outward through the enterprise and supply chain as illustrated in Figure 3-11.

---

* For the reader who does not wish to follow this deep dive into Lean Manufacturing planning and execution, I encourage you to skip chapters 4 and 5 and go directly to Part 2, where we continue the exploration of IT in support of the holistic Lean Enterprise.

	Automate	**Inform**
Lean Operations	Production and Service Operations	Transformation, Planning, and Execution Decisions
Lean Enterprise	Internal Value Streams	Feedback Key Performance Measures to Executives, Manager, and Teams
Lean Global Network	Collaboration and Transaction Activity	Share Information with Global Network Partners to Reduce Friction

Figure 3-12. Automate and inform the Lean constituencies

In the same way, the improvement of information flows may best be started at the core, radiating outward as the three stages of Lean mature. As suggested in the introductory chapter, the two primary roles of an enterprise information system are to automate and inform; within the three stages of Lean these roles are shown in Figure 3-12.

The advancement of Lean Manufacturing practices beyond their repetitive origins, toward low-volume and high-mix operations, is confronting practitioners with new challenges. In these new environments demand is volatile, product and process mix is variable, knowledge content is high, and there are many constraints to be skillfully managed.

Because of the high level of interest, the enterprise software industry has begun to deliver legitimate Lean Manufacturing solutions for both repetitive and nonrepetitive environments. With the emergence of these new tools, perhaps the greatest challenge for practitioners is that no single approach applies to all types of manufacturing operations. Mixed-mode* manufacturing is now commonplace, where several types of manufacturing coexist within the same enterprise, sharing planning, production, and distribution resources. The full continuum of mixed-mode manufacturing is illustrated in Figure 3-13. As many names are commonly used to describe these various environments, in this book we will describe them on a continuum between *repetitive* and *discontinuous* operations.

To help *all* types of manufacturing organizations on their journey to Lean, and to describe specifically where and how IT may add value in each type of Lean environment, we must clearly define the variables and how they differ among these production environments. We must look *deeply* inside the core planning and execution processes of these different Lean Manufacturing environments, for that is where the Lean rubber meets the road. It is within these core operations that the familiar conflicts between Lean and IT are focused, the *Planning* and *Execution* processes that are *traditionally* managed by the MRP II system:

---

* A *mixed-mode* manufacturer has several types (modes) of manufacturing operations within a single plant, company, or enterprise. A *mixed-model* manufacturer produces a variable product mix (several models within the same product family), often using cellular production and level schedule.

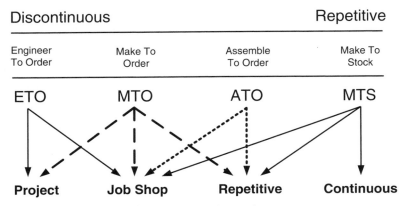

Figure 3-13. The continuum of manufacturing operations

- MRP II is concerned with the planning and scheduling of material and resource requirements.
- Lean Manufacturing is focused on the control of shop floor execution, where material flow is regulated by demand pull.

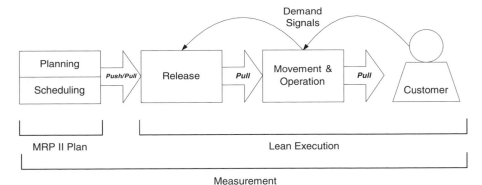

Figure 3-14. Lean Planning and Execution model

There are five basic stages in the planning and execution of work, illustrated in Figure 3-14:

1. **Planning**—anticipating demand and determining the material and capacity requirements to satisfy it
2. **Scheduling**—the prioritization and timing of work, determination of proper batch sizes and job sequencing, and allocation of particular resources

3. **Release**—the timely dispatch of work to the shop floor
4. **Movement and Operation**—the control of resource movement and production operations on the shop floor
5. **Measurement**—the capture of material consumption, resource utilization, quality, and process information, in order to monitor and control processes, measure performance, and guide continuous improvement efforts

In Chapters 4 and 5 we will explore each of these elements carefully, to identify where IT may add value and eliminate waste in a Lean Manufacturing operation. In particular, we will examine the role of scheduling, for it is here that the interaction between Lean and MRP II must be carefully defined, based on the flow and pull characteristics of each particular manufacturing environment along the continuum from repetitive to discontinuous.

To examine the mechanics of scheduling, we must first understand the logic of traditional planning and control systems. Most Lean Manufacturing educational resources in circulation today—books, articles, and workshops—begin with Lean methods. As a result many individuals have been introduced to Lean Manufacturing without a thorough understanding of the basics of inventory, production, and supply chain operations management. This lack of fundamentals may lead to misuse of ERP software, because their MRP II subsystems were originally designed according to traditional theory and practice. Some enterprise software systems have recently been extended to support various Lean techniques, while in other cases stand-alone Lean planning and scheduling applications have been integrated with their ERP hosts. In either case, these new Lean tools coexist with traditional planning and control functions, so practitioners must understand the original assumptions and limitations that form the MRP II bedrock, in order to leverage these tools to enhance Lean performance.

To this end Chapter 4 offers a brief explanation of *traditional* Production and Inventory Management theory and practice, providing the necessary foundation for the understanding of MRP II software capabilities. Chapter 5 then offers a thorough examination of the role of *Lean* planning, scheduling, and execution across the entire continuum of manufacturing operations.

Chapter 4

Fundamentals of Production and Inventory Management

To explain how an ERP system can effectively support Lean Manufacturing operations we must first understand the fundamental design assumptions underlying MRP II, the operational planning, scheduling, and execution component of ERP. This chapter will therefore briefly explain the basic theory and practice of Production and Inventory Management, as described by the Association for Operations Management (APICS):

1. The Product/Process Continuum
2. Inventory Management Basics
3. Bill of Materials (BOM)
4. Material Requirements Planning (MRP)
5. Sales and Operations Planning (S&OP)
6. Master Production Scheduling (MPS)
7. Capacity Planning
8. The Integrated Planning Process
9. The Lean Transformation

THE PRODUCT/PROCESS CONTINUUM

We begin by examining the continuum of manufacturing operations shown in Figure 4-01. At one end, each product is unique and built individually by crafts-

Lean Enterprise Systems: Using IT for Continuous Improvement, by Steve Bell
Copyright © 2006 by John Wiley & Sons, Inc.

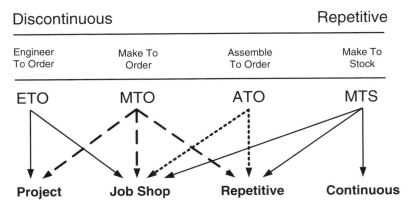

Figure 4-01. The continuum of manufacturing operations

men, often by hand with general purpose equipment. At the other end is mass production, in some cases literally a pipeline or conveyor system where products flow continuously and without variation.

Of course, there is plenty of middle ground within this continuum, and this middle ground is where most manufacturers operate, often with products and processes at several positions simultaneously. At each position along this continuum, different demand, supply, and production patterns occur, so it is important for a company to understand the dynamics of their positioning and where their resulting core competency and competitive advantage lies. There are several generally accepted descriptions for these positions along this continuum:

- **Engineer to Order (ETO)**—Products whose customer specifications require unique engineering design, significant customization, or new purchased materials. Each customer order results in a unique set of part numbers, bills of material, and routings.
- **Make to Order (MTO)**—A production environment in which a product is made after receipt of a customer's order. The final product is usually a combination of standard and custom-designed items to meet the special needs of the customer.
- **Assemble to Order (ATO)**—A production environment in which a good or service can be assembled after receipt of a customer's order. The key components (bulk, semifinished, intermediate, subassembly, fabricated, purchased, packaging, and so on) used in the assembly or finishing process are planned and usually stocked in anticipation of a customer order. Receipt of an order initiates assembly of the specially configured product. This strategy is useful where a large number of end products (based on the selection of options and accessories) can be assembled from common components. This is also called Configure to Order (CTO).

• **Make to Stock (MTS)**—A production environment where products can be and usually are finished before receipt of a customer order. Customer orders are typically filled from existing stocks, and production orders are used to replenish those stocks.[62]

In 1979 Harvard Business School professors Robert Hayes and Steven Wheelwright, published two articles[63,64] describing a pair of continuums: a *Product Life Cycle* based on the marketing and maturation of a product and a *Process Life Cycle* describing the methods of its production. They illustrated the progression of these life cycles and described the positioning and dynamics of various companies, products, and strategies. We will begin by exploring the Process and Product Life Cycles they described separately, and then we'll put them together and draw some important conclusions.

The Process Life Cycle

At one end of the Process Life Cycle is a *Discontinuous Flow* operation with highly customized and often unique production methods, which Hayes and Wheelwright describe as *a fluid process that is highly flexible but not very cost efficient*. Moving toward *Continuous Flow* involves an increase of standardization, mechanization, and automation, requiring greater capital investment and resulting in a higher economic production volume and break-even point combined with reduced flexibility.

As a company moves along the Process Life Cycle toward continuous flow we may be inclined to use the term *maturation*, which implies that one point on this continuum is more advanced than another. This is not the case, and the authors emphasize that each location actually represents a strategic position chosen by a company for a particular product. As a company moves along this continuum, certain management challenges and best practices are indicated in Figure 4-02.

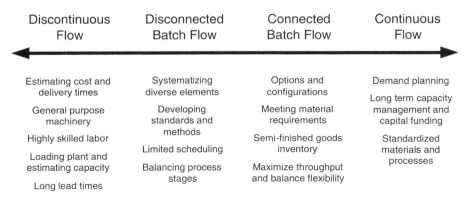

Figure 4-02. The Process Life Cycle

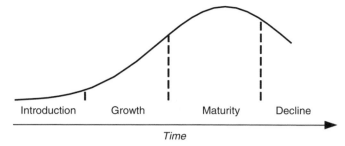

Figure 4-03. The Traditional Product Life Cycle

The Product Life Cycle

Products generally progress through a market life cycle from the initial release through growth, maturity, and decline, traditionally illustrated as in Figure 4-03.

Robert E. Cannon describes these traditional stages in his *Tutorial on Product Life Cycle*:

> **Introduction stage** has recently been termed the product development process, which begins with idea generation and input gathered from customers, users, market research, outside inventors, competitors, other markets, and employees. An idea that survives preliminary evaluation will be passed along for a technical and market evaluation, prototyping and marketing planning, and finally commercialization where production is started and the product rollout begins.
>
> **Growth stage** is where the rising tide of consumer interest lifts the boats of all participants. Costs are declining with increasing volumes and profits are improving, so competitors are attracted to enter the market.
>
> **Maturity stage** occurs when the market has become saturated with commodity products and price competition. Market share becomes the primary focus, and many attempt to differentiate the product, looking for new markets, new applications, more models, and other ways to increase usage or diversity.
>
> **Decline stage** is recognized by the downturn in demand [or oversupply], and may be hastened by the introduction of an innovative new product or changing consumer tastes. There are many appropriate strategies when a product is declining: finding new uses, finding new markets, product variations, extending technology, re-packaging, re-branding, finding avenues for increasing consumption, re-positioning, co-branding, and pricing.[65]

Some products mature slowly over years, whereas others such as consumer electronics may have a life cycle of less than six months. In general, product life cycles have shortened in recent years, one of the key disruptive pressures that manufacturers face today. At the beginning of the Product Life Cycle are emerging products with high variability, often made individually to customer requirements. As a market matures, it may develop into several distinct seg-

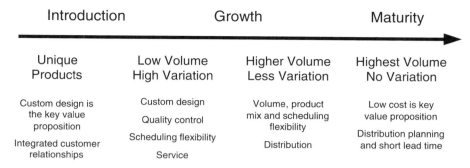

Introduction	Growth		Maturity
Unique Products	Low Volume High Variation	Higher Volume Less Variation	Highest Volume No Variation
Custom design is the key value proposition	Custom design	Volume, product mix and scheduling flexibility	Low cost is key value proposition
	Quality control		Distribution planning and short lead time
Integrated customer relationships	Scheduling flexibility	Distribution	
	Service		

Figure 4-04. The product strategy life cycle

ments, each with its own distribution channels, customer preferences, and pricing structures. As a product matures along this life cycle the tendency is toward a commodity—to increase the volume and limit the variations, so that large volumes can be produced and delivered to the marketplace at a lower cost, and certain competitive strategies are appropriate as illustrated in Figure 4-04.

The Combined Product/Process Life Cycle

When the process and product life cycles are combined, a diagonal line naturally emerges between the two continuums, illustrated in Figure 4-05. The region covered by this diagonal is the *sweet spot* most manufacturing companies pursue, and it is consistent with the ETO/MTO/ATO/MTS continuum described earlier. Two companies may compete with a similar product while occupying distinct positioning along the diagonal—one emphasizing customer choice, the other offering lower cost and faster delivery. A company also may make a strategic decision to move along the diagonal, reacting to (or creating) market pressures.

Moving along the diagonal, however, is not a smooth progression, but rather a series of steps along the product and process life cycles that can be made independently, and thus these steps can be out of synch. A company may manufacture a product that has moved along the market maturity life cycle toward a commodity, but by retaining its traditional Make to Order processes a company positions itself *above the diagonal* and is therefore unable to compete on cost alone. A company may choose such a distinct competitive position above or below the diagonal sweet spot deliberately, but success far off the diagonal must be a clear strategic decision (or a huge blunder) and is rare and difficult to maintain—a *no man's land*. For example, Rolls-Royce individually builds automobiles as a custom process that is not matched with the product maturation of the automobile industry, and their market is extremely limited by the cost/volume factor. Companies can similarly be positioned below the

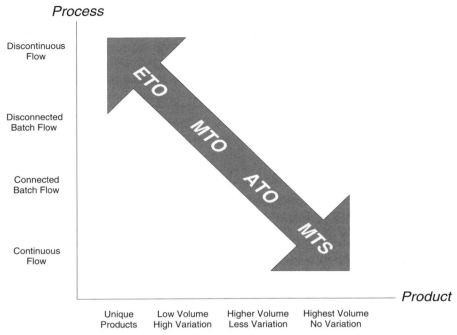

Figure 4-05. Combined product/process diagonal

diagonal when they attempt to standardize and automate a process when the market isn't ready for a standardized product—Hayes and Wheelwright suggest the troubles in the manufactured home industry during the 1970s, when manufacturers invested heavily in mass production techniques as consumers were clamoring for individual touches. These dangerous no man's land regions are illustrated in Figure 4-06.

Movement along the Product Life Cycle alone offers numerous opportunities for cost reduction through product redesign, changes in the distribution channel, and other appropriate marketing strategies. Likewise, movement solely along the Process Life Cycle offers many opportunities for cost reduction through economies of scale and process improvement. However, Hayes and Wheelwright suggest that it is very difficult to move smoothly along the diagonal, matching corresponding movements across both product and process axes. A failure to coordinate movements on these two fronts, however, may result in accidental positioning above or below the diagonal, resulting in a competitive disconnect. The authors termed a smooth diagonal movement the *learning curve*—which requires skillful coordination of production and marketing strategies.

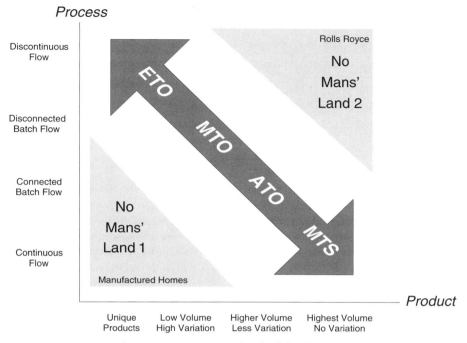

Figure 4-06. No man's land off the diagonal

Market Dynamics Along the Diagonal

When a company occupies a specific position on the diagonal and the market suddenly shifts, what to do? If they're up the diagonal toward custom production, clinging to their inefficient product and process too long means cost erosion. And if a company moves down the diagonal prematurely, sales volume does not materialize to justify the capital investment.

If the market moved down the diagonal in a consistent manner over time, making appropriate strategic decisions might be relatively straightforward—but the market is not so predictable. For example, General Motors achieved a significant competitive position in the early days against Ford Motor Company, when they introduced customer design and color choices against Ford's "any color as long as it's black" strategy based solely on high-volume and low-cost production. So product life cycle maturation isn't a relentless move down the diagonal, and visionary companies can reverse course against their competitors by segmenting the market and creating variety in a standard product, moving up and back along the diagonal.

In fact, the cleverest companies compete against themselves. Two prime examples, Intel and Microsoft, continuously announce and introduce new products, replacing their current ones while they are still in the maturation phase. Early announcement often causes the market to stop investing in the competition's early lead with product innovation, until the giant introduces a similar product sometime later. This is a deliberate strategy for a large company to use their strong market position and financial depth to direct the market, while the competition expends all their resources just keeping up. A dominant market leader can afford the cost of directing the market and technology, and can absorb the enormous costs of R&D and premature product obsolescence as the price of staying on top—an example of an eight hundred pound gorilla behaving as a nimble and aggressive chimp.

The Four Vs

The *traditional* assumption with the combined Product/Process Life Cycle is that a company must choose between custom and mass production, that for each product and process they should deliberately occupy a distinct position on the diagonal. Although it is true that an enterprise may have different products and processes positioned at several locations on the diagonal, the trade-offs for any particular product/process positioning decision must be considered carefully. If they move from customization too late, they suffer cost and market erosion. If they invest heavily in capital equipment and move to mass production too early, they can lose their shirts to more agile competitors and changing market preferences. Therefore a company must develop a balance among the four Vs:

- **Volume**—Is there an economy of scale in production and supply chain operations we may leverage?
- **Variation**—Is the primary value proposition lowest cost or design flexibility? Can the product be segmented and configured within family groups, or is every unit a unique one-off design driven by complex technical specifications? How quickly are market preferences expected to change? Can we direct them, or do they control us?
- **Velocity**—How quickly does the market expect delivery? Is this a standard product line with inventory waiting on a store shelf or warehouse, or is the customer willing to wait a reasonable time for just the right configuration, or for an entirely custom product?
- **Value Proposition**—The king of the Vs: If we're not able to produce the right mix of the first three Vs in alignment with the customer's perception of value, then it is not a viable strategy.

The search for balance among Volume, Variation, and Velocity, to find the right Value Proposition, seemed to require a fundamental trade-off. It seemed to be a law of nature, an imperative that could not be violated, and which has directed the course of the global economy since the industrial revolution.

And then Toyota demonstrated that these were not inviolable laws. In their seminal book *The Machine That Changed the World* based on the Massachusetts Institute of Technology's five-year study of the automotive industry during the 1980s, authors Womack and Jones coined the term *Lean Manufacturing*. Following Toyota's pioneering efforts, manufacturing enterprises worldwide have since been striving for an agile balance among the variables of Volume, Variability, Velocity, and Value.

So what happened to the evolution of enterprise software during this revolutionary period? Software for manufacturing companies emerged in the 1960s and picked up speed through the 1980s—when traditional assumptions guided the design and development of today's most popular enterprise software applications. So before we understand how software can enable this new Lean approach to manufacturing, we must first understand the traditional models, theory, and practice, around which most MRP II and ERP software was originally designed.

INVENTORY MANAGEMENT BASICS

The traditional trade-off has always balanced inventory investment against customer service levels. Too much inventory costs us money, too little inventory costs us customers—period.

The sales department would like everything available in unlimited quantities all the time so they never miss a sale. The finance department wants as little as possible in stock to preserve working capital. And the production department is found in the middle, constantly juggling the two conflicting priorities—needing to maintain sufficient inventory to meet the company's objectives while being held accountable for efficiency and resource utilization. From this natural tug-of-war have emerged many sophisticated approaches to forecasting and managing inventory levels to optimize the balance between inventory and customer service.

Inventory Policy Fundamentals

In general we must identify and buffer the sources of variability in demand and supply—managing lead times, quality problems, scrap and yield losses, batch sizes, supplier delivery and production schedules, as well as countless other contributing factors. The natural first step is to develop *safety stock inventory*, which accounts for four key factors:

- **Demand Variability**—How stable is demand for the product either for customers or for internal use?
- **Frequency and Quantity of Reorder**—How much and how often should we reorder, based on policies and characteristics of the suppliers, products, shipment costs, storage space, and other factors?

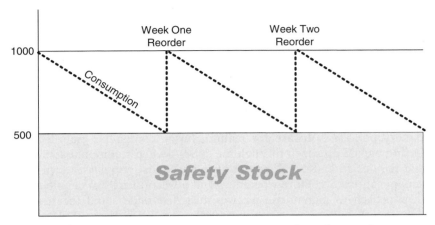

Figure 4-07. Inventory movement pattern using safety stock

- **Replenishment Lead Time**—How long does it take to get the product from our supplier or to make it ourselves?
- **Desired Service Levels**—How quickly do customers expect the product, what percentage of the time must we have it available within that expected delivery time, and how costly are stock-outs measured by lost sales or customers?

Here is an example of a safety stock calculation using a simple *Order Point Replenishment Method*. If we determine that usage of a particular item is 100 units per day, that the replenishment lead time is consistently 5 days, and that we always want 5 days' inventory just in case, then our safety stock must be 500 (5 days × 100 per day). If we reorder once per week, then we must order an additional 500 to cover our weekly requirements. This means that our target stock at the beginning of the week should be 1000, with 500 remaining at the end of the week. This example is illustrated in Figure 4-07.

A common variation of an order point system is the two-bin system,* which is described by the APICS Dictionary as:

A type of fixed-order system in which inventory is carried in two bins. A replenishment quantity is ordered when the first bin (working) is empty. During the replenishment lead time, material is used from the second bin. When the material is received, the second bin (which contains a quantity to cover demand during lead time plus some safety stock) is refilled and the excess is put into the working bin. At this time, stock is drawn from the first bin until it is again exhausted. This term is also used loosely to describe any fixed-order system even when physical "bins" do not exist.[66]

* As you will see in Chapter 5, the two-bin order point is the basis for a product-specific kanban pull replenishment system.

All order point systems work under the same assumption—estimate demand based on some sort of forecast and buffer anticipated demand variability with an appropriate amount of excess inventory. Order point systems can be deceptively simple and easy to manage, and often they do not require a computer. A bucket of bolts with a red line marked halfway down the container is an order point system with one simple rule: When the red line is visible, order another bucket. A silo filled with powder or liquid can be managed the same way—visual order points are very effective as long as the safety stock level is set just right—not too much and not too little. So how do we achieve that balance? By assessing the pattern and variability of demand and lead time, determining the cost of carrying inventory to buffer demand variations, evaluating the potential cost of a stock-out, and thus determining the desired service level we wish to maintain. This tells us where to draw, and occasionally redraw, the red line.

Service levels are best described with simple statistics. If you're like many people, you have the sudden urge to head for the refrigerator or to some other distraction the moment someone says "statistics." Don't worry, I'll keep this simple and I promise it won't hurt a bit—no formulas or calculations are needed for you to understand what's important here. If we investigate the usage of a part over a period of one year, we may develop a chart like the one shown in Figure 4-08. The *Units Consumed* column shows the quantity of inventory used each day, and the *Number of Days* column shows how many days within the year that quantity was consumed. For example, at one extreme 40 units were consumed in each of 30 separate days in the year, whereas at the other extreme 200 units were consumed in each of 25 separate days in the year—quite a significant demand variation. When we plot these figures on an

Number of Days	Units Consumed per Day
30	40
40	60
60	80
80	100
60	140
40	160
30	180
25	200
365	**41600**

365 days x # units per day = 41,600

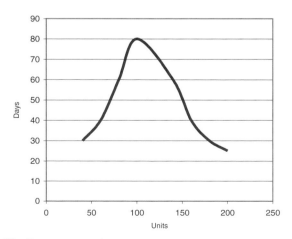

Figure 4-08. Part consumption pattern

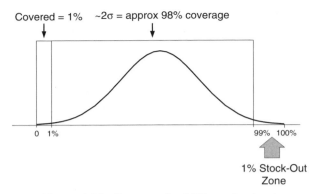

Figure 4-09. Coverage for 99% service level

X-Y axis, we develop a *distribution curve*, illustrated in the diagram. This provides us with a visual and intuitive presentation of the variability of demand for this product. For example, note how demand tails off approaching 200. Most of our demand (the median) is around 100, but 7% of the time (25 of 365 days) demand spikes to 200, and if we want to entirely prevent stock-outs then we must be prepared for this spike, which means carrying much more inventory than we need most of the time.

Using the past to predict the future is always a tricky proposition, and forecasts by their nature are guaranteed to be incorrect. However, by using a statistical technique called *standard deviation** we can determine with reasonable certainty how much safety stock is required to maintain a service level of 95% (resulting in stock-outs 5% of the time), or 98%, 99%, and so on. Generally there is a decreasing marginal return as we approach 100% because we're trying to cover the remote possibility of an extreme but infrequent demand spike. The symbol for standard deviation is the Greek sigma (σ). Note that in Figure 4-09 we use approximately 2 standard deviations that result in covering 98% of demand, the vast majority of the normal distribution of the bell curve: we may refer to this as *Two Sigma*,[†] shown symbolically as **2σ**. Note that to the far left there is a region not covered by the standard deviation, where demand is *less* than the distribution—although this is outside the stan-

---

* In case you're a glutton for punishment, here's how standard deviation works, but don't worry, there are many software programs (including spreadsheets) that can do this for you. Standard deviation is a measure of the dispersion of a population of data. It is computed by calculating the difference (+/–) between the average and each actual observation, squaring each difference, summing the squared differences, dividing the sum by the number of squared differences to determine an average, and finally taking the square root of the average. This technique smoothes the averages and weights the large variations more heavily than the small ones.

[†] Two Sigma (**2σ**) or two standard deviations is generally a reasonable target for variation when planning safety stock. To attempt to plan for safety stock beyond **2σ** means that under ordinary circumstances you're carrying too much inventory, unless the cost of a stock-out is extraordinarily high.

dard deviation range it is covered because we have enough inventory. It is only the region on the far right, outside the standard deviation boundary, where we risk a stock-out, and in this example that is only 1% of the time.

In this way, safety stock levels are set in relationship to the variability of demand as represented by a demand distribution curve for each product. Sophisticated companies often use computers that are carefully monitoring activity and forecasting demand, whereas others use experience and rules of thumb. Whether software or intuition is used, the underlying logic for order point planning remains the same.

Demand Patterns

In a simple environment, safety stock, reorder lead times, and expected demand remain relatively stable and evenly distributed like the smooth bell curve illustrated earlier. However, many companies must deal with demand that is trending, seasonal, or otherwise unstable, and an unchanging safety stock value will not do. We will illustrate two examples that take the same demand distribution that resulted in a nicely bell-shaped curve, and distribute them differently over time.

The first example in Figure 4-10 shows a trend where the daily demand starts low and ends high. There is clearly demand growth occurring, which may require a management plan to boost production. However, safety stock may not require increase, if such a stable and predictable demand growth pattern is expected to continue. Remember that safety stock buffers the *uncertainty* of demand, not the absolute amount of the demand.

Now observe what happens in Figure 4-11 when we reorganize the daily amounts but do not change the total demand, causing the demand pattern to represent a seasonal spike in July.

Jan	Feb	Mar	Apr	May	Jun	Jul	Aug	Sep	Oct	Nov	Dec	Total
1260	1740	2280	2400	2920	3000	3220	4340	4280	4960	5300	5900	41600

Figure 4-10. Growth demand curve

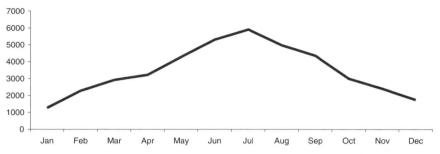

Jan	Feb	Mar	Apr	May	Jun	Jul	Aug	Sep	Oct	Nov	Dec	Total
1260	2280	2920	3220	4280	5300	5900	4960	4340	3000	2400	1740	41600

Figure 4-11. Demand spike curve

Note that in all three examples we have illustrated the quantities of the demand data per day are exactly the same; they have just been redistributed to show different trends. Figure 4-08 shows the distribution unrelated to *sequential* time. Figure 4-10 shows these amounts growing steadily, whereas Figure 4-11 indicates seasonality—or perhaps the sudden increase in popularity of a product followed by a precipitous decline. The point here is that the drivers and trends behind the data must be understood before appropriate inventory levels and replenishment policies may be determined.

Consider a situation where there is a seasonal demand, such as suntan lotion and bathing suits in the spring,* a company may adopt one of two basic approaches: 1) level production or 2) chase strategy.

Level production assumes that the production quantity and the resources (including staffing) required will remain stable throughout the year. This suggests that the company will build inventory ahead during slow periods, called *seasonal safety stock*, and deplete this inventory during peak demand periods. The cost of carrying excess inventory is presumed to be offset by lower costs of level production and stable employment.

A *chase strategy* matches periodic demand with capacity—when demand goes up, production goes up. When demand goes down, so does production. If demand spikes are predictable, the company can plan for excess internal capacity that is used elsewhere during the remainder of the year. If demand spikes are unpredictable (which they often are in even a seasonal business, where fashion manufacturers cannot anticipate with certainty which models, styles, and colors will be popular), safety capacity may be planned with additional plant and equipment that is available on short notice, extra shifts, overtime, and rapid outsource capability. In a situation in which demand quantity may be stable but *product mix* is variable, a chase strategy is preferred, because

---

* Bathing suits and suntan lotion in spring? Of course that's not when they're consumed, but when they may be planned, produced, and delivered into the early stages of the supply chain.

there is a risk of building the wrong model or configuration of inventory ahead of time.

Independent vs. Dependent Demand

Safety stock requires forecasted demand, and this is appropriate for finished goods that are sold to customers. But what about the component products, the raw materials and subassemblies that are used to manufacture those finished parts? We shouldn't have to forecast components, if their demand is *dependent* upon the demand for their parent-level finished parts.

By calculating material requirements for dependent items based up the forecasted demand of the parent items, the overall amount of forecasting and demand planning is considerably reduced. Dependent demand for components is *derived* from the demand of the parent items and should be calculated from the forecast at the parent-item level. To calculate demand of dependent items we need a product structure that defines the relationship between parent and child items. This product structure is called the Bill of Materials (BOM). It is also called a *recipe* or *formula* in process industries where liquids and powders are mixed, such as foods, pharmaceuticals, and chemicals.

BILL OF MATERIALS (BOM)

To explain the BOM we'll use the simple example of a computer system with three components: the computer, the monitor, and the keyboard. Of course, each of these major components may have many subcomponents; for example, the CPU may be composed of a case, motherboard, disk drive(s), power supply, screws, cables, etc. There may be hundreds of items in a single computer that are assembled in stages (subassemblies) and appear on the BOM as hierarchical or indented levels, as illustrated in Figure 4-12.

It's important to remember that the only demand that should be forecast is that for the finished product to be sold. All component inventory stocking, purchasing, and production decisions should be based on dependence on the finished product. Why can't we manage components independently with the order point method described earlier? Let's consider what happens if we try.

We'll start with our simple example of a computer that has only three components. If we set our stocking policy for all component items to a 99% service level, we will ensure that we'll only have a production stock-out 1% of the time, right? *Wrong!* The probability of a stock-out with multiple components is the result of the multiplication of their individual probabilities; this is shown in Figure 4-13. In this case with three components each with a stocking level of 99%, the probability of a stock-out on final assembly is 97% ($.99 \times .99 \times .99 = .9702$). In other words, we have a 3% chance of having a stock-out for the assembly of the parent item when each component has only a 1% probability.

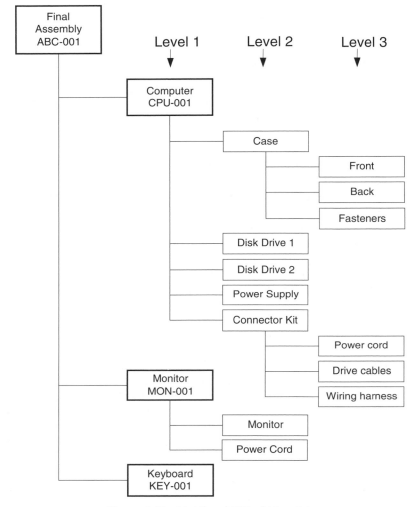

Figure 4-12. Multilevel Bill of Materials

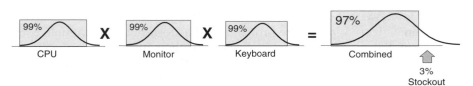

Figure 4-13. Service level of three items at 99% each

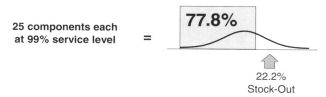

25 components each
at 99% service level   =   77.8%

22.2%
Stock-Out

Figure 4-14. Service level of twenty-five items at 99% each

This stock-out probability *amplifies* rapidly as more components are added to the BOM. Each time a nicely formed demand distribution bell curve is added, the resulting compound demand distribution curve flattens as more variances appear outside the stocking range. It is not uncommon for a BOM to have hundreds or even thousands of individual components—consider an automobile or airplane. Figure 4-14 illustrates an example where there are *just 25 components*, and if each component is managed independently to a 99% service level, the resulting service level is 77.8%—a stock-out can be expected 22.2% of the time!

This clearly demonstrates the disastrous paradox of managing dependent items independently based on an order point method. The amount of inventory carried for each individual component item is unnecessarily high, because safety stock is set to cover independent demand 99% of the time. We hold so much excess component inventory because we're attempting to cover the infrequent and unpredictable demand spikes. We may call this *just in case* rather than just in time inventory. At the same time, the number of stock-outs during final assembly increases dramatically, because demand variations are amplified by the number of components in the BOM. Managing dependent inventory independently simply does not work; this is why Material Requirements Planning is necessary.

MATERIAL REQUIREMENTS PLANNING (MRP)

The APICS Dictionary defines Material Requirements Planning as:

> A set of techniques that uses bill of material data, inventory data, and the Master Production Schedule (MPS) to calculate requirements for materials. It makes recommendations to release replenishment orders for material. Further, because it is time phased, it makes recommendations to reschedule open orders when due dates and need dates are not in phase. Time-phased MRP begins with the items listed on the MPS and determines (1) the quantity of all components and materials required to fabricate those items and (2) the date that the components and material are required. Time-phased MRP is accomplished by exploding the bill of material, adjusting for inventory quantities on hand or on order, and off-setting the net requirements by the appropriate lead times.[67]

Simply put, MRP tells us how much of each item to order (purchase order) and produce (work order) to meet our delivery dates. MRP also alerts the planner when there is a problem needing attention.

How MRP Works

The MRP calculation begins with demand for finished parts, and *time phases* these requirements into periods—usually days or weeks—by the due date of each order. These are called the *gross requirements* for the parent items. MRP then *nets out* (deducts from the gross requirement) finished parts in stock or already scheduled for production within the appropriate time periods. The resulting *net requirements* for finished parts in each time period are then *exploded* down to the first level of components on the BOM, and MRP then calculates time-phased requirements for each component at that level. It then nets out components already in stock, on open purchase or production orders, resulting in component net requirements at that level of the BOM for each period. MRP continues down the BOM for all subsequent levels, performing the same *gross to net calculation* at each level.

When it reaches the bottom level of the BOM for this part, the MRP calculation proceeds to the next parent item and repeats the entire process. When MRP is done with all the parts at all BOM levels, it totals all net requirements for purchase and production and then time phases each part requirement *based on the standard purchase or production lead time value stored in each item file*. This means that if a particular part has a standard 10-day lead time, the purchase or production order is issued 10 days ahead of when it is needed. MRP then suggests the appropriate actions to the planner, who may turn these suggestions into actual purchase orders that are sent to suppliers with expected receipt dates and production work orders that are released to the shop floor at the appropriate time to meet order delivery dates.

The APICS Dictionary describes several types of orders used by the MRP engine:

- **Open Orders**—A released production or purchase order. Using the standard lead time, MRP calculates a scheduled receipt date.
- **Planned Orders**—A suggested order quantity, release date, and due date created by the planning system's logic when it encounters net requirements. MRP suggests a planned order release date and a planned order receipt date based on the standard lead time of each item.
- **Firm Planned Orders**—A planned order that can be frozen in quantity and time. The computer is not allowed to change it automatically; this is the responsibility of the planner in charge of the item that is being planned.[68]

Note the MRP calculation illustrated in Figure 4-15: Gross requirements represent total demand per week, quantity 0 is available in Week 2, and there's

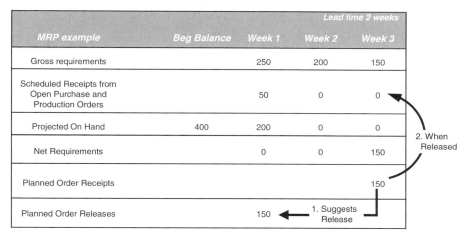

MRP example	Beg Balance	Week 1	Week 2	Week 3
				Lead time 2 weeks
Gross requirements		250	200	150
Scheduled Receipts from Open Purchase and Production Orders		50	0	0
Projected On Hand	400	200	0	0
Net Requirements		0	0	150
Planned Order Receipts				150
Planned Order Releases		150		

2. When Released

1. Suggests Release

Figure 4-15. Material Requirements Planning

a net requirement of 150 in Week 3. A Planned Order Receipt of 150 is created in Week 3, and with a lead time of 2 weeks there is a corresponding Planned Order Release of 150 in Week 1. Once the Planned Order is released by the planner, it becomes an Open Order with a Scheduled Receipt.

MRP can manipulate a staggering volume of transactions with ease, relieving the burden on the human planner by enabling management by exception. One of the key contributions of an MRP system is the capability of generating automatic notifications, which tell the planner what needs attention and when. There are two basic types of notifications: *exception* and *action* messages:

- An exception message indicates where things are not working according to plan.
- An action message advises the planner on the required action to be taken. A typical message generated would be "orders to be released." This means for a manufactured item, the manufacturing order and work pack needs to be produced and sent to the stores for picking so the manufacturing process can commence. For a purchase order, the releasing of an order would entail the planner passing the requirement to the buyer, who contacts the supplier.[69]

You can imagine that with hundreds or thousands of finished parts, each with several BOM levels, and with a moderate transaction volume, MRP can be extremely tedious and calculation intensive. For this reason, not only does MRP require a computer but it also requires a substantial degree of record accuracy to be effective. It is generally accepted that inventory record accuracy must be at least 95%, and BOM accuracy at least 99%, for MRP results to be reliable.

If inventory and BOM records aren't sufficiently accurate then planners and inventory handlers will spend most of their time trying to figure out why the balances aren't correct, resulting in shortages and excesses that manifest as unreliable production schedules and missed delivery dates. When this happens, people often respond by ignoring the results of MRP and reverting to their old methods of planning and purchasing, causing a further decrease in record accuracy, rendering the planning system ultimately a useless investment.

To avoid this failure, the first initiative companies should take when implementing MRP is to focus on inventory record accuracy, which usually leads to *cycle counting* as a method for identifying and eliminating the root causes of inventory record accuracy variances.

By its nature, periodic (monthly or annual) physical inventory counting can cause as many inventory record accuracy problems as it solves. Let's be realistic, it's difficult to imagine that the confusion and disruption caused by the complete shutdown of a plant results in an accurate count. And although the typical monthly physical inventory fiasco may help to identify variances, it is often useless to identify the root causes for those variances, because they're masked by the sheer quantity of inventory that's counted, the volume of transactions that have occurred since the last count, and by the passage of time and loss of memory.

Cycle counting, on the other hand, suggests counting inventory on a regular basis to identify the cause of a variance soon after its occurrence. For example, high-value and fast-moving items, or those more prone to variance, may be cycle counted more frequently. An ideal time to cycle count is immediately after a series of transactions for a particular part, or when replenishment has just been received. Another good time is when a bin quantity is very low—making it both easier to count and more important to identify variances when there is little quantity left to buffer a shortage. Unlike periodic physical inventories, cycle counting is an effective continuous improvement technique because its goal is not just to identify and correct inventory variances but to identify and eliminate their causes.

Safety Stock and MRP

Despite what we've just said about not managing dependent demand independently, there may be dependent items that just aren't appropriate for MRP. For example, we may have a silo of powder or liquid that is replenished by a fixed-quantity truckload—the order for replenishment is based on the visual level of the product, and the truck arrives within a few hours of receiving the replenishment order. Similarly, we may have a bucket of bolts, a vat of adhesives, or a box of rags that is consumed regularly during production, and is of such low value that it doesn't warrant the effort or cost of planning. In cases where lead time is short, reorder quantity is fixed, the relative value of the inventory is low, and the quantity of product can be managed visually, MRP may be unnecessary and safety stock may be managed by an order point method.

Variability of demand and supply will always exist to some degree, so finished parts safety stock (also known a finished goods supermarket) may be appropriate to buffer variable demand. However when holding a finished goods safety stock it is generally unnecessary to also hold safety stock of dependent (component) items for a simple reason—when safety stock is held at the finished parts level it contains within itself additional safety stock quantities of the component items. To also maintain safety stock at the component or semifinished level is therefore redundant and creates excess inventory.*

An alternative approach to finished goods safety stock is to hold inventory with safety stock at the component and semifinished level (also known as a final assembly supermarket). In an environment where many finished items are produced from common components, this results in reduced overall inventory; this approach is variously known as Make to Order, Assemble to Order, Configure to Order, or postponement.

MRP and the Product/Process Continuum

Beyond the basics we have just explored, MRP techniques are driven by the special characteristics of the demand and production environment. Recall our exploration of Engineer to Order (ETO), Make to Order (MTO), Assemble to Order (ATO), and Make to Stock (MTS) environments. Each type of production operation presents unique challenges to the planner and materials manager.

Make to Stock. In a Make to Stock environment, products are standard and built to a forecast or reorder point. This involves a BOM with limited variations that is manufactured repetitively. MTS forecasts demand through the distribution channel, accounting for issues such as distribution lead time and building ahead for seasonality.

Common MRP complications include the following:

- Product development, manufacturing testing, and product rollout require the management of prospective BOMs, purchasing, and inventory for consumption by research and development. MTS, ATO, MTO, and ETO environments share these R&D issues to some degree, although the scope of development research may be particularly acute for MTS and ATO

---

* Safety stock at the component and finished good levels may be appropriate if there is both dependent *and* independent demand for the same parts, such as when they are sold to the customer as replacement or service parts.

environments because they involve the repetitive and ongoing production of predefined models and configurations, and may include collaborative design with OEM (original equipment manufacturer) and distribution partners.

- Engineering change management and version control may require several versions of a BOM for the same item used over a period of time.
- Replacement, service, and warranty parts may require independent demand forecasting and inventory management, especially complicated by a geographically extended distribution and service channel, and where multiple BOM versions must be serviced and supplied over an extended period of time.
- Distribution Requirements Planning (DRP) requires forecasting demand, replenishment, stocking, and movement of materials among multiple warehousing and distribution locations.
- Transportation Management manages considerable inventory that may be in transit at any time, and this must be factored into the availability and lead time planning calculations.

Engineer to Order. There is often a complex presales design and engineering process in an ETO environment where a significant amount of the total lead time may be found and potentially eliminated through concurrent engineering and product life cycle management tools and techniques. By using such techniques, designs may be standardized to some degree, and similar components and assemblies may be reused on several designs, limiting the variety of raw and component materials that must be managed.

Once a design is completed and the sale is booked, the unique design may require a large and complex multilevel BOM using many unique and often custom-built parts that have never before been purchased or manufactured. In many ETO environments, a permanent finished part record is not created in the inventory part master for the end item on the sales order; rather, a job or project record is created, which stores the BOM, routing, and job costing information. This limits the number of one-time finished good part numbers that must be permanently stored in the inventory system.

Supplier relationships and material requirements can be very dynamic in an ETO environment, with new parts continuously estimated and added to the inventory and BOM record only if the sale booked. Procurement decisions are often made with limited or no historical record of procurement lead time and supplier quality performance. Finally, ETO lead time to design, procure, and manufacture can be very long, and is often scheduled as a multiphase project rather than a single process flow. This requires complex MRP calculations across interdependent phases of design, development, planning, and production, over an extended period of time. Material requirements may extend weeks or months into the future, and receipt of these parts should be timed to coincide with their phased requirements. With project management tools such as Gantt charts and the Critical Path method, rescheduling of a single element or phase of the project may cause a cascading MRP reschedule that affects capacity and material purchase and production releases. For large and complex projects this may require special project management, scheduling, MRP, and purchasing interfaces not found within some repetitive MRP II systems.

Make to Order/Assemble to Order. Make or Assemble to Order environments often require numerous BOM core components and options that are *configured* to customer specifications during order processing, which are then fabricated and/or assembled from planned inventory. Make to Order generally refers to a product where a finished part is fabricated from raw materials, whereas Assemble to Order refers to a product where the finished part is assembled from preexisting components, and of course there can be a mix of both.

Figure 4-16 illustrates the typical inventory usage pattern where multiple raw material inputs are planned, purchased, fabricated or assembled, and stocked as core components, options, and subassemblies in anticipation of customer orders. As customer orders are received, the inventory is then configured during a final assembly process and quickly delivered to the customer.

This approach is often called a *postponement strategy*, because we keep our options open to the last minute, purchasing the long lead time inventory in advance while postponing the commitment for final assembly, thus maintaining maximum flexibility for the use of available inventory to commit to sales. Because the core components are standardized, they can be forecast ahead and outsourced (often offshore) to low-cost and capital-intensive repetitive production operations with high volume requirements and relatively long pro-

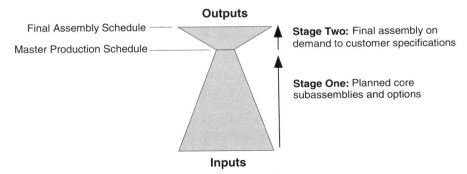

Figure 4-16. Two-stage BOM with MPS and final assembly

duction and transit lead times. The final assembly takes place to order, in a less capital-intensive and flexible plant that may be close to the customer.

When the lead time for production and shipment is long, the core and option items must be forecasted far in advance. Although this may require the maintenance of considerable core and component inventory, this approach requires far less inventory than if the company were to attempt to stock every possible variation of the finished product. In the case of a configurable BOM such as the computer in our earlier example, the customer may have hundreds of options for disk drives, modems and network cards, monitors, and preinstalled software. Despite its underlying complexity, the configured final assembly may represent a single line item on the customer sales order and invoice. In addition to a configured base item, separately packaged add-on items (in the case of the computer, this might include software and peripherals) may also be included as a *kit* included with the base assembly. Although these may appear combined as a single invoice line item with a special package price, the kit represents several items that are picked separately and shipped together. Another example is when an appliance store sells a washer and dryer separately but offers a kit where they are sold together with a special price, and the kit is assembled (picked together) during delivery.

Make or Assemble to Order calls for a *Planning BOM* (also may be called a *Super*, *Pseudo*, *Modular*, *Two-Level*, or *Family BOM*) where the first stage plans for long lead time component requirements based on a forecast and the second stage manages the rapid final assembly process to the customer order.

The first stage of the planning BOM is used to ensure that there are sufficient quantities of the various core and optional components in stock. A planning BOM uses *proportions* of the various options based on past sales history or a planned sales promotion. For example, if in the past we've sold 50% blue, 30% red, and 20% yellow, the planning BOM multiplies these proportions against the expected total requirements of the finished parts according to the forecast, in order to derive required quantities of the various options for inventory.

In fact, a planning bill may factor a little more of each option than needed to account for variability of customer demand for the various options. In this case the total of all options will add up to more than 100% of demand—this is called *option overplanning*. For example, with a forecast of 1000 computers, and based on our sales history and market forecast of model popularity, through the planning BOM, MRP may tell purchasing to order 270 tower cases, 420 midsize, and 370 mini, for a total of 1060 cases. So 6% excess component inventory is built right into the options planning equation, a form of safety stock designed to buffer the variability of customer preferences for the various options. The second stage of the process begins when the customer places an order. Special software called a *product configurator* is often used to aid the customer service representative in selecting the right options according to the customer's desires and specifications.

ABC Inventory Classification and MRP

MRP requires time and effort, so it is not appropriate for every item. One guideline for determining whether MRP is appropriate is by classifying inventory into ABC levels. According to the APICS Dictionary, ABC inventory classification is:

> A classification of a group of items in decreasing order of annual dollar usage or other criteria. This array is then split into three classes, called A, B, and C. The A group usually represents 10% to 20% by number of items and 50% to 70% by projected dollar volume. The next grouping, B, usually represents about 20% of the items and about 20% of the dollar volume. The C class contains 60% to 70% of the items and represents about 10% to 30% of the dollar volume. The ABC principle states that effort and money can be saved through applying looser controls to the low dollar volume class items than will be applied to high dollar volume class items.[70]

The traditional definition of ABC emphasizes dollar volume as the primary criterion for ABC classification, but there may be other factors to consider including relative unit cost, obsolescence or perishability risk, storage difficulty and cost, stock-out risk, and lead time variability. It is often assumed that A-level inventory must be managed carefully using MRP, but this is not necessarily the case. For example, the material that represents over 50% of the volume and cost of a finished product may be a powder or liquid stored in a silo, delivered periodically by truckload, and thus managed by order point method. When the product reaches a certain level in the silo another truckload is ordered, and a future delivery schedule can be determined with a rough estimate of consumption volume. On the other hand, a small, inexpensive, infrequently used, long lead time bolt may delay the shipment of a large assembly, wasting shop floor resources as the job sits idle waiting for the tiny piece. Therefore, a company may wish to develop a comprehensive ABC classification that guides their approach to forecasting and material planning. This approach would factor the usage pattern, cost, risk of stock-out, storage cost, physical characteristics, and supplier lead time of a particular material, to determine the appropriate stocking levels (no, low, or high safety stock) and replenishment method (MRP or order point).

Level A.   When low volume and high value, these products may use MRP with carefully defined lead times, combined with frequent cycle counting, because the value of the inventory is high and the cost of a stock-out is critical. If volume is high, an order point method (such as a two-bin order point or kanban system) may be used effectively. If supplier lead time is short, a JIT replenishment program may be created, reducing safety stock.

Level B.   These products may be of lower value and volume and may use either MRP or order point method with a larger reorder quantity because the cost to carry inventory may be lower and the reorder cost may be higher than A-level items. You may use MRP if there is low volume and high variability and an order point method if higher volume and lower variability. If supplier lead time is short, small safety stocks may be carried with periodic replenishment.

Level C. These products may include consumables—a bin of rags or fasteners, adhesives, packaging, or bulk containers of liquid or powder—where a visual order point is sufficient to trigger reorder. Consumables are usually included on the BOM for product costing, but purchases are not planned by the MRP engine and receipts are often expensed when received and not carried as perpetual inventory. With such a method, transaction costs may be reduced; during production these consumable items are not issued from inventory with the other components on the BOM, but are costed to the finished good with an overhead factor.

SALES AND OPERATIONS PLANNING (S&OP)

During the 1960s and 1970s, many hoped that MRP would be the answer to all their material management challenges. With a sufficiently powerful computer, MRP could recalculate purchasing and production requirements every time a change in demand or supply occurred . . . how wonderful!

However, it was soon discovered that the computer's ability to change the plan was not matched by the shop floors' ability to keep up with those changes. If adjustments were made too often, as these changes rippled through the many items and levels of the BOMs, the interdependent purchase and production schedules created by MRP become *jittery* or *nervous* and impossible to follow. Frequent changes make buyers, suppliers, schedulers, and production staff equally jittery and nervous.

MRP is simply a tool, a calculation engine, a small part of a larger planning process. For it to be useful there must be a *Master Production Schedule* (MPS) to guide MRP replanning, and to minimize disruption and expediting of purchasing and production. But there is still a missing link. Executive management regularly reviews company strategy, devising a business plan that sets forth annual goals and objectives for revenue, expenditures, capital investment, production, etc. With market conditions regularly changing, along with variations in material availability and productive capacity, how are these strategic goals regularly aligned with current market conditions and communicated to the planner so that production (and therefore what the company is capable of selling) remains consistent with the business plan? This is the purpose of the monthly Sales and Operations Planning process.

The Mechanics of S&OP

All companies engage in some form of an S&OP process whether they know it or not. However, many do it in an informal, unstructured, departmentalized, unsynchronized, unpublished, overly detailed manner, using an incomplete picture of demand and with disregard for capacity and financial constraints. And very often the planning department *independently* makes critical resource allocation decisions that affect sales and customer service, in response to changing market conditions, leaving top management wondering who is steering the ship.

The executives of a manufacturing enterprise are continually assessing business climate, evaluating major and minor, long- and short-term adjustments to their business strategy and tactics. This constant awareness should translate into a disciplined long-range strategic planning process that determines new investments in plant and equipment, new directions in industrial technology, product development, distribution, new supplier, and customer strategic relationships. To be effective, these plans must be communicated in such a way as to clearly guide the daily activities of the entire enterprise.

S&OP is a disciplined process resulting in company-wide consensus, ensuring that top management's objectives are realistic and reconciled to the aggregate production plans of the company. The top executives and heads of all functional areas in the company must participate in this process, along with scheduling and marketing personnel.[71] Getting started with S&OP first requires education. The next steps include defining families and formats, preparing pilot data, developing a policy and meeting agenda, and, finally, beginning the monthly meetings.[72] The meetings should involve the active participation of the following constituencies:

- **The Marketing and Sales Organizations**, who are continually assessing demand, including existing orders, forecast, and promotional plans.
- **The Product Development Organization**, who is keeping an eye on the market, looking for ways to develop competitive advantage through innovative product and service offerings. As new products are developed, the life cycle of production, marketing, and distribution are considered, along with an assessment of the impact of a new product on the forecasted sales of the current products.
- **The Planning and Procurement Organizations**, who are constantly evaluating internal and external capacity, material availability, and lead times, to determine how these resources may be optimally utilized to satisfy demand.
- **Production Operations**, who are responsible for delivering the output necessary to balance supply (production output plus available inventory) with demand (the sales forecast and customer orders).
- **Materials Management**, who have a clear understanding of current inventory levels, inventory policies, and the movement of raw materials, WIP, and finished goods.
- **Finance**, who constantly monitor the availability of cash, inventory, assets, risk, corporate accountability, and governance.

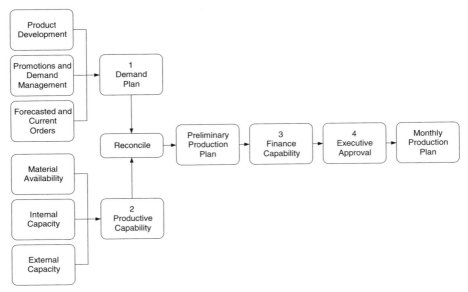

Figure 4-17. The Sales and Operations Planning process

The disciplined S&OP process (illustrated in Fig. 4-17) helps these con-stituencies to collaborate effectively and in a structured, cross-functional manner. Leading up to the final meeting there must be considerable prepara-tion, collaboration, and problem solving. S&OP culminates in a meeting with executive management, where representatives of the critical inputs to the pro-ductive process (demand, supply, production, and finance) meet and resolve final issues, and where the activities of the organization are reconciled to the strategic plan. The final S&OP executive meeting should be completed in less than two hours, focusing at the product family level measured in unit quanti-ties and converted to dollars or other currency, which is the executive-level default unit of measure. If during the S&OP meeting an executive wishes to drill down into the details of any aspect of the operation, the answers should be readily available.

Many companies view S&OP simply as a repetitive monthly meeting to rubber-stamp the production plan, not an ongoing process to reconcile the strategic and tactical plans and to foster cross-functional coordination. They naturally obtain much less value from the process than it offers. The executive S&OP meeting should be clear and concise, not a dull ritual, nor should it be overly detailed or chaotic. This may be the only opportunity executive lead-ership has each month to peer deep inside the detailed operations of the company, so it must be done well. The executive team may also determine that changes are required to the strategic and tactical plans as a result of this monthly reality check. According to Richard Ling, whom many consider the father of S&OP, with integrated enterprise software and supply chain management systems a company can improve its ability to plan and execute consistently:

> In short, a successful S&OP process delivers more predictable financial results. With an integrated planning process, you gain the ability to make better decisions and forecast expected results. In the past, there was typically a lack of true collaboration—both inside and outside the company—because you had access only to static, fragmented data and often could not get the right people involved soon enough in the process. Now, with global real-time access to information, you can collaborate with the right people and, at the same time, collapse the planning cycle. Software enables this collaboration, allowing you to coordinate customers, suppliers, and data.[73]

ERP systems control most of the detailed information that is required to facilitate the S&OP process, with the possible exception of demand management—which may receive inputs from SCM and CRM systems. Unfortunately, demand management and forecasting are still uncoordinated manual processes for many companies. Because of the repetitive nature of S&OP, and because of the layers of complexity when multiple product families and locations are involved, many companies also get bogged down when they try to use ordinary spreadsheets to automate key elements of this cross-functional and collaborative process. As large and complex spreadsheets are e-mailed around the company, undocumented assumptions, spreadsheet errors and omissions, and version overlap often result; at best this creates confusion, and at worst serious planning errors. Organizations that make effective use of S&OP often automate the process through the use of integrated and database-driven forecasting and planning tools that, while they offer spreadsheet-like user interfaces, eliminate the shortcomings of spreadsheet-based collaboration.

The Production Plan

The output of S&OP is an approved *Production Plan*, which describes the rate of planned production *at the product family level*—which again is measured in units and converted to dollars, using average prices. The production plan is the direct linkage between executive management's strategic view of the organization and the monthly operating objectives that guide the rest of the organization. The production plan is delivered to the Master Scheduler, who generates the Master Production Schedule (MPS) *at the parent item quantity level* based on the existing order backlog reconciled with the detailed sales forecast. The MPS considers actual demand from the forecast and committed sales orders, which then communicates finished part requirements to the MRP engine, which in turn calculates the purchasing and production requirements at the individual item, component, and raw material levels.

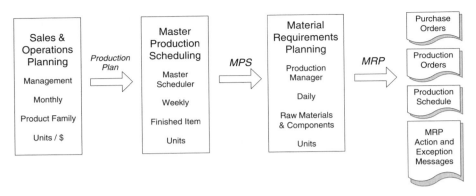

Figure 4-18. The step-down monthly planning process

This is an important sequential process, where S&OP begins at the product family level, stepping down to the next level of detail only once there is agreement, as illustrated in Figure 4-18.

To begin at the detail level, or to move to the next level of detail before reaching agreement, creates confusion and wasted time. Therefore, it is important that the S&OP process is disciplined and well-orchestrated if it is to be performed on a monthly basis and in sufficient time to guide operational decisions.

During the month, if conditions (demand, supply, productive capability, or finance) change significantly, this will immediately become apparent to the planner. If there is little net impact to the overall production plan for the month, the planner may execute the changes himself—adjusting and rescheduling orders appropriately. However, if the changes force a significant schedule, material, or capacity trade-off decision, which in turn causes a significant deviation from the approved production plan, the planner must devise alternative solutions and then escalate the decision to upper management. Depending on the magnitude of the change, this management decision might be a brief hallway discussion or a thorough review of the earlier planning process as illustrated in Figure 4-19. This closed-loop regulation of changes to the Master Production Schedule ensures the strategic and business plans are kept in alignment with the daily activities of the shop floor.

MASTER PRODUCTION SCHEDULING (MPS)

The MPS is the detailed plan of production representing all parent items of the BOM, in other words, all *independent demand*. The MPS is the fulcrum of the traditional planning process, providing a control point from the boardroom to the shop floor.

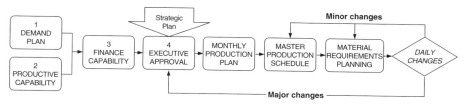

Figure 4-19. The S&OP feedback process

MPS Mechanics

The MPS is arranged by the parent item in the BOM, which may be a finished item or a semifinished item awaiting final assembly to customer order.

In an Assemble to Order environment where semifinished items are manufactured in advance for later assembly into finished goods, there are two sets of parent items that must be planned separately. In this environment the semifinished item production schedule is called the Master Production Schedule, whereas the Final Assembly Schedule (FAS) controls the rapid assembly of the finished items. The MPS demand for the semifinished items is usually driven by a long- or medium-range forecast, whereas the FAS demand is driven by actual customer orders on a real-time basis. The demand inputs to each line item of the MPS thus may be composed of a combination of forecasted demand for semifinished items and actual order backlog of final assembled items, depending on the nature and timing of product demand compared to lead time.

MPS Item ABC-001	Beg Balance	Period 1	Period 2	Period 3
Demand (Forecast and Backlog)		150	250	100
MPS		150	150	150
Projected On Hand	100	100	0	50

Figure 4-20. The Master Production Schedule

The MPS calculation grid (shown in Fig. 4-20) is similar to the MRP grid shown earlier in this chapter, however, MPS is only concerned with the due date of the parent item. It is important to remember that lead times are not applied in the MPS, because this is a schedule of when the product is due—the MRP engine then takes each parent item demand by due date, *offsetting* by the purchase and production lead times of each lower-level component, to calculate the right time to issue purchase and production orders to meet the delivery date of the MPS.

In this example, note that demand varies from one period to the next; however, the scheduler has apparently decided to pursue a *level schedule* of 150 units each period, rather than a chase schedule of producing according to each period's demand. This results in on hand balances at the end of periods 1 and 3—these quantities are Available to Promise because they are not committed to demand. It is helpful for the MPS to make these ATP quantities known to the sales team.

As mentioned earlier, the MPS is the fulcrum of the planning process, and it also provides vital availability information to the sales team. If a sudden change in supply (material availability, production, or purchases) or demand (orders, due dates, or forecast) occurs, the Master Scheduler must evaluate whether the change may be accommodated within the guidelines of the Production Plan, and whether the change is allowed by *time fence rules* (Fig. 4-21):

- **Frozen Zone**—Also called the *Point of No Return*, the Frozen Zone is where materials and capacity have been committed and work has been issued to the shop floor. Within the Frozen Zone changes are highly disruptive and must be tightly controlled, often requiring management approval. Without well-defined rules, executing changes within the Frozen Zone becomes the chaotic triage world of the expeditor, and productive capacity is permanently lost when the smooth flow of production is violated.

- **Slushy Zone**—is bounded by Cumulative Lead Time, which is the longest lead time required to purchase or produce all materials and subcomponents required for the finished part. Within the Slushy Zone, also called the *Trade-off Zone*, jobs of similar routing may be swapped by the Master Scheduler with little impact on capacity, while jobs with similar materials

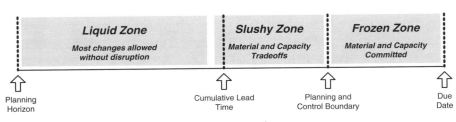

Figure 4-21. Time fences

components may be swapped with little impact on material requirements. However, if trade-off decisions create material or capacity constraints, the decision may require management intervention.

* **Liquid Zone**—Events are fluid and dynamic, and most actions taken do not create negative consequences for the production schedule. The planning software may be permitted to automatically schedule and reschedule planned purchase and production orders, because the execution of most decisions is still far away.

If management escalation is required to make a Slushy Zone trade-off or Frozen Zone expedite decision, then an ad hoc S&OP session should occur to frame the change of circumstances, effectively modifying the Production Plan. This disciplined closed planning and feedback loop ensures that boundaries are in place that will immediately alert the appropriate elements of the organization and activate a prescribed decision process whenever there is a significant deviation from the plan.

Forward and Backward Scheduling

MRP must be instructed how to schedule the start date of a production work-order—as soon, or as late, as possible. *Forward and Backward Scheduling* techniques are therefore used when prioritizing and sequencing work within the Frozen Zone. A *backward* schedule means that the order may be committed to production at the last possible moment to be completed by the due date. A *forward* schedule means that an order is committed to production immediately (based on the existing backlog), possibly to be completed ahead of when it is needed. For example, if it is now January 1, and an order that requires a 10-day production lead time is due January 31, backward scheduling will release the job on January 21, as illustrated in Figure 4-22. Forward scheduling will release the work immediately so it will complete on January 10, causing inventory wait in finished goods for 21 days before the finished part is delivered to the customer. Alternately, the job may be periodically interrupted, waiting in various WIP queues as it is postponed for higher-priority work.

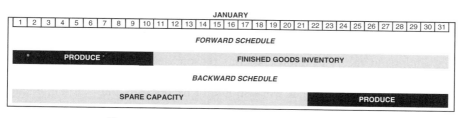

Figure 4-22. Forward and backward scheduling

There are countless ways that a skilled scheduler can manipulate these backward and forward scheduling rules to achieve a desired result; here are a few examples. A job shop that produces a combination of one-of-a-kind jobs (with many unknowns and potential process variation) alongside jobs of a more repetitive and predictable nature may choose to forward schedule the unique orders to allow more cushion for the unknowns, while backward scheduling the more predictable ones.

The scheduler may choose to assign all jobs that are for actual customer orders to a forward schedule basis, ensuring that they'll complete and deliver ahead of schedule, filling up the near-term capacity. Similarly, he may choose to assign all MTS orders (based on a forecast) to be backward scheduled, pushing them into the future. If there is available capacity during the period, this technique utilizes early capacity gaps after completion of actual orders and before start of MTS orders, so a new order that is suddenly received may be inserted without causing a trade-off decision.

After the initial backward and forward schedule calculations, and if no new orders are received, the scheduler has the flexibility to fill this gap by pulling forward MTO work from the Slushy Zone, or MTS orders that are presently backward scheduled within the Frozen Zone, either by selectively changing them to forward schedule or by moving their due dates forward until the capacity gap is filled. Later in the week, if an actual customer order with an imminent due date is suddenly received, the scheduler may shuffle these orders back out again. And perhaps one of these new orders will consume the forecasted demand that caused the MRP system to plan the MTS order in the first place.

Alternatively, customer orders may be backward scheduled initially, which opens gaps in the near term that the scheduler may use to juggle priorities. This approach minimizes the cost of forward scheduling that results in finished goods sitting in inventory long before they are needed. This assumes that orders are received long before their lead time, which is not very common in many industries these days. The risk when backward scheduling is that an event (for example, machine downtime or material shortage) may occur when there is no time left to take corrective action, causing a backward-scheduled customer order to miss the due date. We hate to hear a customer say "You've had this order for two months and you still missed the due date?"

There is often a desire to forward schedule for this very reason, creating a time and inventory buffer against these uncertainties. This results in a wasteful buildup of excess finished goods inventory—some that is eventually sold and some that may sit on the shelf forever because customer requirements may change at the last minute. For this reason, backward scheduling is recommended—incidentally, *this is consistent with the Lean Manufacturing pull principle*, which stresses that work shouldn't be released until an immediate demand signal is present.

CAPACITY PLANNING

There are two primary constraints that must be managed during the production planning process: materials and capacity. Material Requirements Planning is capacity insensitive; it is only concerned with material availability and assumes an infinite amount of capacity is available. However, the Master Scheduler's primary responsibility is to maintain the validity of the schedule, which must consider available capacity to perform the work that is needed.

Most manufacturers expend considerable effort controlling inventory because it's a highly visible asset—you can touch it, you can see it, sometimes you step around it or stumble over it, and it appears every month on the balance sheet. To manage inventory, estimate costs, and plan for production there must be an accurate Bill of Materials and physical controls on the purchase, storage, and consumption of inventory. Although inventory record accuracy may be lacking, most manufacturing companies have at least an elementary grasp of materials management.

Capacity management is another matter. There are far more variables to manage, and capacity is more dynamic and intangible than inventory. Changing circumstances every minute of the day have a significant impact on capacity: machine readiness and performance, labor issues, alternative routings, tooling availability; shifting setup, run, queue, move and wait times, job prioritization and sequencing—all manner of abstract and complex issues affect the productive capability of a plant. Beyond intuitive back-of-the-envelope capacity planning, many manufacturing companies don't try to formally measure or predict capacity. And even if they try, there are so many variables and assumptions involved in capacity planning that anyone may cast doubt upon the whole exercise with a simple shrug of the shoulders.

Does this mean a company shouldn't try to plan capacity? On the contrary, the foundations of capacity planning enhance the disciplines required for proper scheduling and release of work to the shop floor. There is no element in the entire production planning and control process that more directly affects the throughput of the plant, and thus total revenue and profit potential of the enterprise, than astute scheduling, prioritization, and work release. Thus capacity planning may be difficult, but it's always worth doing in one form or another (we'll return to this thought in Chapter 5 when we discuss Lean Manufacturing techniques).

In a traditional MRP II system there are four levels of capacity management that mirror the step-down functions of material planning, providing closed-loop feedback to upper management whenever a significant problem occurs. These four levels are:

1. **Resource Requirements Plan (RRP)**—validating the Sales and Operations Planning process
2. **Rough Cut Capacity Plan (RCCP)**—validating the Master Production Schedule

3. **Capacity Requirements Plan (CRP)**—validating the Material Requirements Plan
4. **Input/Output Control**—controlling the real-time execution events on the shop floor

In this chapter we'll discuss the first three, leaving the discussion of Input/Output Control (execution) for Chapter 5.

Resource Requirements Planning

The *Resource Requirements Plan* (RRP; also called the *Resource Plan*) is used during Sales and Operations Planning to generate a valid Production Plan. Just like the S&OP process itself, the RRP operates at the summary level based on product families. Three rough estimates are required to produce the RRP:

1. **Capacity Required for Each Product Family**—For example, 1 unit of product family A requires an average 2 minutes in the grinding area.
2. **Total Available Capacity for Each Major Resource or Workcenter Grouping**—At this level of planning only critical resources are considered at a summary level; these might include the load placed on bottleneck or other key resources measured in machine or labor hours.
3. **Total Demand for Each Product Family for the Period**—measured in units of production

With these simple and rough figures and a little time (aided by an electronic spreadsheet) an individual may depict the overall workcenter load profiles for the period, identify any obvious workcenter constraints, and determine the overall feasibility of the schedule with a certain degree of confidence.

During S&OP if it is determined that the demand forecast creates an overload on a critical resource, the team considers their options. Do they plan to run another shift? Do they call in some temporary staff, or shift some labor from underloaded workcenters? Do they notify a standby outsource vendor that some work is coming their way this month? Are the extra costs of ramping up additional capacity balanced by the profit of accepting the extra business in the first place? These are management decisions that must be guided by the strategic and business plans; any changes discovered during S&OP that significantly conflict with the business plan should be escalated to and approved by management, so the strategic and tactical plans of the company remain in harmony. Once these decisions are made, the resulting Production Plan is handed down to the Master Scheduler, providing documented assumptions and directives for the capacity that has been allocated for the period.

Rough Cut Capacity Planning

As we explained earlier, the Master Scheduler takes the Production Plan, the open order backlog, and forecast and develops the Master Production Schedule at the parent-item level, scheduled into weekly or daily time periods. Varying demand patterns may cause an uneven or lumpy schedule that must be smoothed through scheduling, but that comes later. *Rough Cut Capacity Planning* (RCCP) is only concerned with balancing demand and capacity for the period, without regard to the specific timing or sequencing of the work. In fact, the typical RCCP may not even subtract WIP already in the process from the net demand, because that requires additional information and calculation effort and RCCP is intended to be a quick validation of the MPS. The following elements are required to produce the RCCP:

- **Bill of Resources (BOR)**—For each finished item, an estimate of capacity is required for each major workcenter or grouping. This is not a routing that identifies the sequence of operations, just the total duration of work required per operation. At the BOR level the difference between setup and operation time is ignored, and the total standard operation time on the BOR is based on a target batch size that takes the setup and run times into account. Because RCCP calculates the *total* demand for a particular product for a period, we are not estimating the number of distinct lots and setups required, so this approximation is necessary.
- **Demonstrated Capacity for Each Major Workcenter or Grouping**—what is shown in actual operation, not theoretical

Note that the amount of detail contained in the RCCP is considerably greater than the RRP. Each item must have its own capacity requirements, and each workcenter must have an estimated available time. With these figures and the aid of a specialized computer program (or several hours with a spreadsheet) an individual may depict the overall workcenter load profiles for the period, identify any obvious workcenter constraints, and determine the overall feasibility of the schedule with a greater degree of confidence than the RRP.

As we noted earlier, the primary responsibility of the Master Scheduler is to ensure that the MPS is feasible, according to existing material *and capacity* constraints. The Master Scheduler has the authority to make many trade-off decisions on what to produce and when, as long as the overall production and capacity boundaries established by the Production Plan are not violated. Any changes to the MPS that would significantly conflict with the Production Plan should be escalated to and approved by management, so that the strategic plan and resources of the company aren't misdirected.

Armed with the RCCP, the Master Scheduler can identify obvious capacity limitations to the MPS. Note, however, that the RCCP is not time-phased, so the scheduling, prioritization, sequencing, and lumpiness of demand and production are not considered, just the overall capacity required compared to what is available for the period.

Capacity Requirements Planning

The Master Scheduler must determine a feasible schedule within certain boundaries, but it is ultimately the responsibility of the production scheduler to sequence and prioritize the work properly according to daily conditions on the shop floor. Once a valid MPS is produced, the MRP engine calculates detailed material requirements and suggests purchase and production order release dates for all finished items, subcomponents, and raw materials with the appropriate lead time offsets. With specific release dates for each work order, the appropriate sequencing of work may be determined and the capability of the shop may be measured very carefully. It is at the *Capacity Requirements Planning* (CRP) stage that distinct differences between operation types come into play.

In a continuous flow environment, where there is a single continuous process like a conveyor or pipeline, the production rate and sequence do not vary by item—there may be a steady rate of production at all times. CRP is not necessary in such an environment. However, to the degree that different items have different workcenter requirements and routing pathways, varying the mix and release sequence of those items may result in dramatically different results. With variable routings, some workcenters may be overburdened at the same time that others are starved. So the sequence of work release, and the order in which work flows from one workcenter to the next, is critical for CRP in a discontinuous flow environment such as a job shop.

The paradox is that as the complexity of the environment grows, so does the value of accurate CRP—but as the complexity grows, the difficulty of computing an accurate and timely CRP also escalates, often to the point of impracticality. This is because there are many precise elements required for accurate CRP:

- Specific routing steps and sequence for each manufactured part, including:
  - Workcenter and operation sequence; this may include alternate workcenter routings considered by the CRP calculation in order of priority and efficiency, if the primary workcenter is unavailable at the desired time
  - Standard setup, run, queue, wait, and move times for each operation
  - Tooling requirements
  - Labor requirements, designating time, skills, and possibly even specific personnel required
  - Inspection and testing requirements

- Demonstrated capacity (measured under normal conditions, not theoretical) for each distinct workcenter, or group of workcenters if they are scheduled concurrently or shared, including:
  - Detailed shop calendar indicating work hours, shifts, and vacation schedules
  - Machine availability, which includes scheduled downtime for maintenance
  - Machine efficiency and utilization factors
  - Human resource availability, which includes individual vacation schedules if specific people are called out to perform certain operations
  - Tooling availability to account for tools used at multiple workcenters
- Demonstrated capacity must then be reduced by requirements for each work order that is already released to production

Capacity Planning and the Product/Process Continuum

The appropriate degree of capacity planning is governed by the complexity of the process, as illustrated in Figure 4-23.

On the continuous flow end of the spectrum, imagine a pipeline that produces 1000 gallons of product each and every hour. You can't force through 1100 gallons an hour, and if your demand is only 900 gallons then you have 100 gallons of spare capacity each hour. That's the capacity plan, period. At the other end of the spectrum, imagine a discontinuous environment like a job shop where the routing pathway of each job may be unique. Although the aggregate RRP and RCCP calculations may indicate that there is available capacity to satisfy MPS demand, when CRP models all the detailed interactions among specific items, routings, lead times, and workcenter capacities, a bottleneck may appear at a particular workcenter, causing the MPS to be invalid.

Judging by the amount of detail required, not only is the CRP calculation process lengthy and iterative like MRP, but the number of dynamic variables

	Resource Requirement Plan RRP	Rough Cut Capacity Plan RCCP	Capacity Requirements Plan CRP
Planning process	S&OP	MPS	MRP
Level	Product Family	Parent Item	All components
Frequency	Monthly	Monthly and Weekly	Weekly and Daily
Workcenter Loading	No	General	Specific
Route sequence	No	No	Yes
Lead Time	No	Yes	Yes
Difficulty	Low	Medium	High
Require a computer?	Maybe not	Probably	Absolutely

Figure 4-23. Comparison of capacity planning methods

that must be considered make CRP in a discontinuous environment one of the most complex software challenges imaginable. For this reason, many companies choose not to implement CRP. We'll explore solutions to this challenge in Chapter 5. For now it is important to understand that from a traditional planning perspective, as we move toward the discontinuous end of the product/process continuum, a production schedule isn't considered valid until it passes both the MRP *and* CRP tests.

Regardless of the type of environment, if a company doesn't make an *appropriate* effort to validate the MPS against available capacity during the planning process, it leaves many unknowns to play out as the daily drama unfolds on the shop floor.

THE INTEGRATED PLANNING PROCESS

The entire closed-loop planning process ensures that decisions at all levels of the organization are coordinated, and that any deviation from the plan is escalated to the appropriate level for troubleshooting and resolution.

Because of the massive amounts of detail that are required at the lower levels of the planning process, software tools for material and capacity requirements planning are usually necessary. The skills to use these tools do not come easily; there is a need for education and training at all levels of the organization.

In summary, the entire integrated planning process (illustrated in Fig. 4-24) works like this:

Step 1. Executive management reviews and revises the strategic plan, business plan, and performance targets at least once per year, setting performance goals for the following year, which are translated into financial and operating goals, particular objectives, financial budgets, and performance measurements. Sales and production targets should be defined at both the company and product family levels.

Step 2. During the monthly Sales and Operations Planning process, management reviews demand and new product release plans by product family compared to material availability, productive capacity (RRP), and financial capability to develop the monthly Production Plan. This is approved by the Executive S&OP team and then delivered to the Master Scheduler.

Step 3. The MPS is the fulcrum of the entire production planning and control system. The Master Scheduler, using the Production Plan as a guideline, generates the Master Production Schedule at the parent-item level. RCCP may be used to summarily validate capacity of the MPS. Changes to the schedule follow time fence rules established as company policy and may escalate to upper management for trade-off or expedite decisions.

Step 4. MRP takes the parent item requirements from the MPS and explodes the BOM to calculate the component gross requirements, netting

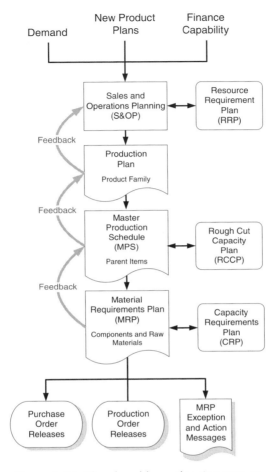

Figure 4-24. The closed-loop planning process

these requirements against available materials, suggesting detailed purchase and production transactions to be executed, and notifying the planner of exceptions that require attention. CRP may be used to validate the detailed release schedule against the current conditions on the shop floor.

Step 5. Purchase and production decisions are executed.

Steps 2 through 5 are repeated as necessary in response to changing conditions. If at any stage during planning or execution a problem arises, there is a clear escalation process to ensure that serious deviations reach the appropriate decision-making level.

THE LEAN TRANSFORMATION

It's eleven-thirty and you're hungry, so you and a friend visit your corner deli for a bite to eat. You ask the person at the counter for a roast beef sandwich; your friend wants ham. The person at the counter goes back to the kitchen for a status report, and a few moments later returns to tell you . . . "Well, the batch of 500 roast beef sandwiches was started on schedule at 8:00 A.M., and since the cycle time per sandwich is 15 seconds, they were finished at 10:06 A.M. and they're ready to eat except they may be a bit soggy by now. Our schedule shows that we have 600 turkey sandwiches in production right now, followed by 400 ham sandwiches scheduled to start at 12:37 P.M.—they're supposed to be ready at 2:12 P.M. But we're having some throughput problems on the pickle line, so it may be closer to 3:00 P.M.—we'll know better when we recalculate MRP sometime in the next hour or so—do you mind the wait?"

Hold on, that's no way to run a restaurant! Perhaps that's no way to run a factory, either. A fast food restaurant is a perfect example of Lean demand flow—an order is configured just the way the customer wants based on a very large combination of possible options, and moments later the finished product comes out fresh and ready to eat. To accommodate this flexibility and speed many specific techniques have been introduced by Lean Manufacturing, including one piece flow, pull, flexible workcenter arrangement, flattened BOMs, mixed model production, cross-trained workers, short changeover times, small batches, and minimal WIP inventory. When they decide to pursue Lean improvements, an enterprise will find that many of these principles are in opposition to the mass production assumptions of traditional manufacturing, whereas others may be complementary, depending upon the nature of their environment and their current position on the product/process continuum.

As companies pursue Lean performance, their specific improvement initiatives depend on where they are starting from. In general, repetitive products and processes strive for flexibility of product mix—moving upward along the diagonal, whereas variable products and processes strive for speed—moving downward. This creates a *Lean Squeeze* toward the center of the continuum, as shown in Figure 4-25.

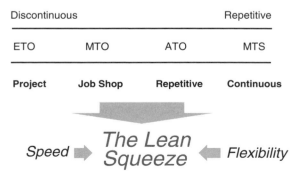

Figure 4-25. The Lean Squeeze

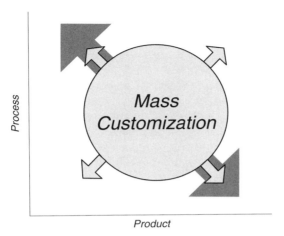

Figure 4-26. Mass customization

All types of production environments are squeezing toward the Lean middle ground, striving to occupy the most real estate along the diagonal, developing the ability to quickly make strategic product and process positioning decisions, rather than being forced into a rigid position by their habitual and institutionalized patterns of planning and production. This initiative has been given the paradoxical name *Mass Customization* (illustrated in Fig. 4-26) because it suggests the blending of two apparently contradictory methods.

Mass customization *minimizes the trade-offs* between flexibility and repetitive production described by the traditional product/process life cycle. Note that I said "minimize" and not "eliminate"—for there are still, and will always be, companies at each end of the continuum that may not benefit from certain Lean techniques. Even so, Lean offers many useful methods for waste reduction and the continuous improvement of *any* production environment.

The Next Step

We have just completed a concise exploration of the APICS *traditional* Production and Inventory Control body of knowledge, and then in this section we briefly contrasted this traditional approach with Lean methods. How do we reconcile the two? What do we keep, and what do we discard, to achieve Lean Manufacturing in any particular environment along the continuum? We turn to Chapter 5 for answers to these questions.

Chapter 5

Lean Planning and Execution

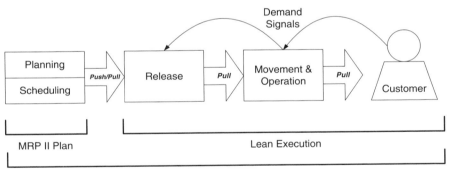

MRP II Plan & Schedule	Release	Movement & Operation	Measurement
• S&OP	• Prioritize	• Routing	• Backflush
• MPS	• Sequence	• Kanban	• Exceptional Event Capture and Notification
• RCCP	• Dispatch	• Supermarkets	
• MRP	• Heijunka	• Material Movement	• Scrap and Yield Loss
• Level Schedule	• Kanban	• Resource Consumption	• Throughput Measurement
• Cycle Time	• Material Issue	• Constraints	• KPI's
• Setup Time	• Work Instructions	• Pacemakers	• Quality and Process Control
• Takt Time	• Drawings		• Lot Traceability
• EPE			• Regulatory Compliance
• Pitch			

In this chapter we explore the detailed relationships between planning, scheduling, and execution across all modes of Lean Manufacturing from repetitive to discontinuous. It is here that the greatest disagreements and misunderstandings between Lean and MRP II proponents occur.

This is by necessity a lengthy and detailed chapter, so it is divided into eleven sections:

1. The Need for Careful Planning in a Lean Environment
2. Flow Production Basics
3. The Lean Planning Model
4. Kanban Essentials
5. The Lean Job Shop
6. Discontinuous Scheduling
7. Theory of Constraints
8. Bringing It All Together
9. Variations on a Lean Theme: CONWIP, SMP, and POLCA
10. Searching for the Right Scheduling Software?
11. The Transition to Lean

This chapter strives to reconcile traditional, Lean, and Theory of Constraints approaches, demonstrating that it is possible to apply the fundamental Lean principles of continuous improvement and waste reduction in any production environment. It also offers a framework to aid in simplifying, scheduling, and controlling any type of operation, with or without the use of software.

THE NEED FOR CAREFUL PLANNING IN A LEAN ENVIRONMENT

Question: Why do we plan?

Answer: Because we don't like surprises, and neither do our suppliers and customers.

We plan in order to anticipate change so that we can keep our commitments. Planning attempts to anticipate and smooth variation to create a relatively stable production environment. In a traditional mass production environment this was accomplished by long-range planning, push scheduling, producing in large batches, and storing large inventories throughout the supply chain. Hence the focus was on efficiency and resource utilization, accompanied by the all-too-familiar mantra: *"Keep all the machines running and all the workers busy; otherwise, we're losing money!"*

Lean Manufacturers have learned that continually running a plant at full throttle can cause significant harm. Excessively high resource utilization does not allow enough slack capacity to respond to sudden changes such as supply interruptions or machine downtime. Too much WIP on the shop floor leads to congestion, confusion, large queues, unpredictable quality problems, long lead times, and missed due dates. A Lean operation needs *protective capacity* to provide for flexibility and responsiveness. The general guideline is to run no more than 80–85% of capacity on a regular basis*, allowing flexibility to react to changing circumstances without overstressing any component of the system. This includes people—spare time may be used for value-added activity such as preventative maintenance, education, training, cross-training, and other continuous improvement activities.

However, with sales and marketing's hand on the throttle making deals, stretching for sales goals and market share, and battling the competition—there must be a governor on the demand engine to ensure that the plant is not pushed beyond the protective capacity threshold. That governor is the monthly Sales and Operations Planning (S&OP) process, which develops a production plan that is consistent with the goals and objectives defined by management.

In an *ideal* repetitive Lean environment, setup times are short and the flow of work is balanced evenly across all workcenters. Work flows steadily according to the rate of customer demand, and a level schedule may be calculated to release a regular product mix to the shop floor. Work is scheduled at only one point in each process and is pulled through production by signals that indicate available downstream capacity. This instant feedback and single point of scheduling and control allows the rate of production to be carefully regulated with minimal effort and complexity, and within a very short time horizon.

In a Lean environment the S&OP process is responsible for calculating the production plan, which may be stated as the target monthly *takt time*—more on this later. To calculate takt time and plan a level schedule, the S&OP process must look ahead for expected demand, *which requires forecasting and planning*. By contrast, Lean *execution* synchronizes release of work to actual demand in real time when requested by a pull signal. So Lean Manufacturing and MRP II converge on the boundary of planning and execution, of forecast and reality. Forecasting and planning are always unreliable to some degree, and in a Lean environment of continuous improvement and rapid response to changing conditions this boundary between plan and actual is very *fluid* (there's the water metaphor again). This is why highly structured MRP II push planning and scheduling software has great difficulty adapting to a fluid Lean shop floor pull execution environment.

Planning to Reduce Lead Time Chaos

Time is the ultimate constraint in all planning, scheduling, and execution decisions. Long lead times require a distant demand forecasting horizon, which in turn requires more planning. The further into the future we must plan, the greater will be the differences between forecast and reality. This uncertainty

---

* Slack capacity guideline does not apply to bottleneck operations, as we'll discuss later.

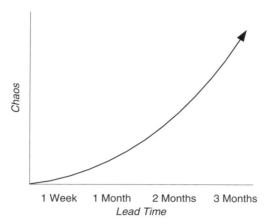

Figure 5-01. The relationship of lead time and chaos

of demand is factored into inventory policy calculations resulting in additional safety stock inventory, creating waste. According to Kevin J. Duggan in *Creating Mixed Model Value Streams*, there is a direct relationship between lead time and manageability of the overall system, which he describes as *chaos* (Fig. 5-01):

> Plotted on a curve, the relationship between lead time and chaos becomes apparent. The longer the lead time, the more chaos enters into the system. Long lead times make it almost impossible to predict what is needed far enough in advance. The shorter the lead time, the more accurate the forecast.[74]

This idea of chaos is particularly useful when communicating the importance of lead time reduction to a new audience, because most people instantly recognize the signs of chaos within their own plant.

Lean Manufacturing reduces lead time and consequently the planning horizon by decreasing *Throughput Time*, which is the elapsed time required for a job to flow through the factory from release to completion. By reducing throughput time we can synchronize the rate of production to actual demand (takt time), whereby lead time may be reduced from months to weeks, weeks to days, or days to hours and minutes. The forecast accuracy of the shortened planning horizon improves, resulting in less inventory waste. As we approach the *ideal* goal where 100% of production is built to order and synchronized to real-time demand, then there is *almost* nothing in raw, WIP, and finished goods inventory based on a forecast. With minimal throughput time combined with supplier lead time reduction, we may hope that the complications of traditional MPS and MRP planning would disappear completely in a Lean environment. Of course there are many real-world reasons why this does not happen entirely.

Suppliers need visibility to our long-range material requirements so they can plan their long-term capacity investments and customer allocation targets. This is especially true with high-technology supplier relationships, where the product life cycles are short, supplier lead times are often long, and demand is volatile and unforgiving. For example, in the late 1990s the entire computer laptop industry was constrained by the supply of flat panel color displays from Japan. Manufacturers placed orders months in advance, timing their product introductions and life cycles to the availability of this constrained component. Companies that didn't plan accurately suddenly found themselves overcommitted to large allocation backlogs, holding excess inventory of obsolete displays as competitors shifted to the next generation of technology. Given this perspective, you can understand why Dell Computers, who Dr. Richard Schonberger describes as the global leader of Lean-ness, improved their inventory turns during a 10-year period ending in 2001 from 4.79 to 63.5—now holding on average less than 6 days of inventory.[75] Two years later Dell was down to 4 days' inventory, with their next goal of reaching 3 days—over 100 turns per year.[76] Careful planning, supplier coordination, and lead time reduction are necessary to achieve this astonishing performance.

In the early days of Lean, Toyota led the way in the development of supplier planning and collaboration techniques. According to Taiichi Ohno in his seminal book, *Toyota Production System, Beyond Large-Scale Production*:*

Toyota naturally makes production schedules—just like other companies. Just because we produce just-in-time in response to market needs [. . .] does not mean we can operate without planning. To operate smoothly, Toyota's production schedule and information system must be tightly meshed. First, the Toyota Motor Company has an annual plan. This means the rough number of cars—for instance, 2 million—to be produced and sold during the current year. Next, there is the monthly production schedule. For example, the type and quantities of cars to be made in March are announced internally early on, and in February, a more detailed schedule is set. Both schedules are sent to outside cooperating firms as they are developed. Based on these plans, the daily production schedule is established in detail and includes production leveling.

In the Toyota production system, the method of setting up this daily schedule is important. During the last half of the previous month, each production line is informed of the daily production quantity for each product type. At Toyota, this is called the daily level. On the other hand, the daily sequence schedule is sent only to one place—the final assembly line. This is a special characteristic of Toyota's information system. The kanban acts as a production order for the earlier processes. In other companies, scheduling information is sent to every production process. In business, excess information must be suppressed. Toyota suppresses it by letting the products being produced carry the information.[77]

Note the important distinctions between planning, scheduling, and execution in Ohno's explanation. Lean Manufacturing does not eliminate the need

---

* Originally published in Japan in 1978 and translated into English in 1988.

for planning but actually increases its importance. With Just-In-Time inventory movement *there is little margin*—of inventory or lead time—for error. For this reason even the Leanest organizations will maintain a safety stock at various stages within their production processes; in the Lean vernacular these may be called *kanbans, supermarkets,* or *constraint buffers.* Regardless of their name, they are carefully planned and managed pools of anticipation inventory and protective capacity to buffer uncertainty.

Lean Manufacturing authorities generally agree that long- and medium-range planning and supplier collaboration is necessary for the many reasons just described. So we must look more closely at the short-range processes described above by Ohno (the daily level), which require methodical planning, daily scheduling, and release of work to the shop floor, to find the critical and often controversial inflection point between Lean Manufacturing and MRP II.

FLOW PRODUCTION BASICS

As we discussed in Chapter 4, the MPS is regulated by time fence rules, which are designed to suppress volatility to maintain a stable and reliable schedule as we approach the time of production. At a certain point, time fence rules declare the production schedule to be "frozen," not allowing further changes. However, time fence rules may be overly restrictive in the fast-paced, demand-driven world of Lean Manufacturing. According to John Dougherty in his article "Managing MPS Changes Despite Time Fences and Frozen Horizons":

> Frozen schedules make cold customers. Fewer changes is not good performance. In fact, over the long term, the more changes the better! What this means is the ability to effectively change and still meet the schedule 95%+ of the time and the customer requirements 98%+ of the time should be the goal. Developing supply chain linkages, manufacturing processes, and planning and scheduling approaches that allow more change, less cost, and higher reliability needs to be seen as the ideal goal.[78]

In the *traditional* planning world, elaborate time fence rules were designed to help schedulers manage chaos. However, in order for these rules to provide a complete solution, a company may have to maintain multiple simultaneous planning scenarios and time fence boundaries for various demand, supply, and production characteristics that exist within the enterprise and its locations and product families. This creates an enormous burden of complexity, data capture, and management. Without powerful software and the skilled resources to perform these tasks, managing this sort of complexity is impossible. Even with good software and great schedulers, it may not be practical and is certainly never perfect. Why?

There is a fundamental flaw in the logic of MRP when applied to scheduling and execution. MRP calculates recommended purchase and production release times based on *fixed lead times* that are stored in the BOM and Routing data files. But in reality lead times vary constantly according to the conditions that exist in the plant and the supply chain at any moment. Because queue time (the time a product spends waiting for the next operation) often represents more than 90% of the throughput time (waste), scheduling short-term shop floor execution based on fixed lead times (which include standard queue times) is fundamentally flawed; in fact, it can create a self-fulfilling prophesy of long lead times. Pushing work onto the shop floor or to the next work-center regardless of its readiness to perform the work creates uneven flow and waste in many forms.

So what is the answer? Better scheduling logic? Data capture devices at every operation to track the input and output of each job in real time? Faster computers? More schedulers? More time fence rules? *How about reducing inventory, batch sizes, setup and throughput times?* This improves the responsiveness of the plant, shortening the duration of the Frozen Zone while reducing the time horizon for forecasting and planning. And while we're at it, let's couple the operations together using pull signaling mechanisms, which drive production release based on the real-time conditions on the shop floor, allowing us to schedule at a single point.

In a Lean operation the lead time may be very short, and thus the duration of the Frozen Zone may be measured in hours or even minutes; it is the time that work is flowing through production and should not be interrupted. The forecast horizon is short and thus extremely accurate, causing fewer surprises (chaos). Short setup times enable rapid and frequent product changeovers, so products may be configured individually to order with a very short lead time. There may be no time to find a manager to authorize changes in the Frozen Zone because jobs are moving through so quickly—and that's exactly what we want. In fact, as the rate of production increases, paradoxically the scheduler has *more* last-minute flexibility.

Consider this: In a repetitive Lean environment, when you have a difficult scheduling decision, you can defer the job one position in the release sequence. If your takt time is eight minutes, then that is all this job's due date has lost by leapfrogging the next job onto the shop floor sequence. This gives you eight minutes, or sixteen, or twenty-four . . . to gather necessary information to make an educated decision. So if you're unsure, defer. Of course, the scheduling and release system, and the demand signaling apparatus that orchestrates the execution of work (such as a heijunka box, described shortly), must be flexible and transparent to the operators so last-minute sequence changes cause no confusion or disruption.

The Lean Manufacturing approach to near-real-time planning and execution, where production is matched to the rate of customer demand, obliges us to reconsider the traditional integrated planning model and time fence rules. To do this we must first understand takt time, mixed-model production, and a level-scheduling release mechanism called heijunka.

Takt Time

Takt time is an essential concept for the development of flow. Womack and Jones elegantly describe takt time in *Lean Thinking*:

> [Takt time] synchronizes the rate of production to the rate of sales to customers. For example, let's assume that customers are placing orders at the rate of forty-eight [bicycles] per day. Let's also assume that the bike factory works a single eight-hour shift. Dividing the number of bikes by the available hours of production tells the production time per bicycle, the takt time, which is ten minutes (sixty minutes in an hour divided by demand of six bikes per hour). Obviously the aggregate volume of orders may increase or decrease over time and takt time will need to be adjusted so that production is always precisely synchronized with demand. The production slots created by the takt time calculation [...] are clearly posted [...] so everyone can see where production stands at every moment. This can be done with a simple whiteboard in the product team area at the final assembler, but will probably also involve electronic displays in the assembler firm and electronic transmission for display in supplier and customer facilities as well.[79]

This is a wonderfully simple illustration, as it describes not only how takt is calculated but also how it visually communicates a steady rate of production across the entire value stream. Of course there's a little more to takt time and flow than this simple example suggests, so let's take a closer look at the mechanics. Takt time is calculated by dividing total effective working time for the period by total demand. If demand is 2800 units per week*, and there are 80 hours of effective work time in the week, the takt time is:

$$80 \text{ hours} \times 60 \text{ minutes}/2800 \text{ units} = 1.71 \text{ minutes per unit.}$$

Effective working time is the available work time minus breaks and *scheduled* downtime. To prevent waste from creeping into the takt time calculation, it's important to ensure that scheduled downtime operations are continuously improved. Keep in mind that we should strive to run the operation at less than 100% capacity (unless it is a bottleneck) so that we can flex for unplanned downtime†, and so we do not overwork our people, allowing slack time for preventative maintenance and other continuous improvement activities. For simplicity, in this example we'll use 100% of available capacity in the following takt time illustrations.

Takt time is limited by cycle time‡, which is governed by the longest operation cycle time (the constraint) within the cell (or the entire value stream).

---

* Monthly demand for the product family volume may be determined by the Sales and Operations Planning (S&OP) process.
† In some environments this planned slack capacity also buffers short-term demand variations, but in the case of a level schedule, the supermarket buffers this demand variation so the takt time may be kept stable.
‡ Cycle time is the time required to perform a single workcenter operation.

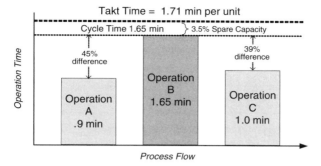

Figure 5-02. Unbalanced line, total work content = 3.55 minutes

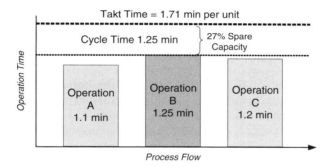

Figure 5-03. Balanced line, total work content = 3.55 minutes

A balanced cell is preferred, where the cycle time of all operations are within 30% of one another.[80] Even in a well-balanced cell, one workcenter will usually have a slightly longer cycle time than the others; this is technically a bottleneck (constraint), and this operation is used to measure the cycle time of the entire cell. Figure 5-02 illustrates an unbalanced cell with operations A, B, and C. Note that operation A with a 45% difference* to the cell cycle time, and C at 39%, are both regularly idle, whereas operation B (the constraint) is running at 96.5% capacity†, beyond the appropriate slack capacity limit.

Compare this with the balanced cell illustrated in Figure 5-03, where the total work content of 3.55 minutes has been redistributed evenly among the operations. Not only is the work more evenly balanced, but the cycle time of the entire value stream is reduced from 1.65 to 1.25 minutes, accommodating a shorter takt time should the rate of demand increase. This cellular arrangement has a cycle time supporting the maximum velocity of one unit every 1.25 minutes (based on operation B, which is still the constraint), so a takt time of

* $1.65 - 0.9/1.65 = 45\%$.
† $1.65/1.71 = 96.5\%$.

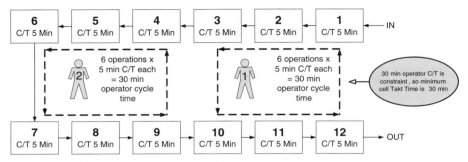

Figure 5-04. Two people in a U-shaped cell with six operations each, supporting a minimum takt time of thirty minutes

1.71 minutes is generous with 27% spare capacity at this time. With a utilization of only 73%* an additional product sharing some characteristics with the rest of the family may be added to the cell without jeopardizing its throughput.

A plant will usually have less than 100% of theoretical capacity available at any particular time because of a variety of factors including machine downtime, slower operators on some shifts, machine performance variations, unscheduled maintenance, material supply interruptions, quality problems, absenteeism, and so on. These variables can all influence the actual cycle time of an operation or cell. Likewise, the rate of demand may also shift suddenly, which can change the takt time. According to Chris Gray and Tom Wallace in their article "Manage It":

> Takt time is the rate of production required to meet customer demand. Operational takt time [which includes factors such as] inventory adjustments, products with seasonal sales curves, plant vacation shutdowns, intermittent large demand shifts, and other factors may require a wider view than pure takt time provides.
>
> Within the context of Sales and Operations Planning (S&OP), pure takt time would be calculated from the customer orders and forecasts; operational takt time would be derived from the production plan, which is the pure demand plus or minus necessary adjustments.[81]

Takt time may be set at a monthly rate according to the production plan, determined through the S&OP process; operational takt time may then be recalculated on a weekly basis if variations are significant, while daily fluctuations are buffered by supermarket inventory. How quickly can takt time adjust to changing demand? Consider the following *ideal* example of a balanced and flexible cell (shown in Fig. 5-04[82]) containing twelve operations, each

* 1.25/1.71 = 73%.

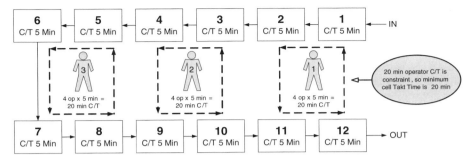

Figure 5-05. Three people in a U-shaped cell with four operations each, supporting a minimum takt time of twenty minutes

with a cycle time of five minutes. Because the cell is designed in a U-shape, workers can move around the cell, performing several operations within their scope of standard work. In this first illustration, there are two workers in the twelve operation cell; each worker performs six five-minute tasks. *The total cycle time of all operations performed by an individual worker represents the cycle time constraint of this cell, so thirty minutes is the fastest achievable takt time in this cell configuration.*

Now let's say that demand increases, requiring a takt time of twenty minutes (three units per hour); what can be done? Another worker is added to the cell, so each is now responsible for four tasks creating a minimum takt time of twenty minutes, as shown in Fig. 5-05.

The absolute minimum takt time supported by this cell layout is five minutes (constrained by the longest operation cycle time within the cell), requiring twelve workers, one at each operation. Note that the *throughput time* (lead time) is sixty minutes (twelve operations at five minutes each) from the time a particular job enters until it exits the cell, regardless of how many workers are assigned to the cell. But the takt time (rate of production) is affected by staffing.

In such a perfectly balanced and flexible cell, takt time can flex simply by adding or removing workers from the cell. Of course, a real plant may not be this agile. Like overzealous planners using MRP, reconfiguring a cell or changing the rate of production too often can cause the entire system to become nervous and chaotic. Frequent staff rescheduling may create havoc for managers and employees, so some degree of stability is desirable; thus a level schedule feeds a supermarket that buffers minor production and demand variations, creating a relatively stable takt time.

The Pacemaker

Once a level schedule for a value stream is calculated, it must be communicated simply and visually to the shop floor, ideally at a single point in the value

Figure 5-06. Product flow analysis

Part #	Product Name	Injection Mold	Stamp	Debur	Paint	Weld	Mechanical Assy	Electrical Assy	Final Assy	Configure and Test	Ship
18392	XS2 Servo Motor	X	X		X	X	X	X	X	X	X
21000	Sensor-Activated Arm	X	X	X	X	X	X	X	X	X	X
19283	Photoelectric Detector				X	X	X	X	X	X	X
19299	Ionization Detector				X	X	X	X	X	X	X
54950	Dust Filter	X	X	X			X		X	X	X
23756	Fiber Optic Visual Sensor						X		X	X	X
34556	Carbon Monoxide Detector						X		X		X
98840	Multi-unit housing	X	X	X	X	X					X

stream called the *pacemaker*. This is complicated by the fact that an enterprise usually manufactures several products and/or product families*; these families may be organized into separate value streams within the plant, each with its own takt time and pacemaker. An enterprise must logically evaluate and organize their products and flows. There are two natural stages in this organization process:

1. **Transformation**—Making logical and physical changes in the production process based on product family groupings, similarities in product movement, similarities in materials and operations, cellular design, bottleneck characteristics, and other flow characteristics of the value stream. Transformation is an ongoing process of continuous improvement, because a change in any of these conditions may require the redesign and rebalancing of the cell or line, or recomposition of the product groups. Although sophisticated design and modeling software tools are available, in general, information technology should take a back seat to creative team-based problem solving during the physical transformation process.

2. **Operational Management**—Forecasting demand, identifying the expected demand composition among product families and options, calculating takt time on a periodic basis, sizing kanbans and supermarkets, leveling the schedule, and then communicating the anticipated and actual release schedules to suppliers and the shop floor. Operational management is an ongoing process that usually requires integration with the MRP II planning and scheduling systems.

To group products that share common flow patterns into families and value streams, it is helpful to use a product flow analysis such as the one illustrated in Figure 5-06.[83]

---

* Ideally the organization of these families should be consistent with the Sales and Operations Planning process, where the product families from a demand management perspective are aligned with their respective level schedules and production cells.

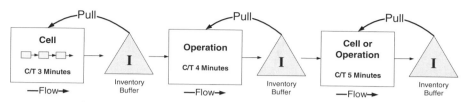

Figure 5-07. Buffer pull signals uncouple variable flow rates

Once commonalities among product families are established, a single value stream for each family may be designed, comprised of several sequential operations and/or cells. An operation is a distinct activity, whereas a cell is a combination of balanced operations. A value stream is comprised of a series of cells and operations. There are usually places where an operation or cell flowing at one production rate meets an operation or cell of a different rate. These distinct flows are joined by a buffer of inventory which decouples the two production rates. This relationship between flow and pull is illustrated in Figure 5-07, where a value stream is comprised of cells/operations with three-, four-, and five-minute cycle times.

Takt time is derived from the demand for the entire value stream; by contrast, each cell or operation can produce at a different rate. The slowest cell or operation is effectively the constraint of the entire value stream; in Figure 5-07 it is five minutes. This relationship between takt time for the entire value stream and the production rates and capabilities of the various cells and operations complicates scheduling and control. Traditionally, many plants tried to schedule each operation or cell separately, which is not only complicated but adds no value. A basic principle of Lean Manufacturing is that we should only schedule at one point in each value stream; this is called the *pacemaker*.

As Ohno explained earlier, Toyota communicates the schedule to final assembly and pull signals regulate all other upstream operations and cells. Focus on the pacemaker greatly simplifies scheduling, operational feedback, and control tasks in a complex environment. Because the pacemaker naturally regulates the throughput of the entire value stream, to schedule and control the flow of the process at any point other than the pacemaker is *a waste of time and effort*. This is an extremely important point whether we are scheduling visually, manually, or electronically.

The pacemaker synchronizes production throughout the entire value stream and is placed as far downstream as possible, close to the point where demand signals are received from the customer, so that pull signals can cascade backward through upstream operations and cells. Takt time, representing demand, is the drumbeat that drives the pacemaker. Once we have organized our operations and cells, calculated the takt time of each value stream, and

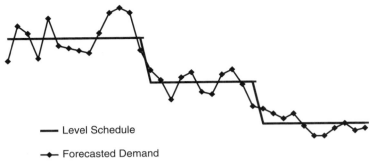

— Level Schedule

—◆— Forecasted Demand

Figure 5-08. Level schedule

identified the pacemakers, how should we determine how much of each product within each family to schedule, and in what sequence? There are two basic approaches to scheduling and sequencing a mixed-model value stream. First, the process may be designed so that numerous models and options within a product family may be changed over from one piece to the next without causing any delays. This is *one-piece flow*, where production is pulled directly from customer demand. In many cases, however, setup times between model and configuration changeovers do not permit individual piece flexibility, requiring the use of batching. These batches usually replenish a finished goods supermarket, whose inventory (safety stock) levels are carefully planned to satisfy forecasted demand and buffer demand variability. The calculation of batch size and frequency to replenish this supermarket at a steady rate is called level, or *heijunka*, scheduling (Fig. 5-08).

Heijunka Scheduling

Heijunka is a Japanese term meaning "to make flat and level." The heijunka method is also referred to as *Rate-Based*, *Level*, or *Campaign Scheduling*. Heijunka breaks down the total volume of orders for a given planning period (two months, one month) into scheduling intervals (weekly, daily). A heijunka calculation then defines a repetitive production sequence for that scheduling interval, which dictates the model mix scheduled on a given line. That schedule is put into operation through the use of kanban cards or signals for the mix of products.[84]

Heijunka works as a shock absorber, buffering variations in supply and demand, providing the shop floor and suppliers with a stable short-term production plan. It is very important to understand that a heijunka schedule requires a forecast of quantity and mix for each value stream, so it is applicable only if demand is relatively stable. To buffer the inevitable differences between forecast and reality, heijunka uses a supermarket. With a level schedule, customer demand is fulfilled from a finished goods or final assembly super-

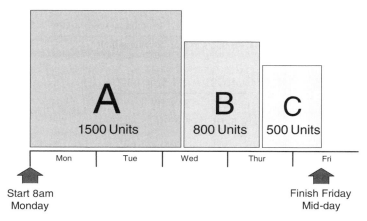

Figure 5-09. Traditional batch and queue work order schedule

market. When the product mix is so variable that it is not practical or economical to maintain a finished goods supermarket, then an Assemble or Configure to Order method may be favored, where the final assembly process is served from a final assembly (semifinished goods) supermarket. A hybrid of these two approaches may be used, where a standard product mix is level scheduled to supply the finished goods supermarket, while individually configured orders are inserted into the scheduled mix on demand.

Let's begin by illustrating the traditional workorder scheduling approach for a value stream and then contrast this with heijunka scheduling. Assume that we manufacture three products in a particular family, and we expect a weekly demand of 1500 units of Product A, 800 units of Product B, and 500 units of product C, for a total of 2800 units per week. Setup time is 15 minutes, cycle time is 1.5 minutes per unit, and there are two eight-hour shifts.

In the traditional batch and queue mode we would schedule each of these jobs (A, B, and C) as a single work order to minimize setup costs (Fig. 5-09). Each batch would move slowly through the production process, overloading some work centers while leaving others starved and waiting for work. If we started Monday morning, Product A would finish midmorning Wednesday, Product B would finish Thursday afternoon, and Product C would finish midday Friday. If everything goes as planned, a customer ordering Product C on Monday would have to wait until Friday, unless we maintain a substantial safety stock to anticipate the demand that cannot be satisfied by such a protracted and lumpy work order-based production schedule.

There simply isn't enough time available if we attempt to satisfy this demand with one-piece flow. Running each unit individually, using an impractical assumption that there is a setup after each unit, requires 770 hours* in

---

* 2800 units × (15 minutes setup + 1.5 minutes run) = 770 hours, assuming a changeover after each unit.

Effective working
time in period − (Run time x Quantity for Period)

――――――――――――――――――――――――――――――――

\# Products in Mix x Setup time

$$\text{EPE} = \frac{4800 - (1.5 \times 2800)}{3 \times 15}$$

Figure 5-10. EPE formula example

an 80-hour two-shift workweek. Obviously some batching is required, but how large should the batch sizes be to satisfy the demand? As small as possible, but how is that calculated? Note that takt time calculates the effective working time per unit of demand, but *does not factor in setup times*. Therefore, takt time alone cannot help us to determine the optimal batch size and changeover frequency when we have multiple products running on the same cell. To match demand to the available time for production (including setup time) we must calculate a *changeover interval*, which is the period of time required to produce one full cycle of a product family. A shorter interval creates a shorter lead time so that production can be more closely matched to demand. Setup time reduction is the key to reducing the interval. The interval is calculated with a method called the *Every-Part-Every* interval[85] (EPE, EPEI, or EPE*x*). Total effective work time available in the period is reduced by the total *run* time required for the period. The remainder is the available time for setups, which is divided by the number of products in the mix multiplied by the setup time*. This results in the interval—the number of product family changeovers possible during the period. Figure 5-10 demonstrates the EPE calculation using the example of Products A, B, and C.

Note that the effective working time is 4800 minutes (5 days × 8 hours × 2 shifts × 60 minutes), assuming for simplicity that there is no downtime. Based on 4800 minutes, this calculation suggests that 13.3 intervals, each representing *360 minutes of total demand* (4800 min/13.3 = 360) should be produced for Products A, B, and C before the next changeover occurs, as shown in Figure 5-11.

By running smaller batches of each product over shorter periods of time, a factory is able to maintain smaller inventories while being more responsive to changing demand. To prevent frequent recalculation of this schedule and disruption of the natural rhythm of production, a supermarket buffers minor variations.

In this example we have assumed that all 113 units of Product A are identical, followed by 60 identical units of B, and so on. But consider an automobile plant using heijunka scheduling to determine the interval of model

―――――――――――――――――――――――

* This assumes that setup time is consistent for each product in the family; otherwise, a weighted average setup time may be used.

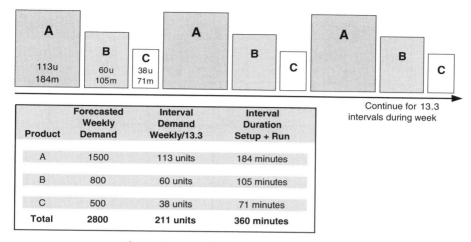

Product	Forecasted Weekly Demand	Interval Demand Weekly/13.3	Interval Duration Setup + Run
A	1500	113 units	184 minutes
B	800	60 units	105 minutes
C	500	38 units	71 minutes
Total	**2800**	**211 units**	**360 minutes**

Figure 5-11. Heijunka schedule calculation

changeovers. Each individual automobile within a model/family may require a slightly different configuration of options, either according to a specific customer order (configurable demand pull) or based on a forecast of configuration frequency planned in advance (Planning BOM). This requires not only heijunka scheduling but also a release mechanism to ensure that the proper materials and configuration instructions (drawings, BOM, work instructions) are communicated (using paperwork or electronic displays) as each unit is pulled through the shop floor.

The release of work (as well as other documentation and work instructions) may be communicated to the pacemaker through the use of a *Heijunka Box*, also called a *Load-Leveling Box* or *Post Office Box Scheduling Board*, as illustrated in Figure 5-12. When kanban cards are used to control production, they are inserted into the appropriate time slots. In this illustration the sequence is AABBCDAABBCD, resulting in an ABCD production ratio of 2:2:1:1. The earlier EPE calculation suggests an ABC production ratio of 113:60:38, so the heijunka box would be loaded with 113 A cards, followed by 60 B cards, followed by 38 C cards; then the sequence would be repeated.

When the batch size is one (one-piece flow) it is simpler to change an individual card, altering the sequence to accommodate a sudden demand shift, or inserting a special configuration for a custom order. If the cycle time between units is short, however, it may not be practical to distribute 113 separate cards to the same workcenter either individually or all at once—we certainly wouldn't want to make a job for someone just creating and distributing these cards all day! To limit the number of cards and the resulting wasted motion, the heijunka slot could contain a single card to produce 113A, followed by a single card to produce 60B, and so on.

Shift 1	8:00	8:20	8:40	9:00	9:20	9:40	10:00	10:20	10:40	11:00	11:20	11:40
Shift 2	4:00	4:20	4:40	5:00	5:20	5:40	6:00	6:20	6:40	7:00	7:20	7:40
Product A	A	A					A	A				
Product B			B	B					B	B		
Product C					C						C	
Product D						D						D

Release sequence AABBCD Pitch 20 minutes

Figure 5-12. Heijunka box

In some environments this may cause a challenge with the irregularity of the job start times for each card, complicating the smooth milk run movement of cards from the heijunka box to the gateway* workcenter, and between cells. This may be addressed by having a card represent a standard batch size (kanban card or container quantities) so each heijunka time slot must represent a standard quantity. For example, if takt time is 30 seconds, and a reasonable time to deliver one round of cards and materials around the shop is 20 minutes, then batch size is 40 (20 min/.5 min per unit = 40 units). This release frequency (interval between time slots in the previous illustration is 20 minutes) is called the *pitch*, which is calculated simply by multiplying the takt time and the batch quantity. This ensures the delivery of a consistent quantity of material to the pacemaker on a regular schedule. There is usually an individual (*water spider, runner, material handler*) responsible for making regular trips around the plant, gathering and distributing heijunka and kanban cards, moving materials according to the schedule and kanban signals, and watching carefully for signs of interruption anywhere in the flow.

The rate of production of a particular value stream may be visually monitored by observing the regularity of the pitch delivery to the pacemaker operation—this is often compared to a train schedule, and if a product misses the train then we know there is a problem within the duration of one interval. Pitch sets the value stream in motion, then measures its performance.[86] Alternately, performance against the pitch rate can be measured by the rate of delivery from the end of the line to the supermarket. In either case, measuring the input or output unit flow provides instant and visual feedback that the entire process is running according to the takt and pitch time.

---

* Gateway workcenter is the first operation in a routing, which initiates the work and first pulls raw materials from inventory.

Because takt time changes according to the rate of customer demand, and because pitch is derived from takt time, the pitch interval may also periodically be recalculated. However, the entire plant may develop a rhythm according to this pitch time, which must also allow time for the material handlers to make their rounds, problem-solving as they go. Frequent changes to the takt and pitch times should therefore be considered carefully. From a mechanical perspective of the heijunka box, the pitch time interval across the top must be easy to change, so a magnetic or greaseboard schedule may be used. Alternately, an electronic heijunka box can automatically calculate and display the schedule with the appropriate intervals, release sequence, and quantities.

Mixed-Model Complexity and Software

This overview of cycle time, takt time, cell balancing, and heijunka mechanics is offered to make a point from an information systems perspective. On the one hand, do not underestimate how sustained improvement efforts using these techniques can simplify and fine-tune a plant. On the other hand, do not underestimate the complexity that may arise in a dynamic environment, with many variables to skillfully manage including:

- Production volume
- Production velocity
- Variability of product mix
- Maintenance and issue of configuration and work instructions for each job
- Lead time of the various components of the product mix
- Variability of supply lead time
- Number of option components in final assembly
- Interrelationships among configurable options contained in multilevel Planning BOMs and specified by an order entry product configurator
- Volatility of demand
- Short-term responses to competitive actions, such as pricing and promotions that dramatically and suddenly shift product volume and mix
- Changes in the S&OP and MPS
- Cell balancing
- Kanban container quantity and capacity
- Variability of actual production rate (cycle, takt, and pitch times)
- Integration with real-time changes in order processing and inventory control systems
- Shifting supermarket levels for raw materials, final assembly stock, and finished goods
- Managing schedules for multiple plants, value streams, product families, and cells

The sheer volume of calculations and recalculations often requires software assistance of some kind. If you're interested in learning more about the mechanics of mixed model heijunka scheduling, *Creating Mixed Model Value Streams* by Kevin J. Duggan includes a CD-ROM that contains a number of helpful spreadsheet templates to get you started.

Although many companies may initially attempt to manage heijunka schedules using spreadsheets that manually interface to their MRP II systems, they may reach a point where the spreadsheet (and the planner driving it) cannot keep up with the complexity, volume, and frequent changes. To avoid burdensome manual calculations, takt and pitch time and the resulting heijunka schedule may be calculated monthly (or even less frequently); however, this may result in larger than necessary supermarket quantities to buffer the short-term variations. Fortunately there are several software publishers that offer dynamic heijunka scheduling capabilities that are either built in or integrated with ERP/MRP II systems.

Although heijunka smoothes the flow of mixed-model production, it can also create an overwhelming number of transactions in the MRP II system if we insist on capturing data on materials and labor in the traditional manner. Mixed-model production and the ideal goal of one-piece flow requires a new way of thinking about work orders, data capture, transactions, and cost accounting, which we'll explore in Chapters 9 and 10.

Heijunka scheduling may be appropriate in mixed-model situations where there is some visibility to future demand patterns through an extended supply chain. An ideal example is the automotive industry, and heijunka is a key element of the Toyota Production System, where demand is planned months in advance and volatility is buffered by an immense and multilayered distribution inventory channel. A variation of the level schedule approach allows for a different configuration to be inserted into the level production schedule according to an individual customer order, while preserving a stable rate of production. For example, Dell Computer uses a "sort of" heijunka by loading up work into two-hour increments (pitch) that are issued to the shop floor.[87] This dynamic Configure to Order level scheduling is likely to require more computer interaction than a stable and repetitive product mix; we'll explore the mechanics of this approach later in this chapter.

A final word on heijunka—it is considered by many to be the highest aim of Lean execution, and heijunka may not be practical in many environments. There are many basic transformations throughout the enterprise that must be accomplished first—stabilized demand, shortened lead times, flattened BOMs, product family organization, cellular layout, load balancing, heijunka box design and signaling mechanisms, and so on—before heijunka is attempted. Participants in our seminars and workshops are often dismayed by the challenge of heijunka, because many of the basic foundation requirements are not in place within their own environment. They make the mistake of hoping to get from A to Z in one step. Lean transformation is a long journey, consisting of patient and relentless progress. Wherever you are, that's where you begin.

THE LEAN PLANNING MODEL

Now that we understand the mechanics of mixed-model production, how do we coordinate with our suppliers, production staff, and customers? For instance, suppliers may be expected to respond immediately to kanban signals for inventory replenishment, so they must plan ahead. Because there is little room for error with JIT delivery schedules, it is important for us to communicate our anticipated material requirements from the forecast horizon until the time of release.

In this section we explore an integrated planning model for a Lean Manufacturing operation. This model is identical to the traditional model during the long-range and monthly Sales and Operations Planning process, but it deviates considerably as we approach real-time demand-driven production.

The Lean Integrated Planning Process

The Lean Integrated Planning Process is shown in Figure 5-13 and is comprised of the following steps:

1. The S&OP process creates the Production Plan, which in turn becomes the Master Production Schedule (MPS) as described in Chapter 4.
2. The MPS is used to drive the Material Requirements Planning (MRP) calculations, which trigger the release of long lead time purchase orders to suppliers.
3. The MPS is used to derive takt time. This process is roughly equivalent to the traditional rough cut capacity plan in that takt time is the demand for capacity, and cycle time (representing the theoretical capability of the line/cell/value stream to produce) represents capacity. To the extent that takt time is in line with cycle time, from a rough cut capacity standpoint the MPS is a valid schedule.
4. A preliminary release schedule is determined. This may be a heijunka schedule in a rate-based environment, or it may be anticipated kanban releases in a pure demand-pull environment. In either case, the anticipated release is based on the net period demand expected for each specific item, which is calculated by adding or subtracting each item's forecasted demand by quantities in the finished goods or final assembly supermarket that are above or below the target order point replenishment level for that item.
5. The preliminary monthly release schedule (heijunka or anticipated JIT kanban release) is communicated to suppliers, so they can plan their short-term production and release schedule accordingly. This schedule may be communicated actively, by sending out MRP or blanket purchase order/release transactions to suppliers by mail, fax, e-mail, or EDI. Alternately, a supplier portal website may passively display the

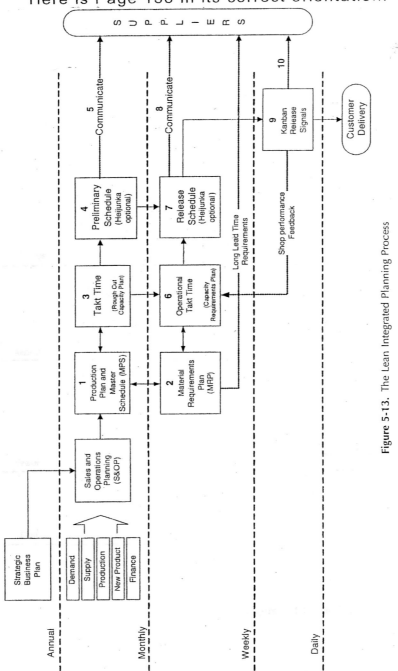

Figure 5-13. The Lean Integrated Planning Process

anticipated release schedule, which the supplier may view at any time. The portal may even provide a data interface so the supplier can import the information directly into its planning system.

6. As we approach the real-time release of work, sudden changes in demand or supply may cause us to revise takt time to an operational takt time, rebalancing production resources, recalculating the pitch, and reorganizing the heijunka box. However, these rapidly cascading changes may not be practical in some environments, so inventory buffers must pick up the slack. If the newly calculated operational takt time is in line with the constraining cycle time, the schedule is considered viable; this is equivalent to the traditional Capacity Requirements Plan, which compares actual productive capability against demand.

7. Takt time (original or operational, whichever is used) and net period demand by item are used to generate a final release schedule (heijunka schedule or anticipated JIT kanban release). If demand and production parameters do not change significantly, it is not necessary to recalculate the EPE interval and relevel the schedule each week. Part of the EPE calculation requires current demand information, which accounts for replenishment of the supermarket stock levels. If EPE is not recalculated regularly, then there should be an exceptions warning mechanism that indicates when a supermarket stock level falls below a critical replenishment threshold because of an unexpected surge in actual demand.

8. The short-term release plan is communicated to the suppliers actively or passively as described earlier. If there is no change from the preliminary schedule, a protocol may be established with the supplier that no communication is transmitted. However, a missed signal could be accepted by the vendor as no signal, leading to miscommunication and potential supply interruption.

9. Finally we are ready for actual production; kanban and/or heijunka signals initiate the flow of materials.

10. Raw material kanbans signal suppliers for replenishment.

Unlike the traditional push scheduling environment, a Lean pull environment can respond very quickly to changes in production rate and mix. In a kanban environment, pull signals ripple back to the gateway workcenter, indicating the next item to be produced; this signal instantly indicates item-specific demand and available production capacity. In a heijunka scheduling environment, change rules can also be nearly instantaneous. Until the very moment the next job is released to a pull signal, the scheduler *may* change the item or configuration to be produced as long as this can be communicated to production (using a manual or electronic heijunka box) and there is no trade-off cost such as a materials constraint. There can be no expediting once a job has been released, nor is any needed, because the work flows quickly and

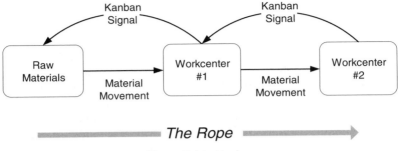

Figure 5-14. Kanban

smoothly to completion. This is the beauty and simplicity of *demand pull and material flow*, which leads us naturally to a discussion of the kanban pull mechanism.

KANBAN ESSENTIALS

Kanban describes a variety of *pull* mechanisms that draw work through a plant (and in fact across an entire value stream within a Lean Enterprise and across the entire Lean Network) according to demand and downstream capacity. Imagine a rope pulling all items along at the same rate so every movement is synchronized. In the early 1900s Henry Ford used this technique by tying a rope to the front and back bumper of each car, pulling them through the assembly line. It was simply impossible for one vehicle to move any faster or slower than the one in front of it. This is the principle of kanban—because the workcenters are connected by a signaling mechanism. As shown in Figure 5-14, the immediate downstream workcenter-available signal indicates that *all** downstream workcenters are synchronized and available.

Kanban containers of specific inventory items may be staged along the production process so they are immediately available when needed. The kanban quantity is a blend of safety stock and WIP, and the total amount is calculated by the amount of inventory per kanban container, multiplied by the number of kanban containers.

Qty per container × Number of containers = Total kanban inventory

Looking back to Chapter 4 you will recognize that kanban is an order point inventory replenishment technique. In an environment managed by kanban there may be hundreds or thousands of mini-reorder points all around the factory,[88] and each must be carefully planned and managed. Like any other

---

* This emphasis on "all" downstream workcenters is an important point we'll return to later when we contrast the use of kanban in repetitive and discontinuous environments.

order point method, the calculation of kanban container size and the number of containers requires demand forecasting. In a consistent or seasonal demand situation this forecasting may simply involve a review of history to guide future reorder point level decisions; nonetheless this is a form of forecasting.

Through the calculation of container size and quantity, kanban *generally* creates a carefully regulated safety stock in several locations:

- A supermarket of finished inventory available for immediate consumption to fill demand
- A supermarket of semifinished inventory awaiting final configuration and assembly
- Stores of WIP for intra- and intercell movement on shop floor
- Stores of components and raw materials whose consumption signals supplier replenishment

It is helpful to think of a kanban system as a loop of information and materials circulating between two stores of inventory. When the downstream workcenter consumes inventory it sends the upstream workcenter a replenishment signal, which in turn initiates the next job by pulling material from its own store. Replenishment information flows upstream and materials flow downstream in a continuous loop. A scenario with multiple stores of inventory at various stages of production and distribution can be illustrated as a series of kanban loops, as shown in Figure 5-15. This triangle-and-loop symbology (representing the interaction of pull and flow) will be used throughout the rest of this chapter to illustrate the common idea underlying various techniques.

Replenishment pull signals may be in the form of one- or two-card systems, physical squares, containers, baskets, pallets, bins, golf balls, paper documents with bar codes, faxes, e-mail, electronic signals, buzzers, bells, lights, EDI releases, etc. Practically any visual, physical, or electronic trigger may be creatively used as a kanban signal. In addition to the types of signals used, there are many subtle types of kanban flow patterns. How do the cards move? Who picks them up? How often? How are flow interruptions identified and communicated? Virtually every practical application of kanban in a real-life production environment has its own creative twist—the flexibility and creativity of kanban is perhaps its greatest virtue.

Figure 5-15. Kanban as loops of information and materials

There is one particular style of kanban that deserves specific mention, the *Priority Kanban*, which allows a batch operation to be inserted within a flow process. Also known as a *signal* or *triangle* kanban, it is used when the upstream operation has a long setup time, thus requiring large batch sizes. An inventory buffer downstream from this batch operation allows units to be consumed individually, triggering replenishment to the upstream operation *when demand for the minimum batch size is reached*. John Bicheno describes a priority kanban system in *The New Lean Toolbox*:

> When there is a changeover, a signal kanban is used. As parts are withdrawn, kanban triangles are hung on the board under the appropriate product column. A target batch size is calculated for each product and the target is marked on the board. When a sufficient number of kanban triangles have accumulated to reach the target, a batch is made. This gives a visible, up-to-date warning of an impending changeover. In normal circumstances the batch is made when the target level is reached. If there are problems, kanbans may accumulate beyond the target level. This would indicate higher priority. Normally, a batch is made to cover all the kanbans in the product column. In very slack periods, a smaller batch may be made to cover only the cards on the board.[89]

A priority kanban can also be used as a heijunka box variation, where the pitch defines the time interval for the next job to be released, but where the priority of the next release is determined by which product has the greatest accumulation of triangles relative to its target batch size.

The Real Value of Kanban

With the many creative options for signaling, kanban may be used effectively in any environment where there are relatively predictable patterns of workflow. Later in this chapter we'll discuss kanban approaches that may be effective where workflow is discontinuous. Regardless of the environment in which it is used, the benefits of kanban include the following:

- Visual control and flow of work, which encourages teams to visualize the problems and find their own solutions, a foundation of continuous improvement.
- Kanban signals indicate demand, so overproduction is not allowed.
- Workcenters cannot be overloaded as long as the kanban size is calculated properly, and movement is based on availability of the downstream workcenter.
- The quantity of individual kanban containers regulates the batch size, and incrementally reducing the container size is a simple method to force batch size reduction.
- The number of kanban containers in use regulates the amount of WIP and inventory buffers on the shop floor.

- As the number of kanban containers is reduced, vacant floor space emerges and workcenters may be grouped closer together (often into cellular arrangements) requiring less move time and effort, while netting more productive capacity in the same physical space.
- Kanban quantities offer a simple method of WIP reduction and continuous improvement, called *space denial*.[90] As operations improve, gradually remove kanban containers—this will cause constraints to appear, so they can be identified and eliminated.
- By tracking the quantity of kanban containers over time, you can simply measure the trend of WIP reduction.

Kanban is clearly a useful approach to simplify and manage the flow of production and control inventory. But don't lose sight of the fact that kanban *generally* builds up small pools of inventory in anticipation of a future requirement; in this regard it's just an inventory replenishment technique. If a plant attempts to implement kanban before basic elements of flow are in place—the transformation process described in the previous section of this chapter, which may include batch size reduction, product family rationalization, flattened bills, cellular design, and so on—a misguided kanban effort can cause more harm than good. Kanban is a double-edged sword: on the one hand it's a very useful tool, but when used inappropriately it becomes just another cause of waste. In *The Toyota Way*, Jeffrey Liker shares his insights on the real value of kanban:

> It is fascinating to watch this work, with so many parts and materials moving through the facility in a rhythm. In a large assembly plant like the one in Georgetown, Kentucky, there are thousands of parts moving about. Alongside the assembly line, there are small bins of parts and small bins are being moved from neatly organized stores. It is hard to imagine how a computer system could do such a good job of orchestrating such a complex movement of parts. When you find out the computer is not doing the orchestration, but rather small, laminated cards moving about, it is shocking. Yet Toyota Production System experts get very impatient and even irritated when they hear people rave and focus on kanban as if it is the Toyota Production System. The challenge is to develop a learning organization that will find ways to reduce the number of kanban and thereby reduce and finally eliminate the inventory buffer.[91]

Two Basic Types of Kanban

It has been twice stated that kanban *generally* creates safety stock, and there is an important reason for that qualification. There are actually two distinct types of kanban systems, although most literature focuses on the *Product-Specific Kanban* method—not surprisingly because it is suited to a repetitive operation, which supports our ideal notion of what Lean Manufacturing is supposed to be. Uday Karmarkar of the UCLA School of Decisions, Operations and Technology Management urges in his *Harvard Business Review* article "Getting Control of Just-In-Time":

> JIT enthusiasts should realize that when a kanban system is implemented in an environment full of variations in supply and demand, it is even less likely than MRP to operate in a stockless manner—that is, without a burdensome amount of WIP.[92]

If there are a large number of distinct products manufactured in the cell, or a high degree of configurability or customization of the finished goods, then it is impractical to store even a single unit of each unique item in a finished goods supermarket. Does this mean that kanban cannot be used to control a less repetitive or high-mix operation? Not exactly, but it requires a special type of kanban that does not store inventory in advance, which at first seems contrary to the very idea of kanban.

Product-Specific Kanban. In a *Product-Specific Kanban* (also called *repetitive, product, name brand, type 1, and type A kanban*) environment, predefined kanban items and quantities are stored at each stage of production, awaiting the next demand pull signal. Let's consider a simple example where there is a sale of an item from a finished goods supermarket and where the items in the supermarket are stored in kanban containers with predefined quantities of *specific* items. The moment a product is picked for a customer, the kanban replenishment signal is sent to the preceding (upstream) workcenter. This is authorization for the kanban container to be refilled with the *same product and quantity*. This consumption and signaling process cascades backward through each successive upstream workcenter, until it reaches the gateway workcenter, which sends a signal pulling a predefined container of raw materials from stock, which in turn sends a signal to the supplier for replenishment of that item. The rope is attached from bumper to bumper. It is important to understand that in Product-Specific Kanban the next part/job released to the shop floor is identified by the part assigned to the empty kanban container. There is no predefined release schedule; the next kanban container that arrives at the gateway workcenter defines what inventory is next issued to production. Visualize this process as a backward wave cascading through the shop, as illustrated in Figure 5-16.

The essential rule in any kanban system is that a workcenter must remain idle until it receives a signal from a downstream workcenter. In Product-Specific Kanban that signal indicates that a downstream workcenter needs a kanban quantity replenished with a *specific product*. Note the many information/material loops at the bottom of the diagram in Figure 5-16; this illustrates the potential disadvantage of Product-Specific Kanban, which is the number of distinct inventory buffers (represented as triangles) at each stage in the flow. Product-Specific Kanban replenishment is appropriate when there is a repetitive demand and low variability, because there is predefined inventory safety stock waiting at each move step. As a form of order point replenishment, the

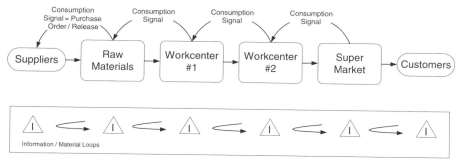

Figure 5-16. Product-Specific Kanban pull

kanban quantity and replenishment interval of these specific items is determined by a forecast, which assumes that demand is relatively predictable. By regulating the capacity and quantity of kanban containers, the amount of WIP for each specific item can be strictly controlled.

Generic Kanban. A *Generic Kanban* (also called *nonrepetitive, capacity, type 2,* and *type B kanban*) system supports low volume, volatility, and high variability of product mix and, in fact, displays some characteristics of a push system. Generic Kanban is typically used in mixed-model and Assemble or Configure to Order environments where the routing of operations remain consistent but the configuration and material requirements vary by individual job. When material requirements are variable it is impractical to stock specific kanban items in anticipation of customer demand. This explains Uday Karmarkar's comment about the creation of "a burdensome amount of WIP" when Product-Specific Kanban is applied to variable demand.

A vital distinction is that with Generic Kanban, the signal received from the downstream workcenter isn't for replenishment of a particular item, because the kanban containers are not product specific. The Generic Kanban merely signals that there is available *capacity* at the downstream workcenter, and that it is ready to receive the next kanban *with whatever materials are appropriate to the next job.* Specific material, configuration, and work instructions often accompany each job—either as a paper traveler document or as a computer device that displays the instructions as the job moves into the work area. The flow of Generic Kanban is illustrated in Figure 5-17.

With Generic Kanban the release schedule is driven by the backlog, containing customer orders for particular items and configurable options. Kanban pull signals regulate the flow based on downstream capacity. Unlike Product-Specific Kanban, where an item is pulled from finished goods stock in a downstream supermarket, with Generic Kanban a new job for the particular item is *pulled into the gateway workcenter based on the prioritized backlog of customer orders ready for dispatch to the shop floor.*

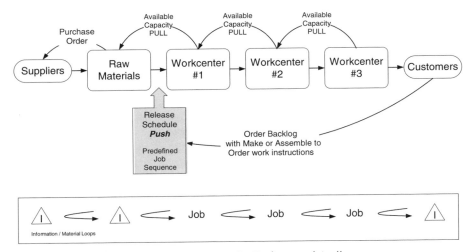

Figure 5-17. Generic Kanban push/pull

Another vital distinction in a pure Generic Kanban environment is that there is no idle WIP on the shop floor awaiting consumption. As illustrated by the information/material loops below the diagram in Figure 5-17, all Generic Kanban quantities are issued to a specific job from raw materials, so there is no idle inventory on the shop floor.

Generic Kanban is a combined push *and* pull system. It is like a push system because a specific job is introduced to the shop floor according to the predefined release sequence driven by customer orders. Unlike a traditional MRP push system, however, work is pulled by capacity-available signals from the downstream workcenters. A downstream capacity-available signal only indicates that the workcenter is ready to receive work and does not signal demand. If the backlog is empty, then no demand signal is issued to the gateway workcenter, so no new work is released to the floor. Thus the demand pull rule of kanban is satisfied. Stated another way, Generic Kanban is a pull-capacity signal, and the backlog is a push-product signal; they *must* work together. If there is pull capacity but no push product then the backlog is empty and work stops because there is no demand. If there is push product but no pull capacity then the job is not started until the downstream operation is available.

Heijunka Scheduling and Generic Kanban

Mixed-model heijunka scheduling requires a special application of Generic Kanban. The EPE determines the interval and repetitive batch quantity for each product in the mix, which in turn determines the quantity, mix, and sequence of work released to the cell. Because the specific item and quantity

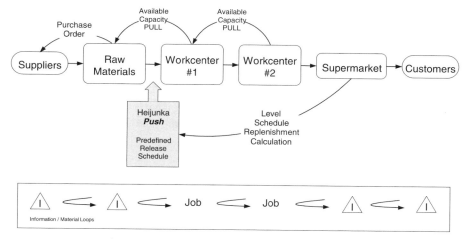

Figure 5-18. Heijunka Generic Kanban push/pull

sequence is determined by the heijunka schedule, and is released to the gateway workcenter when a kanban capacity-available signal is received from the next downstream workcenter, this is a Generic Kanban system. The important distinction with a heijunka schedule is that the demand signal is issued from a supermarket, not directly from the customer. Thus *heijunka is a Make to Stock push/pull scheduling and execution technique*, in which the supermarket buffers demand variation. Heijunka scheduling works as a shock absorber, creating a sense of stability by allowing the plant to produce at a stable rate and level mixed-model sequence buffered by a supermarket, rather than the volatility of producing each unit directly to a customer order.

Heijunka is illustrated in Figure 5-18; note that this and the previous Generic Kanban illustration are nearly identical, except that there is now a finished goods supermarket represented by another inventory triangle in the information/material looping box. Although heijunka communicates a steady and predefined sequence of work, the actual product mix and quantities of kanban containers may be varied (subject to material availability) moments before release. With a heijunka box it may be practical to replace the card in the next pitch slot just before it is distributed to the shop floor, causing no disruption as the shop floor doesn't know what the next card is supposed to be. In an environment where kanban signaling is performed electronically, rather than with physical signals such as containers or cards, the software can dynamically recalculate kanban/pitch sizes, takt time, EPE interval, and the kanban-heijunka release sequence. This provides the plant with a remarkable degree of flexibility to change the release sequence, to change the configuration of an individual unit to a customer order specification, or even to change the insertion of a particular model with another, without affecting the level rhythm of production.

For example, although Toyota plans its level schedule months in advance, they periodically replan that schedule until the day of production, when they can change the product configurations to match incoming customer orders. Some Toyota suppliers have achieved better than 100 inventory turns per month, with the ability to produce a wide mix of products within a few hours of receiving a kanban replenishment signal. To maintain this flexibility, they must obtain reliable product mix planning information from Toyota on a regular basis. According to Alan Cabito, Group Vice President of Toyota Motor sales:

> The Toyota system's not a build-to-order system. It is a *"change to order"* system. We pick a car on the line, any car, and change it. There's a lot of complexity to changing color—you have to change virtually all the accessories. And the way that gets managed is on the allowance of how much change can take place. There will be a limit to the number of green, leather-interior Siennas we can make in the same day.

> We place a single month's order three times. We'll order it four months out, three months out, two months out. During that time, they will set up all the components and suppliers. For July production, the final order will be placed in May. So your order's out there 60 days in advance. Then every week we can change the order in the U.S. plants. Every week we can modify anything that's unbuilt, except for the basic body type.[93]

Toyota manages their kanban buffers carefully so they have flexibility to suddenly change the next unit in the production sequence, as long as aggregate changes are within their WIP change allowance. Note the progression of planning stages Cabito describes, which allows Toyota and their suppliers the flexibility to anticipate and accommodate daily product mix changes. Although Toyota's horizon is several months into the future, and is buffered by an extended distribution network, the process he describes is consistent with the integrated planning process illustrated in Figure 5-13.

As they continue to refine their Lean methods, and as the complexity of their operations and the demands of the market increase, even Toyota has learned to rely more heavily on IT, creatively interwoven with their visual systems. According to Liker:

> These days Toyota is increasingly using computer systems for scheduling. For example, when ordering parts from suppliers, Toyota is moving to electronic kanban rather than sorting and sending cards back. In this case, it does not have to be either/or. Toyota will often use a computer system for scheduling some operations, but then use manual cues like cards or white boards to visually control the process.[94]

Hybrid Kanban Systems

It is important to note there is also a *Hybrid* Kanban system (also called *type C* and *type 3 Kanban*), which is a combination of Product-Specific and Generic

Kanbans. This may be used when there are both standard line-side materials that are consumed by all jobs (stocked and replenished with Product-Specific Kanbans) and job-specific materials using Generic Kanbans.

According to Stephen Moncrief of Parker Hannifin, there is a push/pull relationship between MRP and Lean, where they ideally work together in a hybrid scenario:

> The MRP system, through the kanban control center, controls the number and type of kanban signals on the shop floor. The MRP system specifies what final assembly is to be made. The pull system controls when the production will take place. The [Parker Hannifin] plant can be described as an Assemble to Order facility where a relatively low number of manufactured parts feed into thousands of possible end items. We schedule the final assembly using our MRP system, and we produce many of our lower-level components using kanban cards. The push pull system provides the mechanism to introduce kanban control to the job shop production floor.[95]

This is similar to a situation at the General Motors medium truck division plant, where each truck arrives at a workcenter with an electronic display of work instructions for the configuration of each specific job. According to John Ninotti, director of medium truck production operations for General Motors:

> It's important that we communicate to the operators how to build these vehicles, because the possible combinations are in the trillions. To that end, each operator has a computer screen describing which options dictate what parts to be picked for that vehicle.[96]

Of course General Motors also has many high-volume standard components that are stocked and replenished by the standard Product-Specific Kanban method. As a job moves through the plant, when a Product-Specific Kanban item (such as a standard truck chassis) is consumed from floor stock, the container is sent to replenish with another identical unit. Ford has another name for a similar method: SMART (Synchronized Material Availability Request Tickets), where an operator presses a button when a reorder point is reached for a fast-moving, heavy, or expensive part where floor space is limited on the line.[97]

In the same process alongside Product-Specific Kanbans, seldom-used components may be called out by a Generic Kanban signal. So, in fact, hybrid kanban systems using a mix of Product-Specific and Generic Kanbans combining push and pull techniques are appropriate in many situations. This level of complexity and volume is where a Lean shop may, in the words of *Industry Week* technology editor Doug Bartholomew, "scream out for technology help."[98] Although a simple Product-Specific Kanban environment may require no computerized scheduling or control, a Generic Kanban is quite different and often requires a computerized order entry, configuration, and backlog management system, feeding into a scheduling system for prioritization, sequencing, and release orchestration, producing a paper or electronic traveler with work instructions to follow the order through the production steps.

MRP and Kanban

MRP and kanban must work together when planning inventory buffers. It's important to remember that kanban is an execution signal, but the level of kanban and supermarket inventories must be planned in advance (Product Specific) and unique kanban replenishments also must be planned ahead to account for lead time and material availability (Generic). The APICS Lean Manufacturing Workshop session on Lean Scheduling suggests a framework for material planning and sourcing using a combination of MRP and kanban, which has been adapted in Figures 5-19 through 5-21.[99] Figure 5-19 suggests the common patterns of information (planning, kanban, and heijunka) and material flow.

When a group of products represents a large and stable volume of throughput, these may be called *Runners*, suggesting that a repetitive and possibly dedicated cell be arranged for them. When a group of products share some common elements but are also subject to variation, these are called *Repeaters* and may be arranged into cells allowing for mixed-model manufacture and final assembly. The final group may be called *Strangers*, where every job requires different resources or processes.

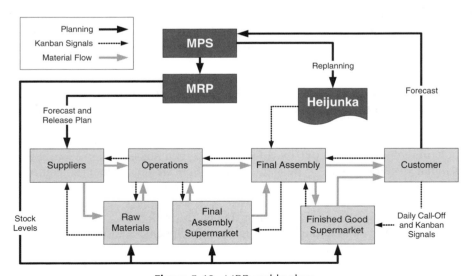

Figure 5-19. MRP and kanban

	Runners	Repeaters	Strangers
A	Tight Kanban	Tight Kanban or MRP	MRP
B	Tight Kanban	Loose Kanban or MRP	MRP
C	Two Bin	Two Bin	Two Bin or MRP

Figure 5-20. RRS, MRP, and kanban

By combining the concept of Runners, Repeaters, and Strangers (RRS) with an ABC inventory classification approach (discussed in Chapter 4), we arrive at the following general guidelines for inventory replenishment planning policy, as shown in Figure 5-20. Tight and loose kanban refer to Product-Specific Kanban with varying degrees of safety stock. *Tight kanban* suggests an expensive item that must be managed carefully with a minimum of safety stock, whereas *loose kanban* may contain more safety stock to allow for higher variation. Note that during the implementation of kanban (and as an approach to inventory reduction) you generally start loose, carefully tightening up inventory levels over time to identify and eliminate waste.

The general point of the matrix in Figure 5-20 is to suggest that we should have stock on hand (Product-Specific Kanban) to supply Runners and Repeaters. A- and B-level inventory for Runners should be managed carefully because their high level of consumption could be amplified into considerable excess inventory. C-level inventory (consumables such as bolts, rags, and lubricants) may be managed with a physical two-bin system, or replenishment by visual measurement of a bin or silo level, because the potential inventory waste of excess low-cost consumable inventory does not justify more careful management. A-level inventory for Repeaters may be managed by tight kanban, although MRP may be preferred if the value is high. B-level inventory for Repeaters may be managed by loose kanban simply because the cost to manage tightly may be greater than the value of waste reduction; alternatively, it may be managed by MRP. A- and B-level items for Strangers should generally be managed with MRP, because their infrequent demand does not justify carrying stock on hand. In some cases where MRP is called for, Generic Kanban may be used, where the supplier anticipates the need for a variety of materials based on MRP and a Generic Kanban signals for that supply only when it is needed.

Finally we may introduce the element of lead time into this matrix, as illustrated in Figure 5-21.

Without describing every intersection of the grid in Figure 5-21, note that those items that have a short lead time may be managed very tightly with

		Runners	Repeaters	Strangers
A	Short LT	Tight Kanban	Tight Kanban	Tight Kanban or MRP
	Long LT	Loose Kanban	MRP	MRP
B	Short LT	Tight Kanban	Tight Kanban	MRP or Two Bin
	Long LT	Loose Kanban	MRP	MRP
C	Short LT	VMI or Two Bin	VMI or Two Bin	VMI or Two Bin
	Long LT	Two Bin with Safety Stock	Two Bin with Safety Stock	Two Bin with Safety Stock

Figure 5-21. RRS, MRP, kanban, and lead time

kanban or even Vendor-Managed Inventory (VMI), where the supplier is responsible for monitoring and replenishing regularly consumed materials that are often supplied in bulk. Long lead time items require looser management, with more safety stock to buffer variation. When this additional inventory is not practical, MRP planning may be appropriate. These are general guidelines, not hard and fast rules. It should be clear, however, that the variety of demand patterns, and the flexibility of MRP and kanban design, offers a manufacturing enterprise many opportunities to consider for reducing waste.

THE LEAN JOB SHOP

Job and project shops come in endless variety, and we will refer to them collectively as *discontinuous* operations. With a discontinuous operation, the material requirements, routings, and operations may vary substantially from one job to another. Because each operation cycle time may vary, and because jobs are often released in large batches, at any moment some workcenters may be overloaded while others sit idle. These characteristics may lead many to conclude that a job shop must redefine itself to a more repetitive model before it can consider itself "Lean." This is mistaken, and in this chapter we will carefully explore the challenges presented by a job shop, concluding with some definitive steps that a job shop can take to become Lean.

Principles of a Job Shop

Although not ideally suited for high volume production, job shops are a vital component of most supply chains. In his book *Speed to Market: Lean Manufacturing for Job Shops*, Vincent Bozzone comments:

Job shops and made-to-order custom manufacturers are the unsung heroes and backbone of U.S. industry. Without the specialized skills and on-demand services these companies provide to larger enterprises, industry would not exist in the United States as we know it today.

Job shops are the most difficult of all types of manufacturing operations to manage. They are infinitely variable. They also tend to be comparatively small— the vast majority have fewer than one hundred people with annual revenues of less than $10 million. Many are only marginally profitable with any available reinvestment capital generally being spent on production equipment and machinery, rather than on the people and organizational side of the business.[100]

A job shop is typically a Make to Order, or sometimes an Engineer to Order, environment where each job is different from the last and potentially unique. The plant is often arranged in a functional layout, where all machinery of a common type is grouped together. One reason for this arrangement is that because each job may be different, equipment is often designed for general purpose, requiring operations that are specific and sometimes unique to each job. Operators are therefore skilled and adaptable in the use of particular equipment, so the equipment and talent are grouped together. Also, several machines may be able to perform any particular operation, so grouping offers the greatest schedule flexibility to perform the work on whatever machine happens to be available at the appropriate time.

Another difference between a job shop and a repetitive environment has to do with batch sizing. One of the primary objectives in repetitive Lean Manufacturing is the reduction of batch sizes approaching one-piece flow. Batch size reduction in a job shop may not be practical or even beneficial. There is an important distinction to be made between process batch* size and transfer batch size. Because a job shop makes to order, the customer order†quantity determines the *process batch* size—the amount of work that is released to the shop floor at one time. By contrast, the *transfer batch* size is the quantity of goods that may move from one operation to the next.

For example, there may be a bottleneck operation with a long setup time that requires the entire job to be run in a single batch, whereas subsequent operations may be performed in smaller batches as they exit the bottleneck operation. This technique where a job flows downstream in smaller transfer batches is called *operation overlapping. Transfer batch size* refers to the quantity that should move to the next workcenter and may be governed by the processing capability of the downstream workcenter or a physical container size

---

* Often the terms "batch", "job", and "lot" are used interchangeably. However, job and lot usually refer to a customer order or total production run quantity, whereas batch represents the quantity that is processed at a particular operation at one time. For example, several production batches may combine into a single lot for traceability.

† In some cases the customer order quantity multiples are guided by minimum pricing, or price break rules, which are designed to encourage the purchase of the product in unit quantities that are appropriate to package size, shipment weight/volume, or optimum production batch size.

such as the capacity of a forklift. In some cases, using transfer batches that are smaller than process batches is not practical because the entire lot must stay together as it goes through the production process—this may be dictated by characteristics of the process itself such as chemical, color, or flavor specifications, space limitations, quality, and compliance requirements.

In some cases, there may be no practical benefits to process batch size reduction in a job shop environment. According to Fogarty, Blackstone, and Hoffman in *Production and Inventory Management*:

> The JIT approach advocates reducing setup time until a shop can afford a process batch of one. If the transfer batch has been lowered to one unit to speed the order through the shop, what additional benefit is to be gained by reducing the process batch down to one unit? Since the order will be shipped to the customer as a batch, does it ever make sense to have the process batch be less than the order size? Unless there is some clear benefit to be obtained by continuing to reduce the setup time beyond that needed to support a process batch equal to the order size for all orders, JIT is creating wasted effort by forcing the transfer batch to equal the process batch.[101]

This argument seems at first to oppose the Lean emphasis on batch size reduction and one-piece flow, but there is much more to Lean than these specific techniques. If there is no need to reduce batch sizes in a particular situation, and the batch sizes are reduced anyhow, this causes unnecessary setups that are wasteful. Maintaining large, intact process batches is consistent with the EPE interval calculation discussed earlier in this chapter, which suggests that the optimal batch size in a mixed-model Lean environment may be greater than one. With that said, a job shop will always benefit from setup time reduction even if it does not lead to smaller batch sizes, because shorter setup times generally enhance flexibility and reduce waste.

The Advantages of a Discontinuous Environment

Job shop scheduling can be quite complex, and a skilled scheduler can make or break the success of an entire operation. Furthermore, because of high process and product variability, Murphy is a frequent visitor to job shops, disrupting even the most careful plans. Considering these challenges, why shouldn't discontinuous operations strive to become repetitive ones? Of course, this is not always practical, nor is it desirable, because discontinuous operations offer several competitive advantages:

- Design and engineering talent focused on solving unique customer problems
- The ability to produce small runs
- Agile production skills and a flexible shop organization capable of satisfying changing customer demand

However, job shops present many challenges that are not addressed by popular Lean Manufacturing techniques. According to Leon McGinnis, professor of manufacturing systems at Georgia Institute of Technology:

> Toyota Production System and Just-In-Time were successful in the automotive industries because the process itself is repetitive, and you can identify and target sources of variability, so improvements can be institutionalized using hardware mechanisms (i.e., poka-yoke) and through the development of standardized work processes and operator training. In a job shop what you make from one day to the next is different, and it is much more difficult to institutionalize systemic improvements.
>
> There aren't many large companies left running job shops where they shouldn't be. They have successfully aligned high volume products in high volume plants, lines, and cells. It is important for a manufacturing company to do those things that are smartest to do first—rationalizing flows, organizing and segregating these repetitive processes. What you are left with are processes that are not high volume and products that are more dynamic, and these should be segregated into more focused environments. This requires a different approach to planning and scheduling, and is more suitable to scheduling software.[102]

This rationalization of product lines by the larger companies is a good example for smaller job shops to follow and suggests that a job shop may group some of its product families and processes, moving along the product/process continuum toward repetitive production. But by their very nature, many job shops will remain mostly discontinuous, because that is their chosen and appropriate competitive positioning along the product/process continuum.

Project Shop versus Job Shop

Project shops are at the far end of the product/process continuum and have information system requirements that are quite different from the typical job shop. Projects are often of extended duration, coming together in a final assembly or single customer order or contract. Often labor and equipment move to a stationary final assembly site, such as the assembly of an aircraft inside a production hangar. This may include on-site installation and field service. Government contracting usually creates a project manufacturing environment, because of the rigorous control and reporting requirements.[103]

Project-based companies may take different approaches to organizational structure. Some companies form project-driven, self-directed work teams, where specialists from all functional aspects of the business are united under a common team dedicated to a single project or program (a family of projects). In the focused team approach, most planning and scheduling is done within the confines of the team. In matrix-type organizations, on the other hand, project managers direct shared resources across multiple projects. Many larger project-based manufacturing companies will have a special organization called the Program Management Office (PMO) that coordinates business processes and schedules resources across all projects. Smaller job shop operations (i.e., metal shop, machine shop, electrical shop, electronics shop) often exist within a project manufacturing organization, fabricating subassemblies that move to the final assembly site.

Planning and scheduling in a project shop requires a mixture of Master Scheduling, MRP, and Project Management software to schedule materials, facilities, and resources. Critical Path Analysis (CPA) is a common scheduling algorithm used to determine the longest path from start to finish that contains no slack time; the tasks along this pathway represent the critical constraints that must be managed carefully, because they directly determine the project lead time. Because people and equipment may be shared across several projects and locations at the same time, the project management software also may need to schedule and load resources across multiple projects concurrently.

A project manufacturing organization should look for software that can hybrid schedule projects and jobs interactively, where project and job costing interact, and where the project management component (work breakdown structure, tasks, timelines, and resource requirements) interacts with material and capacity planning capabilities. Integrating stand-alone project management software with a job costing and scheduling system may not meet the sophisticated requirements of a large project manufacturer.

Discontinuous environments generally have a greater need for data capture, control, and feedback than a repetitive operation, because there is greater variation in the individual product and processes. This is especially true for a project manufacturer. For example, the Boeing Company's assembly database contains every specification—from engine type to carpet color—of every aircraft it assembles. With four terabytes of storage (4 million megabytes) it is one of the world's ten largest transaction databases; 24,000 shop floor mechanics, engineers, procurement managers, and other Boeing employees have access to the database. It holds each plane's bill of materials, which other Boeing applications use for procurement, production scheduling, manufacturing engineering, and shop floor operations. Engineering drawings, schematics, and other database space hogs reside in a separate repository connected to the transactional database.[104]

The Challenges of Lean in a Discontinuous Environment

The primary flow and pull challenges in a typical job shop are caused by:

- Irregular and low volume
- High variety
- Frequent engineering changes

Element	Job Shop	Repetitive
Plant Layout	Group Technology	Product Flow
Work Authorization	Work Order	Flow Rate
Process Routings	Variable by Part Number	Fixed
Cost Accounting	Job Cost	Process Cost
Load Leveling Factor	Labor	Material
Operation Balancing	Limited	High
Material Handling	Irregular	Standardized
Plan and Control	Lots	Rate

Figure 5-22. Comparison of Job Shop and Repetitive environments

- Queues
- Variable production routings
- Moving bottlenecks

The differences between Job Shop and Repetitive production are summarized in Figure 5-22.[105]

In light of these issues, a push-oriented system may be more practical for a discontinuous operation. The fact that a job or project shop cannot implement flow and pull to the extent of a repetitive manufacturer does not mean that a discontinuous operation does not have many other opportunities to eliminate waste. Nevertheless, the majority of the Lean literature focuses on repetitive operations, so when a job shop manager reads about one-piece flow, takt time, and pull signals, he or she may shake their head and say, "That won't work here, so we must not be able to implement Lean." That is simply incorrect, because there is plenty of waste to be eliminated in a job shop, although the opportunities to implement flow and pull are naturally more limited.

Skillful application of push scheduling in a job shop may greatly improve performance, and at the same time Lean waste reduction and skillful pull techniques can also add value. With this flexibility in mind, we must understand that pull methods are complicated by two primary obstacles in a job shop: Takt time cannot synchronize the value stream, and routings are variable.

Takt Time Cannot Synchronize the Value Stream. In a repetitive operation, the routing is consistent and the line is balanced, so production may be synchronized according to takt time. The demand signal from a downstream work-center indicates there is available capacity to accept more work. The flow

Figure 5-23. Variable cycle times

naturally cannot issue a pull signal if there isn't downstream capacity all the way through to completion.

In a discontinuous operation, these rules generally do not apply. The mathematical takt time calculation is practically useless because the actual cycle time consumed for each job at each distinct workcenter may vary significantly from the average takt time. Consider a simple example where workcenters 1 and 2 have a cycle time of 10 minutes, but workcenter 3 has a cycle time of 40 minutes. Because each job may have a different routing and work content, and certain routing sequences may occur infrequently, balancing of work is not always possible.

Routings are Variable. In Figure 5-23, although the line isn't balanced, kanban will still work because signals do not depend on takt time. WC2 should not signal for more work from upstream WC1 until it has received a capacity-available (Generic) pull signal from downstream WC3. That's fine as long as flow is *linear,* as this first example illustrates. But *variable routing* means darned near anything can happen, and this can invalidate the simple linear capacity-available signaling method used by ordinary kanban pull systems. For example, multiple jobs can share an oven or other communal resource. A job can split to multiple resources, with some parts receiving a certain paint, finish and set time, while others do not, with all parts flowing back together for final assembly. Or consider this common situation, where three common upstream workcenters feed three common downstream workcenters as illustrated in Figure 5-24. Any upstream workcenter can feed work to any downstream workcenter, while any downstream workcenter can pull work from any upstream workcenter—it's the classic *many-to-many relationship* that drives software programmers crazy. Now let's say that only downstream workcenter 5 has available capacity, while upstream workcenters 1, 2, and 3 all have a job waiting to release. Which job goes first?

More importantly, *who decides* which job goes first?

Even if we simplify this relationship by placing a buffer between the upstream and downstream workcenters as shown in Figure 5-25, there is no clear guidance for what job should be issued to the next downstream workcenter, unless we use a prioritization technique such as the relative available time remaining in production of each job, or sequence by due date, which may

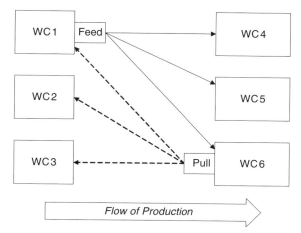

Figure 5-24. Variable routings

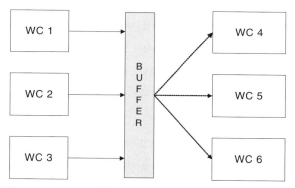

Figure 5-25. Variable routings buffered by supermarket

complicate a simple pull system and in some complex routings may cause sub-optimal scheduling results further downstream.

For example, an order that has a shorter routing duration may be released later and still finish first. Local pull signals cannot see the big picture when variable, branching, looping, and generally chaotic (discontinuous) routings across multiple workcenters are involved. A scheduling system is often needed to make sure the right jobs are released to the gateway workcenter in the proper sequence, and to manage the sequence of jobs in the queue of each downstream work center.

Generic Kanban may be used in a simple job shop, sending capacity-available signals to upstream workcenters. But if there are more than a few routing steps and branches, complex queuing logic similar to a railway switching yard may be needed to sequence and release the right job to the demand signal, to keep the pathways flowing optimally. Variable routings and cycle times introduce bottlenecks that must be managed; later in this chapter we'll explore practical scheduling and kanban pull approaches for this sort of environment.

The degree to which two factors—workstation cycle time and routing pathways—vary in a plant indicates the degree of need for an interactive job flow scheduling system to optimize the throughput of a discontinuous operation. During a visit to a particularly challenging job shop, with numerous setup time, bottleneck management, and prioritization issues, we were told "That's what the foreman is for." But in a complex environment, even a skilled foreman may not grasp the entire solution. And even if the foreman is that skilled and intuitive, should an organization place the control for optimizing throughput in the hands and head of a single human being? If that knowledge can be captured and institutionalized, it then becomes the intellectual property and competitive advantage of the organization. If not, it's a temporary advantage but a longer-term risk.

How a Job Shop Can Become Lean

Although scheduling issues can be extremely challenging to a job shop, paradoxically this creates a perfect opportunity to develop competitive advantage through lead time reduction. And here is the surprise: When they perform an analysis across their entire value stream, many job shops discover that the majority of their total cost is determined, and most of their lead time is consumed, *before the job ever reaches the shop floor!* This certainly calls into question the perceived significance of scheduling and flow.

According to the *Design for Six Sigma* methodology, the earliest stages of product development cause the greatest impact in the outcome of the product design, quality, and cost, as shown in Figure 5-26.[106]

And according to Vincent Bozzone in *Speed to Market: Lean Manufacturing for Job Shops*:

> Job shops already work on a pull system—nothing is produced until an order is received. The objective of Lean Manufacturing in a job shop or in a custom manufacturing environment is to cut lead time.

> Flow is achieved by eliminating delays in the total business process—from the conversion of RFQ's to orders, orders to shipments, and accounts receivable to cash. Companies that can bid and ship an order quickly will realize a competitive speed advantage and an increase in sales. When RFQ's are issued, buyers typically have a delivery date in mind or specified. The clock starts running when

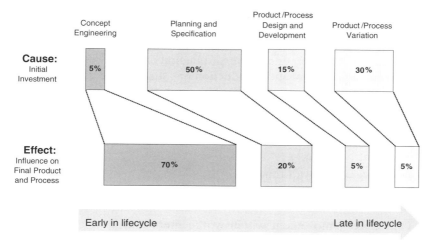

Figure 5-26. Design for Six Sigma quality

the RFQ is issued, not when the order is won. Often, the lost time in estimating and quoting must be made up later on the shop floor. This can become very expensive.[107]

The majority of product and process cost may be predetermined, and the majority of the lead time may be consumed before a job reaches the shop floor. Disorganized presales design, engineering, estimating, and quoting processes waste valuable lead time. Furthermore, the preproduction operations of planning, manufacturing engineering, and purchasing may also contribute significant time waste. These initial processes require clearly defined and standardized procedures that support rapid decision-making. Our experience suggests that the inability to make accurate decisions *on a timely basis* is the primary cause of excessive lead times in preproduction activities. In a knowledge-heavy environment, the improvement of these sales and prepro-duction processes will certainly delight the customer, creating a competitive advantage that is difficult to overcome.

Draining the Swamp

When planning a Lean implementation, many job shops find it difficult to get started. Here are two approaches to break the ice: Simplified Market Pull Scheduling (SMP) and Group Technology.

In *Learning to See*, authors Rother and Shook emphasize that you should flow where you can, and pull where you must. In a pure flow environment where operations are physically coupled (such as a continuous conveyor) one operation immediately initiates the next, so there may be no buffers of time or inventory. Excess time and inventory waste are only introduced when there are gaps and uncertainties in the flow of a process. Now apply that notion to a job shop where production is performed to demand. Because material is issued to the shop floor only to satisfy a customer order, the challenge isn't inherent excess inventory due to overproduction or erroneous demand signals but the confusion of flow once production has started. The *congestion* caused by the functional organization of the shop floor, combined with variable routings and the erratic sequence of jobs, leads to:

- Excess WIP inventory pooled at certain workcenters
- Transient bottlenecks
- Excessive queue wait times
- Extra transportation between workcenters because of routing complexity
- Unnecessary movement within workcenters caused by shifting one job aside to work on another
- Defect and rework waste caused by the general confusion often found in a discontinuous environment
- Delayed recognition of and response to quality problems

Job shops often release too much work to the floor because they see individual machines sitting idle. By releasing too much work at once, however, they may keep certain workcenters busy temporarily, but excess WIP is created, hindering the smooth flow of work in general. The pattern of flow in job shops may be erratic, and as a result there is a *congestion effect* where large queues of inventory and production bottlenecks may appear and disappear spontaneously and unpredictably as job routings converge along random pathways.

It is proven that there is a nonlinear relationship between capacity utilization and delays, and systems overload at less than 100% capacity.[108] This is easy to prove in your own experience, because traffic congestion appears and disappears suddenly and without apparent cause, long before the highway is 100% full. Traffic engineers have shown that an accident or other disturbance can cause a wave motion, like ripples in a pond after a stone is dropped, to propagate along a highway long after an accident is cleared. This same sort of prolonged phantom disruption can be propagated along the pathways and queues of a discontinuous shop floor.

Nonlinearity of capacity and delay explains why it is devilishly difficult to control the throughput of a job shop with push scheduling software alone, because the assumption of fixed lead times and their relationship to capacity are invalid. Traffic engineers have learned that very simple control mechanisms can prevent congestion; a good example is the signal light that limits the number of cars entering a freeway. Even if the freeway is moving along at much less than 100% capacity, the introduction of new vehicles should be carefully regulated so as not to slow the rate of flow on the main artery. If too many cars pull in at once and slow the flow of traffic, the backward rippling congestion will last for several minutes. This approach is known as *traffic calming* and can be applied to a job shop as well as a freeway.

This is analogous to the simple pull rule—don't insert another job into the flow until the downstream workcenter is ready for it! Of course, in a cellular operation with simple routings this is easy to determine, but in a discontinuous environment with variable routings the availability of the immediate downstream workcenter does not ensure further downstream availability, so more sophisticated communication and control mechanisms may be required to limit the release of excess work to the floor. A good first step for the improvement to such an environment, which does not require sophisticated scheduling, is to *drain the swamp*—drawing down the excess inventory that is the primary cause of this congestion and confusion.

This philosophy led to a straightforward method called Simplified Market Pull Scheduling (SMP), which may be used to establish a beachhead of control in an environment with variable routings and *excess capacity*.* SMP is a derivative of Drum-Buffer-Rope (DBR), the mechanism for applying the Theory of Constraints, which we will explore later in this chapter. Whereas DBR is focused on the identification and management of bottleneck (constraint) resources, SMP assumes that there are no constraints that *should* restrict total throughput because the plant is operating at less than full capacity due to insufficient demand†—nevertheless, congestion causes the plant to be unable to keep up with the demand that does exist.

SMP suggests simple rules for releasing work to the plant at the right time and dynamic priorities for moving jobs through the plant for on-time completion. To be effective, however, SMP requires reduced WIP inventory and shortened lead times. According to Dave Turbide in his white paper *Simplified Market Pull Scheduling*:

> Most manufacturers today are operating at something less than full capacity. Ironically, many still struggle to get product out the door on time. It is a fact that the actual active production time is a small fraction of the overall lead-time to produce. The rest of the time, the work is waiting in queue or waiting while other pieces in the batch are being processed. One of the first steps to implementing SMP is to drain the majority of WIP inventory out of the shop. The company will find that the excess capacity that was already there will now become visible.[109]

Here are the steps Turbide recommends to implement SMP:

1. Stop releasing work to the shop immediately.
2. Continue to quote your regular lead times. When nonbottleneck workcenters are starved for work, temporarily reallocate the workers to the

---

* This is also the basis of CONWIP, which is discussed later in this chapter.
† This is a fine point in the Theory of Constraints that we'll revisit shortly.

constrained workcenters if that will increase their throughput. Judiciously release new work to the floor only to feed critical bottleneck workcenters. You may also carefully release jobs that do not pass through the existing bottlenecks, while being careful not to create new bottlenecks or congestion. Release of too much work at this stage will defeat the SMP effort. At the end of each week you will have purged a week's worth of lead time from the shop floor.

3. Begin releasing new work that has been accumulating in the backlog once it reaches roughly half of the originally quoted lead time. Pay careful attention as the new work settles in behind the old that is being flushed out, making sure to allocate resources dynamically to smooth out the bottlenecks—this should be done simply and visually by the workcenter operators who have been coached on this exercise. Sequence the initial release and queue management according to the amount of lead-time buffer that has been consumed, similar to the critical ratio* used in traditional production control.

4. Once new orders are moving through production more quickly and regularly according to the shortened lead time release schedule, excess capacity will begin to reveal itself. Reallocate this capacity to reduce the remaining backlog of WIP. Soon there won't be piles of WIP in front of every workcenter. Work sequence and priorities will become clear to the operators.

5. Over a period of time, release work to the shop floor progressively closer to the due date, until you have reduced production lead times by 60% or so. You are now at the point where you should be very careful with lead times, since you are running Leaner and with less margin for error. As you continue fine tuning the release schedule and workcenter priorities, you will begin to clearly understand the natural patterns of flow constraints and disruptions which occur in the plant.

6. As your performance improves consistently, you may begin searching out opportunities where clients place a premium value on faster delivery times. If you have an opportunity to be aggressive, for example the chance to lure an important customer from a competitor, you may quote an exceptionally short lead time with the comfort of knowing that this order may be "expedited without expediting" by simply releasing it early, since you have eliminated excessive queue time delays in your job shop.

SMP begins with the assumption that although there is excess capacity there are still temporarily bottlenecks and flow problems due to congestion, and so its focus is on WIP reduction. SMP may be an appropriate *first* step for any

---

* Critical Ratio is calculated as hours of work remaining divided by available hours before the due date and prioritizes the jobs that have the most remaining work.

discontinuous production environment where there is excess capacity, because we're not physically rearranging the shop floor, changing the flow of work, or implementing complex scheduling mechanisms. To treat SMP as a final objective, however, would miss the point. With SMP you will reduce excess WIP, adding flex to the shop floor, subduing the chaos—or at least revealing it—while shortening lead times and improving on-time performance. The new Lean job shop, with shorter lead time and more reliable deliveries, may be able to gain market share over less agile competitors and thereby increase the demand to the point where internal production bottlenecks appear. At that point, the company can switch to full-blown DBR to identify, exploit, elevate, and manage those constraints—more on this shortly.

Group Technology

Regular patterns of workflow commonly exist in even the most variable job shop. *Group Technology* suggests that products and processes that share characteristics may be combined into product families and cells to smooth the flow of work. This is very similar to the product flow analysis described earlier in the chapter (Fig. 5-06), where product families sharing similar processes are identified and assigned to cells in a mixed-model environment.

With group technology, products within a discontinuous environment may be grouped according to size, design characteristics, materials used, etc. Processes may be grouped according to machines used, types and sequencing of operations. To begin group technology analysis, you may use your intuition to categorize products and processes. If you have a large number of products and processes it may be helpful to perform a product flow analysis. Key product characteristics may also be stored in the MRP II database, represented as characters or segments within the part number itself, or stored in special user-defined fields in the item table. By storing these characteristics along with other vital BOM and routing information, sorting and reporting software tools may aid in the identification of common manufacturing operations, grouping the information on planning and production reports.

Analyses of potential group technology patterns in a job shop may lead to cellular organization of some processes, rather than organization by function or machine type. In many job shops, most of the plant may remain in a functional layout because of the highly variable nature of production (Strangers) but if a handful of key products (Runners and Repeaters) can be organized and level-scheduled on a dedicated and flexible production cell, this may result in dramatic improvements in throughput and reduced WIP inventory and lead time.

In addition to the benefits of process improvement, group technology may lead to the standardization and reuse of some materials and subassemblies. This may in turn lead to inventory and lead time reduction, level scheduling of some processes, and the creation of final assembly supermarkets for some components—at first unimaginable yet possible in some job shop environments once there is a focus on standardization of design. When design and engineering strives for maximum reusability of designs and components, this also shortens the presales design and tooling processes.

In fact, there are wide-ranging benefits resulting from group technology:

- Reduced engineering cost
- Accelerated product development
- Improved costing accuracy
- Simplified and accelerated estimating and quoting
- Reduced tooling lead time and costs
- Reduced setup times
- Extended staff training and skills
- Improved product consistency and quality
- Improved serviceability due to parts standardization
- Reduced warranty service costs
- Simplified process planning
- Flattened BOMs and routings
- Simplified cellular scheduling
- Simplified material planning and purchasing
- Reduced raw material, component, and WIP inventory
- Increased available floor space
- Reduced lead times
- Accelerated production throughput

A Lesson in Product Rationalization

Job shops may find themselves further up the product/process diagonal than they should be. This mismatch is often due to a failure to distinguish between the best value for the customers' needs, what customers think they want if given limitless choices, and what salespeople and product designers think the customer needs and wants.

Recall when our firm was hired to design a centralized purchasing system, described in Chapter 3 (Fig. 3-04). This company manufactured several models of complex equipment, but they could have been a fairly repetitive job shop using standardized options; in fact, they were an Engineer to Order operation because virtually every job contained a unique design element or material.

Regular delays were caused by the difficulty of procuring unique parts for each job, and raw material inventory was overflowing with one-off parts that would never be used again. Product design consumed a large amount of time, extending lead times well beyond the industry average. The cost of ordering individual parts drove up the product cost, dashing the hopes of developing an efficient central purchasing process. BOMs were rarely accurate, and each job was unique in subtle ways, jeopardizing planning accuracy, product consistency, manufacturability, and serviceability.

We ultimately identified the source of the problem, the company's earnest desire to please their longstanding customers. But did these uncontrolled product variations really add value? Was the company's willingness to do whatever the customer wanted a competitive advantage, or their Achilles heel? Well, it was both. In many cases their flexibility did add value to meet a customer's special requirements, whereas other times it just added waste. Most importantly, this company lacked the discipline and judgment throughout the presales design process to make appropriate configuration recommendations with each order. The old saying "the customer is always right" should be questioned if excessive product variation does not add value to the product; otherwise, inappropriate positioning on the product/process diagonal may result, leading to competitive disadvantage.

Summary of Lean Job Shop Tactics

The primary value proposition of a job shop is flexibility. However, too much flexibility can create chaos, so a job shop should strive for a balance of simplification and standardization of products and processes. Group technology is but one method of enhancing the Lean performance of a job shop through an emphasis on standardization; here are others:

- Remove waste during the presales process. Provide sales and engineering teams with Product Life Cycle Management (PLM) tools and the encouragement to work together. Rigorously measure the performance of these processes, and look to customers for improvement suggestions.
- Implement a product configurator system to automate and standardize the repetitive aspects of design and validation during the sales process. Record, institutionalize, leverage, and protect the intellectual property of the product and process.
- Invest in collaboration tools that enhance value to the customer. Become a trusted and integrated partner in their product development process, vital to the development of their future requirements and strategy. Build an unassailable competitive advantage into every aspect of the customer relationship and the product life cycle from concept through delivery.
- Emphasize product design for manufacturability and quality, which includes standardization and maximum reuse of designs, materials, tools, and processes.
- Focus on rapid and accurate estimating, costing, and quoting. Develop material and operation standards, costing and pricing models. Constantly measure and refine these standards so they reflect the realities of production capability and cost.

- Enhance the effectiveness of preproduction planning and purchasing processes, such as S&OP, MPS, and MRP.
- Rethink and reorganize the plant to reduce move waste, even where group technology or cellular organization isn't practical, For example, consider putting smaller machines on rollers so they may suddenly assemble around a larger stationary machine to create a spontaneous work cell.[110]
- Emphasize rigorous process and value stream mapping. This may be difficult in a discontinuous operation, because the flows are naturally less coherent than a repetitive environment. Nevertheless, the effort to map and understand the value streams and distinct processes should lead to moments of lucidity, where patterns emerge from the chaos and new ideas for organization and improvement appear.
- Search for isolated areas where kanban pull may be introduced.
- Practice 5S housekeeping and workplace organization techniques to create an orderly environment, encourage individual and team discipline, and develop thoughtful and standardized behaviors. Pay close attention to the wasteful movement of operators searching for materials, tools, and work instructions.
- Tighten up supplier quality and lead times.
- Relentlessly pursue setup time reduction.
- Scrutinize process and transfer batch sizes. Emphasize small batch sizes and flow wherever practical.
- Encourage customers to place smaller orders more frequently.
- Drain the swamp by carefully reducing work release to the shop floor.
- Establish scheduling methods and systems (manual and automated) to release, route, and prioritize work so that it flows through the shop with minimal interruption, inventory, time, and movement waste (explored in the following section).
- Identify and manage constraints, explored in the section on the Theory of Constraints later in this chapter.
- Thoughtful transformation enabled through group technology, cellular reorganization, and other Lean techniques may ultimately guide the strategic focus and realignment of the enterprise to more profitable products, processes, and markets.

DISCONTINUOUS SCHEDULING

A manufacturing enterprise needs visibility of the production schedule, so individuals can answer the same important questions that seem to arise each day: What is the backlog? What is on the schedule? What is the rate of actual production? When will this order be finished? We need this information to communicate with customers and suppliers on existing jobs, and to make reliable commitments for additional work. It is therefore important to publish the schedule in a format that is easy to understand. According to Mitchell Millstein in "Putting an Eye on the Scheduling Function":

> Non-visual scheduling results in wasted time or movement looking for the next order or checking the status of an order waiting to be produced. Many facilities have one expert who knows where all orders are in the scheduling queue or in the manufacturing process. If that person is gone—be it vacation, sickness, resignation, or a bathroom break—the system shuts down. And even when this expert is in, he or she is inundated with requests from sales and manufacturing to pinpoint the location or status of an order. If customer orders are made visible and the scheduling process standardized, the facility will operate more productively by allowing everyone greater access to order status. The expert could spend more time doing pro-active work in the facility or with customers.[111]

Scheduling in a discontinuous environment is particularly important because it is so unpredictable. In a repetitive environment with a level schedule, if you know the heijunka schedule at the beginning of the week you have a good idea of the expected rate of throughput. Combined with an understanding of the predefined product mix, backlog contents, and supermarket levels, you can make realistic promises to your customers. Not so with a discontinuous environment, where a single order can suddenly change the capabilities of the entire plant. So it is important to schedule a discontinuous environment as best you can, communicating the status clearly and regularly. In a small shop, that may mean a greaseboard in plain view of the shop floor. In a larger shop, electronic schedules are often necessary—not only to crunch the sheer volume of data required but also to disseminate the information to individuals across the entire organization, on computer screens and printed reports, in the specific format appropriate for each purpose. Although electronic schedules may be automated to some degree with spreadsheets, they have practical limits in terms of complexity, scalability, and integration.

The scheduling system should only suggest the release of *workable work* to the plant. It is counterproductive to release scheduled work until there is available capacity, where equipment and tooling are performing within specifications, and sufficient materials are available and of acceptable quality. If work is issued when any of these basic conditions are absent, the job stalls on the shop floor and must be moved aside to make way for another. Worse yet, if materials are scarce, the incomplete job may have consumed inventory that could have been used to complete another job. Ensuring the release of workable work requires interaction between the schedule, material planning and control, and capacity planning.

A capacity planning system considers resource availability and detailed routing information when calculating the schedule, looking for gaps in the schedule where jobs (or portions of jobs) may be inserted. The more detailed and accurate information available to the scheduling system on capacity, routing steps, and start times, the more precisely it can calculate expected due dates, maintaining a valid release schedule. In even a small job shop, however, it is usually not practical to capture this amount of constantly changing detailed production information required to develop an exact schedule, nevertheless a rough cut schedule may be sufficient to aid in many practical scheduling decisions. A reasonably accurate rough cut schedule can also be an early warning system. Exception reports help us sift through enormous volumes of data to identify situations that require an immediate decision— due dates that are in jeopardy, expected purchase receipts that are late, and so on. Many systems also provide notification alerts that can send an e-mail or pager message, warning of conditions that require immediate attention.

And finally, the planning system can use the schedule information for simulations, answering what-if questions like:

- What if we add a third shift for the next two weeks?
- What if we purchase rather than make this component for this one order?
- Machine #1 just went down. What happens if we shift this job to the other machine?
- What will our labor schedule look like if we accept this big order?
- When can this customer expect delivery of the order we just started yesterday?

Types of Scheduling Systems

We will briefly examine four basic types of scheduling systems: infinite-capacity, finite-capacity, APS, and simulation systems. We will then point out their critical weaknesses, exploring what can be done to minimize these weaknesses.

Infinite-Capacity System. As its name suggests, an Infinite-Capacity Scheduling System (ICS) does not calculate capacity limits for each resource and will therefore schedule 16 hours of work into an 8-hour shift. Although this seems overly simplistic, ICS can be very useful. Although you must set up the various routings, lead time elements, and transfer batch quantities for ICS to work properly, it does not require a high degree of precision and can therefore be set up quickly. Loaded with simple routing information, ICS will still suggest reasonably valid sequencing and priorities, although the actual start and stop times may not be accurate. ICS is the most forgiving entry point for a company that is new to scheduling, allowing them time and feedback to refine their workcenter and routing standards before switching to finite mode.

Finite-Capacity System. Finite-Capacity Scheduling Systems (FCS) establish a limit to the capacity available at each workcenter. FCS may use various *heuristics* to suggest the best schedule. These include forward and backward scheduling capabilities, plus sequencing and queuing rules such as First Come First Served (FCFS), Largest Lot Size First (LLSF), Smallest Lot Size First (SLSF), Shortest Processing Time (SPT), Earliest Due Date (EDD), Earliest Operation Due Date (EODD), and Critical Ratio.

These heuristics share a common theme—they are based on due date, job duration, or remaining time sequencing and may be used to derive relatively feasible schedules. But can they derive an optimal schedule? No, because they're *heuristics*:

> **Heu-ris-tic:** A rule of thumb, simplification, or educated guess that reduces or limits the search for solutions in domains that are difficult and poorly understood. Unlike algorithms, heuristics do not guarantee optimal, or even feasible, solutions and are often used with no theoretical guarantee.[112]

Advanced Planning and Scheduling (APS) Systems.

Also called a *Constraint-Based Scheduler*, an APS system extends the capabilities of finite scheduling with the ability to model an entire process, and the interaction of multiple processes, highlighting particular constraints or pacemaker operations that govern the throughput of the entire enterprise. An APS can establish a constraint operation as an anchor, backward or forward scheduling from it to control the flow of work through the entire value stream. Many APS systems are capable of supporting rate-based (level) scheduling. APS systems are often accompanied by sophisticated data capture mechanisms at key points in the process, which monitor process performance and notify the scheduler when there is an event that will cause the process to become out of control.

With all of this sophistication, many APS systems make simple assumptions about capacity, disregarding the nonlinearity of capacity and delay we discussed earlier. Invistics, a Lean software company, is currently working under a grant from the National Science Foundation, developing mathematical modeling, simulation, and scheduling technology with advanced queuing and scheduling techniques adopted from the telecommunications industry. Tom Knight of Invistics explains that these techniques emphasize reduction of the time work must wait in front of the next operation—typically the source of 90% or more of the lead time in a job shop operation. He explains that the National Science Foundation recognized this as a significant issue, and suggests that "If we can make the same type of impact in job shops that have been made in repetitive operations, we'll have a huge impact on the economy."[113] Pioneering work such as this may take time to find its way down market in cost and usability to smaller job shops. It is important to note, however, that many APS scheduling systems exist today with more than enough power and flexibility to schedule and control job shops as long as they skillfully minimize complexity, using techniques we'll describe later in this chapter.

Simulation Systems. Simulation tests the outcome of likely solutions to a problem, comparing each result based on specific evaluation criteria and suggesting the optimal solution. Evaluation criteria in a scheduling simulation may include shortest move distance, least processing time, or best due date performance—when defining the simulation you must choose one for the simulation to optimize, or a combination of factors to balance. Using sophisticated mathematical modeling techniques, these powerful simulation programs may be able to run a large number of iterations to solve for the optimal solution; however, these tools are generally beyond the economic and practical reach of small and medium-sized job shops.

So What's the Catch?

If scheduling software can help manage all these details, then what's the catch? Practically speaking, scheduling software is completely effective only in the most straightforward repetitive operations, where material requirement and routing information is precise and stable and the production rate is steady. Beyond this extreme case, variability and uncertainty can make scheduling more of an art than a science. There are three distinct limitations that make the use of scheduling software very challenging in a discontinuous environment.

Scheduling Software Limitation #1: Care and Feeding of the Software. Parameters for batch sizing, purchasing and production lead times, workcenter and operation standards, shop schedules, and time fence boundaries must be carefully entered into the software by human beings. But many job shops have a limited planning staff with important shop floor responsibilities, and even with a large staff it is often not practical to change thousands of parameters as frequently as actual conditions change on the shop floor. As a result, the accuracy and usefulness of the system may quickly degrade. Scheduling software must store standard operation times (a combination of queue, setup, run, and move times) for each step on the routing. Before improvement, the time consumed by Non-Value-Added (NVA and NNVA) production activities may represent over 95% of the total throughput time. As a plant becomes Lean many of these NVA lead times are reduced or eliminated. However, because these planning factors are embedded in complex, multilevel BOM and routing tables for hundreds or thousands of parts, they cannot be changed quickly. So they become self-fulfilling, as push scheduling imposes irrelevant standard operation times on the process, perpetuating waste and hindering continuous improvement efforts.

Scheduling Software Limitation #2: Complex Routings. To make appropriate sequencing decisions at the time of release, and as a downstream workcenter chooses jobs from an upstream queue, the software requires accurate information on capacity, cycle times, alternate routings, and constraints. Here are some common challenges:

- Machine capacities and configurations must be known, including specific rules governing the sequencing and batching of work. In our experience, virtually every manufacturing environment has at least one perplexing schedule challenge; here are two common examples:
  - Allergen and flavor sequencing in food and pharmaceutical operations and color sequencing in food, chemical, and textile operations require jobs to be run in a particular order depending on the interaction of specific physical or chemical properties of the formulation, or else a lengthy cleaning must be performed.
  - Concurrent (shared) resources, such as a paint booth or oven, where several batches must queue up for a single cycle, and where the cycle time often depends on physical characteristics of specific parts included in the batch, as well as the total volume and density of the loading. There may be rules that regulate what parts can be cooked together, where there are chemical interactions among some products, or where the longest cook time item in the batch forces the cycle time to exceed the maximum allowable cook time for a particular item.
- Rework cycles may be difficult or impossible to predict. This is particularly troublesome as unanticipated rework attempts to jump to the front of the queue and disrupt the entire production schedule.
- Because many job shops are organized functionally, they may contain loosely defined clusters of general-purpose machinery and skilled labor, so that several possible combinations of resources may perform the same operation. This creates a logical and software challenge, requiring the definition of alternate routings and the definition of conditions and order of preference where those alternates are suggested. For example, there may be a backup manual operation using older equipment that is activated if a bottleneck appears, and the scheduling system must know when it is appropriate to suggest the activation of this resource.

Scheduling Software Limitation #3: Timely Feedback.

For scheduling software to make accurate recommendations it must know the current status of every workcenter and resource, every job, and all materials available for production. Traditional theory suggests *input/output controls* on each workcenter that monitor workcenter performance and trigger replanning when actual throughput does not meet the schedule. To provide this real-time feedback information to the software, significant investments must be made in real-time data capture systems, along with substantial human effort to maintain the system and manage the exceptions. When poorly implemented, not only are these invasive data capture and interactive scheduling systems a great nuisance, but they can introduce more waste and confusion than they eliminate.

Placing input/output controls on *every* operation in order to control a discontinuous operation is a fantasy that has outlived its usefulness in most environments.

Can Scheduling Software Really Work for You?

These are three sobering limitations. With all the sophisticated scheduling tools available, and the burdensome maintenance required to keep them running properly in a discontinuous environment, you may feel overwhelmed and despair that scheduling software can never really help you run a Leaner factory. After all, there's no room in the budget for a plant full of rocket scientists, unless of course you're making rockets.

A large ERP vendor recently confided to me that less than 5% of the APS tools they had sold were actually in use. We suspect that many customers became excited during the sales demonstration, seeing the promise of APS, so the scheduling software was bundled in with the purchase price to close the deal. For whatever reason, the implementation of the scheduling software was then delayed indefinitely, or at least until the ERP foundation was in place, which often takes longer than expected.

An integrated scheduling system can and will work, as long as you don't require it to be perfect. And you shouldn't try to schedule *everything*. Although a discontinuous environment may appear complicated and chaotic, there are usually underlying patterns if you know how to find them. So ask yourself these questions: What would deliver the most bang for the scheduling buck in each value stream? What single operation can regulate the entire process? What deserves the primary focus of the scheduler? In a repetitive process it is called the pacemaker, and in a discontinuous operation it may be called the *primary constraint*.

THEORY OF CONSTRAINTS (TOC)

> Focus on everything, and you have not actually focused on anything.
> Dr. Eli Goldratt
> *The Haystack Syndrome*[114]

Dr. Eliyahu M. (Eli) Goldratt, a physicist by trade, was asked by a friend to help solve a production scheduling challenge for the manufacture of chicken coops.[115] Unencumbered by traditional production and accounting thought habits, he introduced a revolutionary approach in his 1984 business novel, *The Goal*. With several million copies of *The Goal* sold worldwide, a variety of other titles promoting different aspects of TOC, and a multitude of consultants providing seminars and workshops, the Goldratt Institute has become a thriving intellectual property franchise. Many people have told me that, after first reading *The Goal* many years ago, they changed their entire approach to manufacturing management.

Some people may make the mistake of underestimating the subtlety and sophistication of Lean Manufacturing because of its simple message about the elimination of waste. Similarly, many may underestimate TOC because of its apparently straightforward emphasis on the elimination of constraints. This may lead to many individuals having no more than a sound bite understanding of the true significance of TOC. Although many people equate TOC solely with the Drum-Buffer-Rope method of constraint-based scheduling, at a higher level Goldratt emphasizes TOC as a disciplined, scientific thought process, describing such esoteric tools as evaporating clouds, current reality trees, and the "thinking process."

In this section we will limit our exploration to the essential elements of TOC that pertain to shop floor control, so that later in this chapter we may explore a solution to the challenges of discontinuous scheduling. If the reader is interested in more detail on the shop floor aspects of TOC, consider *The Race* by Eli Goldratt, from which much of the following content is derived. And if the reader is interested in an introduction to TOC as a disciplined thought process, consider *It's Not Luck*, also by Goldratt.

The philosophy of TOC is deceptively simple and extremely useful to the management of a discontinuous operation. Every value stream has a primary bottleneck (constraint) that limits its ability to reach its goal. The rate of *throughput* of that value stream (total production that is *sold*) is governed by that constraint. The rate of nonconstraint operations should be tied to the primary constraint, and if a nonconstraint operation has no work then it should stop. This is similar to the rule of kanban—the key distinction, however, is that kanban assumes a balanced line, and when one workcenter stops, they all stop. TOC assumes an unbalanced line (the nature of a discontinuous operation) and demands that the bottleneck operation should *never* stop, unless there is insufficient demand, in which case the constraint has shifted from the production operation to market demand.

Logically then, the speed of the primary constraint operation (henceforth we'll call this the Capacity-Constrained Resource, or CCR) governs the throughput, and thus the total revenue of the entire value stream. *An hour of lost CCR production equates to an hour of lost revenue for the value stream*, and an idle hour on a nonconstraint operation is meaningless as it does not represent any lost revenue. This suggests a radical reevaluation of the concept of production cost.

Drum-Buffer-Rope

To manage constraints, Goldratt devised a scheduling mechanism called *Drum-Buffer-Rope*. The production rate of the CCR drives the value stream, so the CCR is the *Drum*. The Drumbeat is similar to the takt time of a balanced line, governing the production of the upstream workcenters according to the consumption rate of the CCR. To ensure that the CCR never stops while there is demand, we must protect it from supply interruption. The CCR is

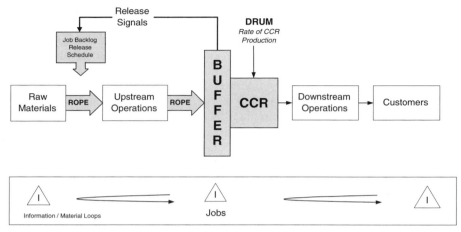

Figure 5-27. Drum-Buffer-Rope

performing operations on a variety of products with various batch sizes and cycle times, so this requires a *Buffer* ahead of the CCR which is measured in *time*, rather than inventory quantity represented by a typical Product-Specific Kanban buffer. This is an important distinction. According to Goldratt, "The concept of revolving inventory in the buffer is vastly different from the usual understanding of safety stock as a constant inventory level for each part."[116] The buffer should store an amount of work to be consumed by the CCR during the probable duration of a supply interruption—in other words, the length of time before more parts arrive at the CCR. Whether that duration is four hours or four days, it determines the momentary size of the CCR buffer.

The *Rope* is essentially a kanban signal pulling work toward the buffer. The rope signals when the next job is to be released to the gateway workcenter according to buffer management rules we will discuss shortly; the next job to release is determined by the backlog. The similarity to Generic Kanban is evident as you see in Figure 5-27, except that the jobs aren't pulled by capacity signals from downstream workcenters but are released to production based on signals issued to the gateway workcenter depending on the status of the CCR buffer.

What is a Constraint?

A constraint is a factor that keeps an organization from reaching its goal. In the context of production, it is a process bottleneck. TOC describes four types of constraints: policy, market, materials, and operation (CCR). Constraints should be identified and eliminated in that order.

Policy constraints are often the greatest source of waste in an organization. Whenever you hear the phrase "We've always done it this way," that's your cue that a policy constraint may be lurking about. Machine efficiency and utilization objectives, unnecessary inventory stocking policies, and end-of-month sales incentives are all common policy constraints. They are causes of irrational behavior leading to waste and should be eliminated.

Market constraints are limitations of market demand. A plant may have a theoretical CCR on its production operations, but if there is not enough demand to load the CCR to maximum capacity, then it is not the current primary constraint. In the case of a market constraint, a company should emphasize demand creation and at the same time anticipate that additional sales will occur, and continue improvements to a potential CCR because it will become the primary constraint once the plant is again under sufficient demand.

Material constraints occur when materials are unavailable to support production. In a situation where materials are available in the supply chain but unavailable to production, this may be a failure of the forecasting, MRP, and purchasing processes and indicates a need to improve planning or increase buffer levels (safety stock).

Finally there is the *CCR*, the primary constraint on the production operation.

Three TOC Buffers

TOC suggests a value stream may require up to *three* buffers: the CCR buffer, the final assembly buffer, and the shipping buffer. We have already described the CCR buffer. The final assembly buffer protects against short-term demand variability. This supports the practice of postponement, where the CCR feeds a buffer of semifinished goods that are used to quickly configure or assemble a finished product to a customer order. The shipping buffer services customer orders from finished goods inventory and buffers variability of supply from the upstream operations. The downstream final assembly and shipping buffers prevent shortages of nonconstraint materials, which might delay the assembly and shipment of a product that passes through the upstream constraint.

The final assembly and shipping buffers (supermarkets in Lean terminology and safety stocks in the traditional vernacular) exist to protect throughput and due date performance. There may be operations that bypass the CCR, and these may be pulled directly into the final assembly and shipping buffers with ordinary Generic and Product-Specific Kanban replenishment rules, as shown in Figure 5-28.

The Five Rules for Managing a Constraint

TOC suggests five simple rules for managing constraints: Identify, Exploit, Subordinate, Elevate, and Search for the next.

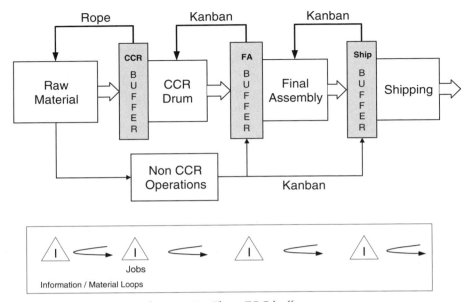

Figure 5-28. Three TOC buffer types

1. Identify the Constraint. Constraints are usually easy to find, though it may not be simple to identify the *primary* one. Just look for large stockpiles of WIP, backlogs, and frequent expediting. If there is an operation around which arguments frequently occur, that may be a constraint. Often the most expensive piece of equipment on the shop floor, purchased specifically to improve throughput of a particular process, becomes a constraint at peak times—to the chagrin of those who made the decision to invest in it. Lean practitioners call these *monuments.*

It is essential to find the primary constraint, because efforts on lesser constraints will have no impact on throughput. Value stream mapping may help to identify and quantify the primary constraint and its buffer.

2. Exploit the Constraint. To exploit a constraint you introduce a new policy or technique to provide more throughput without changing the physical design or capacity of the process itself. The first step to exploiting the constraint is to place a buffer in front of it, so that it is not starved by a sudden interruption of work. After the buffer is introduced, you may consider other methods:

- Rigorous setup time reduction efforts.
- Eliminate unnecessary setups and, when practical, complete an entire process batch before setting up the next job.
- Stagger breaks and shift changes so the constraint is never idle.

- Redirect work to nonconstraint equipment.
- Provide training to improve efficiency.
- Do not allow poor-quality products to be input to the constraint.
- Emphasize quality processes so a poor-quality product is never output from the constraint.
- Introduce a rigorous preventative maintenance program on the constraint to avoid unplanned downtime.

3. Subordinate Everything to the Constraint. The revenue of the entire organization depends upon the primary constraint, so activities must be coordinated to optimize its throughput. Continuous improvement efforts should focus on the primary constraint within each value stream.

4. Elevate the Constraint. If exploiting the constraint and subordinating all other processes to it does not relieve the bottleneck, then you must elevate the constraint by an investment in its capacity. It is important to remember that a minute gained or lost at the CCR is an extra minute of throughput for the entire value stream (or possibly even the entire plant!), so traditional Return On Investment (ROI) assumptions must be reconsidered. Methods to elevate the constraint may include:

- Reduce setup time through investment in tooling.
- Run extra staff, shifts, and overtime on the CCR.
- Use aggressive compensation policies to ensure optimum performance.
- Reactivate retired or obsolete equipment during peak load times.
- Outsource work planned for the constraint.

5. Search for Next Constraint. TOC emphasizes continuous improvement, so the elimination of a constraint is the signal to immediately identify and eliminate the next one. Unlike incremental improvements suggested by Kaizen activities, however, breaking a constraint may have dramatic effects; in fact, it may change the entire competitive situation of a company. According to Goldratt, "In the 'cost world', changing one or two items does not change much. In the 'throughput world', changing a constraint changes everything."[117]

It is therefore important to eliminate constraints in the order suggested: policy, market, material, and then CCR. Once a lower-level constraint is eliminated, it may cause a higher-level constraint to appear, so once a constraint is broken we must start at the policy level and reevaluate the entire situation. For example, a policy constraint such as an inappropriate volume pricing requirement may appear to be a market constraint. An attempt to correct a phantom market constraint without addressing the underlying pricing policy will only lead to trouble. Alternately, we may purchase a new machine that breaks the CCR, enabling us to increase volume of the bottleneck operation. If at that time we don't revisit our market penetration strategy and pricing policies, we may artificially limit throughput of the new CCR by generating insufficient demand.

Scheduling with Drum-Buffer-Rope

Here are the basic steps to implement TOC as a scheduling system:

1. **Identify and Quantify the CCR**—Although *The Goal* predated the introduction of value stream mapping tools and techniques by Rother and Shook, they are very useful when identifying, quantifying, and modeling constraint-based scheduling scenarios. Distinguish between flows that utilize and those that bypass the CCR, then determine whether a final assembly or finished good supermarket is required downstream from the CCR. Don't proceed to step 2 until you can prove that you have identified the primary constraint.

2. **Exploit the CCR**—Taking the steps explained earlier, subordinate the nonconstraints, and if necessary elevate the CCR. Attempt to break the constraint if practical and economical.

3. **Establish the Drumbeat**—Develop a pace for the CCR based on its throughput capacity. Work must be released in the proper sequence and timing so the CCR buffer is not depleted. This often requires backward scheduling from the CCR buffer to the release schedule of the gateway workcenter(s). Monitor the throughput rate of the CCR as the constraint is exploited and elevated, adjusting the drumbeat accordingly.

4. **Design the Rope**—Determine the appropriate material handling and pull signal methods for upstream flows feeding the buffer. For example, pull signals from final assembly or shipping buffers that bypass the CCR may be executed with Product-Specific Kanban for repetitive items and Generic Kanban for unique items. If the routing pathways from work release to CCR are simple, this scheduling decision may be performed manually through kanban signals on a First Come First Served (FCFS) basis. If routings and cycle times are more complex, then more sophisticated prioritization rules may be necessary, and the scenario may require backward scheduling to the gateway workcenter to ensure that jobs are released with the proper sequence and timing.

5. **Quantify the CCR Buffer**—The buffer is sized (time not quantity) according to the pace of the CCR drumbeat and the likelihood and duration of upstream supply interruptions. For example, if there is the potential for a three-day interruption from the feeding operations, then the duration of the CCR buffer should be three days. However, there may be multiple operations and routings feeding the buffer, and other workcenters could take up the slack if one is lost, or the remaining upstream workcenters may be able to speed up their pace, adding resources and shifts to make up for the shortfall, so that the CCR buffer is not con-

sumed by the loss of a single upstream operation. The point is that the calculation of the CCR buffer duration is not a simple mathematical process but must consider a number of interrelated factors involving potential capacity, material availability, and quality disruptions. The exercise of mapping, identifying, and quantifying these factors may help to develop robust value streams that maximize throughput while minimizing inventory in the CCR buffer.

6. **Manage the CCR Buffer**—Because buffers are managed by time, and not unit quantity of production, TOC suggests that the buffer should be segmented into three time zones of equal duration. Like a traffic light, we'll call these the Red, Yellow, and Green zones. The buffer is then carefully monitored, and appropriate responses are determined, based on where gaps appear. Figure 5-29 shows an example of a buffer status board.

CCR Buffer Status Board
Buffer Target = 850 Minutes

RED = 275 Minutes, 32% of Target

JOB	CUSTOMER	TIME	ITEM
000163	Baker Industries	100	ABC-001
000165	Childrens Hospital	50	DEF-002
000171	Duffel Corporation	125	GHI-003

YELLOW = 335 Minutes, 39% of Target

000172	Edwards Industries	200	ABC-002
000174	Firestone Enteprise	50	DEF-003
000175	Grado Plating	85	GHI-003

GREEN = 155 Minutes, 18% of Target

000177	Huber International	155	XYZ-789

Figure 5-29. DBR Three-zone buffer status board

The Red zone represents the work that is due to arrive at the CCR during the first third of the elapsed time for the buffer. A gap in the Red zone is very serious because it indicates that the feeding operations have fallen behind and the CCR may soon be starved for work when the buffer is consumed. The Yellow zone is the middle third of the buffer time period, and a gap appearing here indicates that there is a potential supply problem that should be investigated. The Green zone is the final third of the buffer time period, and gaps are acceptable; in fact, they are desired. If the Green zone is always full, the buffer is too large. If the Green zone suddenly fills, this indicates that either the CCR is falling behind schedule—an extremely serious condition—or too much work is being released to production—a troublesome but not critical condition.

Just like any other visual signal on the Lean shop floor, a graphical display of the buffer condition is helpful to quickly identify problems, as indicated in Figure 5-30.

By closely monitoring the status of the CCR, the system can orchestrate release signals to upstream workcenters and quickly alert a supervisor when there is a problem. This is the sort of critical and noninvasive visual scheduling feedback that is required to keep everyone in the plant focused on the key constraint operations, investing little attention to nonconstraint scheduling information.

TOC and Continuous Improvement

Lean and TOC are compatible, offering complementary problem-solving approaches to eliminate waste. From the very beginning of Lean, Ohno emphasized the reduction of inventory as the primary tool for improvement, because its existence disguised other forms of waste. TOC shares this emphasis on inventory with two critical measures: throughput-dollar-days and inventory-dollar-days, suggesting that a plant should not be judged just on the level of inventory it holds but also on how fast the inventory is moving:

- **Throughput-dollar-days**—the dollar value of late shipments multiplied by the number of days the shipment is late
- **Inventory-dollar-days**—the value of inventory multiplied by the number of days it stays under the plant's responsibility

The plant's prime measurement should be to reach zero throughput-dollar-days (which indicates accurate promising, on-time delivery, and satisfied customers). The secondary measurement should be to do this with as few inventory-dollar-days as possible.[118]

Although they both improve performance, Lean and TOC take slightly different approaches. Lean balances TOC through an emphasis on team orien-

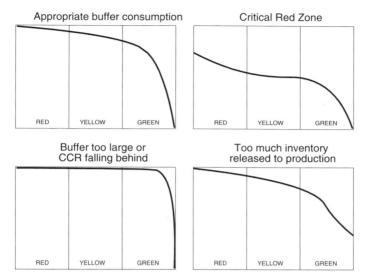

Figure 5-30. CCR buffer zone status patterns

tation and cultural change, relying heavily on intuition, trial and error, and incremental problem-solving. By focusing on all non-value-adding activity, Lean identifies tasks and activities that should be eliminated regardless of the existence of a constraint. This intuitive approach creates weakness when a problem is caused by interactive factors that make problem resolution complex,[119] where a more rigorous and empirical method like TOC or Six Sigma may be more appropriate.

TOC methodically optimizes a discontinuous process by specifically managing the constraint while subordinating all other processes. This helps to focus problem-solving in a complex environment. When the constraint is eliminated, TOC immediately searches for the next one.

Lean continuous improvement techniques are generally broad, team-based, and intuitive, whereas TOC is an empirical and focused initiative. TOC may be used in conjunction with other continuous improvement techniques, by focusing on a particular constraint, directing kaizen teams to creatively problem-solve how best to exploit, subordinate, and elevate the constraint.

TOC suggests focusing on the primary constraint within each value stream, whereas Lean emphasizes the elimination of all waste.* For this reason TOC

---

* Value stream mapping may identify a constraint and prioritize continuous improvement efforts toward it, but Lean does not stop at the constraint, seeking to slowly and incrementally eliminate *all* waste.

may be criticized for its single-minded focus on the constraint, to the detriment of more general waste reduction. However, *Throughput Accounting**, an offspring of TOC, emphasizes the elimination of operating expenses anywhere in the organization, once you have focused on the increase in throughput and reduction of inventory. Any waste that creates an operating expense should therefore be eliminated according to TOC, whether or not it is related to a constraint.

Furthermore, TOC can help to identify general waste and inefficiency through its rigorous investigation of policy and market constraints. TOC's focus on waste reduction is therefore more balanced than it may appear, for example:

- Policy constraints include traditional workcenter utilization and efficiency measures that may overload a nonconstraint workcenter and create unnecessary inventory, treating it as a CCR when in fact it does not constrain throughput.
- Policy constraints include poor prioritization and sequencing techniques that result in overloaded workcenters and frequent expediting, creating the appearance of a moving CCR.
- Ineffective quality policies cause interruptions and discontinuity of flow that may appear as a moving CCR.
- A large transfer batch size policy may cause downstream backups and uneven workcenter loading. Effective application of Lean flow design (smaller transfer batch sizes, workcenter load leveling, pull) may eliminate what appear to be moving CCRs.
- An unprofitable product mix may result from the combination of an ineffective marketing policy with a market constraint, where the market is not willing to cover the cost of certain products. Consider this example, where there are two products, one requires one hour of CCR, the other requires five hours. Traditional cost accounting absorbs only the incremental cost of the CCR activity. But Throughput Accounting imputes the time cost of the entire value stream on the CCR, making the production cost of the CCR-heavy product considerably higher. If the market isn't willing to pay this premium for the second product, then the company should discontinue it.

This last point is clearly demonstrated when a repetitive operation focuses on a single product family, and is therefore able to offer a low price because they have no constraint on volume production. Contrast this with a

---

* Throughput accounting, first addressed in *The Goal* and explored more deeply in later works, emphasizes the value equation of Throughput minus Inventory minus Operating Expense (T – I – OE) *in that order*, whereas traditional cost accounting generally emphasizes the reduction of operating expense as the first priority while recognizing inventory as an asset, not a liability to be reduced.

job shop, having a constraint on volume (in favor of flexibility) and therefore unable to compete on price of a specific item. This illustrates different positioning on the product/process diagonal, where a job shop may seek to differentiate itself by adding more customer-specific value and variety to the product.

Why it is important to optimize throughput rather than focus on the most profitable product?
Two products: A and B
A sells for $100 and costs $25 to make—contribution margin $75
B sells for $100 and costs $75 to make—contribution margin $25
Obviously we want to sell more As than Bs, right? But what if there is a constraint?
The CCR makes one A each hour
The CCR makes ten Bs each hour
How much should we sell, and therefore make, of A and B?
Our profit is $250 per hour by making ten Bs and only $75 by making one A.

Although Lean Manufacturing and TOC are complementary, there is a fundamental difference we must reemphasize. Repetitively focused Lean texts commonly imply that a balanced line should be attainable, whereas TOC accepts the fact that there will be a naturally unbalanced line, a realistic assumption in a discontinuous operation. In a repetitive operation, Lean may use Product-Specific Kanbans to store buffers of inventory at *each* step along the production process, whereas TOC requires a buffer only ahead of the CCR (note that both Lean and TOC suggest buffers at final assembly and shipping).

In general, popular Lean Manufacturing techniques tend to favor repetitive operations, whereas TOC offers the greatest advantages to discontinuous operations by encouraging a sharp focus on the primary constraint. However, these two approaches are not as different as they appear, because emphasis on the pacemaker in a balanced cell is similar to emphasis on the constraint in an unbalanced job shop. In either case, focus is essential to realize the benefits of scheduling software and techniques while avoiding unnecessary complexity.

BRINGING IT ALL TOGETHER

Now it is time to combine the diverse concepts and techniques we've explored throughout this chapter, applying them to several mixed-model scheduling situations.

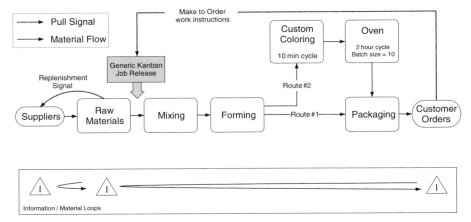

Figure 5-31. Simple job shop

A Simple Job Shop

Let's begin with a simple example of a job shop where some but not all products require a cycle in a shared oven, as shown in Figure 5-31. There are two routing pathways in this example. Route #1 moves directly from forming to packaging with no time in the oven, and route #2 receives a custom color that must be cooked and cooled in the oven before moving to packaging.

This scenario may look relatively simple, but from the scheduler's point of view it contains several variables:

- The oven can only accommodate a limited number of jobs during a single cycle and is a potential bottleneck.
- Jobs must be queued up at the oven in a specific sequence, as some jobs may not combine with others because of chemical interactions during the heat cycle.
- The cycle time of the oven is longer than the preceding workcenter, and the capacity of the coloring cycle is greater (but limited by customer-dictated color and order size), so the coloring process can outpace the oven under most circumstances.
- An order with an earlier due date and a shorter routing duration (route #1 not requiring the oven) may be released to the shop floor later than an order with a later due date that requires an oven cycle (route #2) and still finish first.
- Each job is potentially unique, so a Generic Kanban signal issued from the customer order may pull work from the gateway workcenter, communicating specific product configuration and work instructions.

The scheduler must juggle these issues, coordinating the timing and sequencing from job release through the pathways leading to the shared resource, to ensure the oven is loaded properly and optimum throughput is achieved.

TOC Reduces Apparent Complexity

Now let's consider a more complicated example from a Theory of Constraints point of view. We've determined that the oven is the primary constraint, with a fixed heat/cool-down cycle time of two hours and a capacity of 10 batches per cycle. The preceding operation, custom coloring, has a cycle time of only 10 minutes, so during the two-hour oven cycle the custom coloring workcenter can queue up 12 batches (120 minutes/10 minute cycle time) while the oven can only process 10. The custom coloring workcenter can outpace the oven by two batches each cycle, and the pace must be regulated or else excess inventory will accumulate. But even if an inventory buffer accumulates in front of the oven, if it's not the right mix of jobs, the oven will still run at less than capacity.

If we introduce a jumbled flow leading into the custom coloring operation, then a job could follow any number of pathways through mixing, forming, molding, and finishing. This creates an additional challenge for the scheduler to determine the initial release sequence, and prioritization guidance as jobs pass through the jumbled flow of upstream workcenters, to ensure they arrive at the CCR at the right time and with appropriate buffering of product mix to sequence and load the oven at full capacity. Group technology and product family sequencing may help to simplify this jumbled flow, but because this is a low-volume and high-mix environment it will not completely eliminate the challenges.

Finally, we'll monitor the status of the CCR buffer, using APS scheduling software with a three-light buffer management andon signal to alert operators of a potential sequencing or throughput problem feeding the CCR.

As illustrated in Figure 5-32, (A) a Generic Kanban signal issues from the customer order back to the CCR buffer, and in turn (B) a Generic Kanban signal is then sent from the CCR buffer back to job release, indicating the appropriate release sequence and work instructions—this may require backward scheduling and sequencing software assistance to maintain the CCR buffer properly. (C) Kanban replenishment signals issue to material suppliers: These may be Product-Specific Kanban signals for regularly stocked items and Generic Kanban signals for items that are specific to the job and not regularly stocked. (D) Some jobs bypass the CCR and move directly to packaging. These may be high-volume, low-mix items regulated by a Product-Specific Kanban, high-mix items regulated by a Generic Kanban, or a combination of both.

Does it matter if the CCR operation is at the front, in the middle, or at the end of the production routing? The CCR Red/Yellow/Green zone buffer management logic remains the same, but release signal sequencing and backward

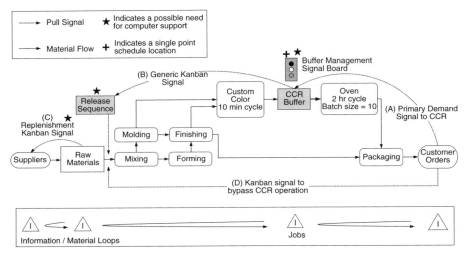

Figure 5-32. Job shop with CCR buffer management

schedule calculations are likely to be more complicated as the CCR is placed further downstream, because there are more upstream pathways and potential interruptions to consider when loading the CCR buffer properly.

It is important to note that a simple buffer management situation may be handled with spreadsheets, but as more complex scenarios arise, integrated APS scheduling and buffer management tools may be required. As complexity increases in a discontinuous environment, we may never completely avoid the use of an APS scheduling software system, but we can minimize its complexity by focusing on key operations such as constraints and pacemakers. As complicated as this scenario appears, notice that the focus is clearly on scheduling and controlling the constraint, which in turn regulates the upstream operations.

Mixed-Model Cellular Production

Now for an example where demand pull signals issue from a final assembly cell back to the CCR. Note that the latter portion of this scenario is mixed-model production, using takt time and a pacemaker operation to provide quick delivery to the customer.

Recall from earlier in this chapter that there are two general patterns of production that may be used in a mixed-model environment: Assemble or Configure to Order and assemble to a heijunka schedule. In both cases it is very important to note that there are two distinct processes: 1) production by the shared resources (constraints, batch-oriented production) that are building to one or several final assembly supermarkets that in turn serve the final assembly of one or several product families (these shared resources build

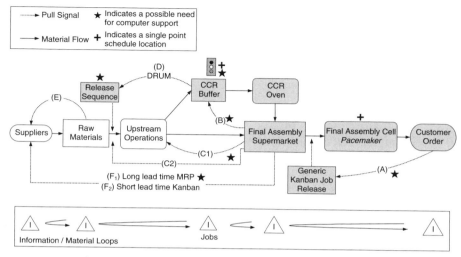

Figure 5-33. CCR feeding Assemble or Configure to Order

ahead to the final assembly supermarket and are scheduled on a batch basis to minimize setups) and 2) the final assembly process (with takt time and a pacemaker) quickly serving individual customer orders by drawing from the final assembly supermarket.

This second stage is where the distinction appears between Assemble or Configure to Order with no finished goods supermarket, and assemble to a heijunka schedule, where customers pull from a finished goods supermarket. The choice between these two approaches depends on the balance among product variation, inventory levels, and customer lead time expectations. First we'll consider the Assemble or Configure to Order scenario, shown in Figure 5-33. This scenario assembles highly variable products directly to the customer order with no finished goods supermarket.

This is an example of the postponement approach, where the demand signal is created when the customer places an order. Materials are pulled from a final assembly supermarket and are assembled to the specifications of that particular order. Because each finished product is assembled to order, it is impractical to have a finished goods supermarket. Nevertheless, the final assembly process deals only with variations within a product family, so it has been balanced and the pacemaker operation runs according to takt time.

Note that there are two distinct points of scheduling in this scenario: the pacemaker and the CCR drumbeat, since preassembly and final assembly are separate processes.

(A)   Generic Kanban signals are released to the gateway workcenter of the final assembly process. Product configurator software may help the salesperson make the right design, engineering, configuration, and pricing decisions. Available to Promise software helps to determine the possible delivery date. Intelligent product configuration and work instructions are sent to job release and follow the workorder through production; these may be a combination of physical and electronic documents.

(B)   The Final Assembly Supermarket sends pull signals to the CCR. These pull signals can be a hybrid of the two kanban types. Product-Specific *Priority* Kanbans issue in target batch quantities from the final assembly supermarket to the CCR to minimize setups. Generic Kanbans for nonstandard parts are issued to the CCR in customer order quantity, although they may be grouped with similar products, using a priority kanban to minimize setups. Computer assistance is likely required to determine and regularly adjust the supermarket buffer size according to a forecast, calculating material requirements with MRP and a Planning BOM.

(C1)   Product-Specific Kanban signals issue from the final assembly super-market (bypassing the CCR) to the upstream workcenters for items that are regularly stocked.

(C2)   Generic Kanban signals issue from the final assembly supermarket (bypassing the CCR) to the gateway workcenter for unique or custom items required by the order. Intelligent product configuration and work instructions may be sent electronically to the gateway work-center and follow the job through production.

(D)   The CCR buffer issues kanban signals and backward schedules to the gateway workcenter according to the buffer status and sequencing requirements. Computer support may be required to determine the backward schedule if pre-CCR routing pathways are complex, and to operate a buffer management andon.

(E)   Raw material replenishment signals are sent directly to suppliers. These may be Product-Specific Kanbans for regularly stocked parts and Generic Kanbans for unique or custom items required for a particular customer order.

For materials that are stored and replenished directly to the final assembly supermarket, or line-side materials stored within the final assembly cell, not passing through a separate storage facility:

(F$_1$)   MRP issues purchase orders for long lead time components based on forecasted requirements and the Planning BOM. This usually requires computer support.

(F$_2$)   Kanban replenishment signals for short lead time components are sent to suppliers. These may be Product-Specific Kanbans for regularly stocked parts or Generic Kanbans for unique items required by the customer order. The supplier may require long-term requirements planning information in order to plan for quick response when a kanban signal is received.

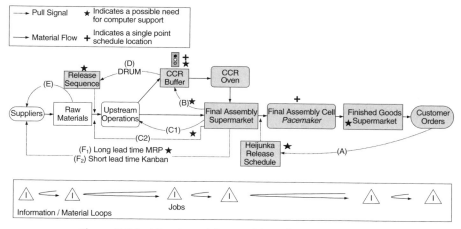

Figure 5-34. Mixed-model assemble to heijunka schedule

Final Assembly to a Heijunka Schedule

Finally we introduce a level schedule replenishing a supermarket of finished goods pulling from a CCR, as shown in Figure 5-34.

There is little upstream change from the previous example. Far downstream, customer orders are fulfilled from a finished goods supermarket, which is replenished by the heijunka schedule. This is possible because there is limited finished item variability, and so this is a Make to Stock push/pull level schedule based on forecasted finished goods supermarket levels. The heijunka schedule on the final assembly cell calculates the takt time, pitch, interval, product mix, and sequence based on demand and cycle time of each product family and cell; this is indicated as push/pull signal (A). Dynamic heijunka schedule calculations may require computer support because of volume, complexity, and rate of change.

An enterprise may employ a hybrid of Assemble or Configure to Order and heijunka scheduled final assembly processes, combining the two prior scenarios into a single integrated process. When this is the case, final assembly is orchestrated according to the heijunka schedule, while a customer-configured order may be dynamically inserted into the predefined heijunka sequence at any time, as long as it is within a change allowance. This scenario was described earlier in this chapter by Toyota as "change to order."

Lean and TOC Working Together

It is important to note how Lean and TOC techniques cooperate in the last two examples. TOC describes three buffers: CCR, final assembly, and shipping; Lean describes two supermarkets, final assembly and finished goods, which pull from upstream shared resources that may be constraints. Although the terminology from these two perspectives may be different, it is clear that they are describing the same general buffering approaches:

- Lean pacemaker (where the line is generally balanced but nevertheless constrained by the longest cycle time operation) = TOC drumbeat
- Lean final assembly supermarket = TOC final assembly buffer
- Lean finished goods supermarket = TOC shipping buffer

Discontinuous operations may be predisposed to think in TOC terms by focusing entirely on the primary constraint, running the entire value stream to the drumbeat of the CCR. This greatly simplifies an otherwise complex scheduling and control situation. If a job shop produces every job to order, and there are no commonalities among operations downstream from the CCR, then the drumbeat does regulate throughput properly according to demand—the CCR is effectively the pacemaker of the entire value stream.

However, if a job shop rationalizes its product/process strategy, determining that some postconstraint operations may be grouped into a mixed-model final assembly process, then the CCR may feed the final assembly supermarket with a priority kanban. This optimizes throughput on the constraint by running larger batches, while maintaining flexibility to quickly respond to the customer's individual pull signals. These are the underlying mechanics of a job shop moving down the product/process diagonal from Engineer or Make to Order, toward Assemble and Configure to Order.

Who regulates the value stream when a CCR *and* a pacemaker are involved? The relationship between the CCR and the final assembly pacemaker can be demonstrated clearly when there is one CCR feeding multiple final assembly cells. In a mixed-model environment this is a common arrangement; while each product family may have its own dedicated final assembly cell, they may all be fed by a shared resource such as a metal fabrication operation. Some shared operations may be capacity constrained while others are not. This arrangement is illustrated in Figure 5-35 with an example of an outdoor furniture manufacturer with three product families. Keep in mind that each final assembly cell operates at a takt time unique to its cycle time and product family demand, creating an uneven demand for components that pass through the CCR. The final assembly supermarket buffers this variable demand, protecting the CCR by aggregating individual pull signals into larger batch sizes with a priority kanban.

In this example the pacemaker operations drive the shared CCR drumbeat. This would seem to contradict the TOC imperative, where the CCR drumbeat controls the throughput of the entire value stream. But this rule only applies when there is sufficient demand on the CCR to cause the constraint. If the pacemakers in this example suddenly experience a drop in demand, they will stop sending kanban signals back to the CCR, *then the CCR is no longer a constraint.* The constraint has shifted to the market because of lack of demand. So in this scenario the CCR drumbeat is *subordinate* to the pacemaker takt

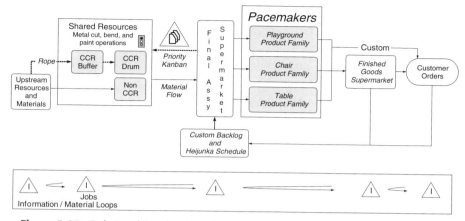

Figure 5-35. Relationship of a shared CCR to multiple final assembly pacemakers

time, because the pacemaker takt time is derived from the market demand for each product family.

VARIATIONS ON A LEAN THEME: CONWIP, SMP, AND POLCA

In this chapter we have focused on two fundamental approaches to describe the scheduling and control of operations: Lean Manufacturing according to the Toyota Production System, and Theory of Constraints. Finally, we look at three less widely known variations that are especially appropriate for a discontinuous environment: CONWIP, SMP, and POLCA.

CONWIP—Constant Work In Process

Introduced by authors Hopp, Spearman, and Woodruff in their 1989 article "CONWIP—A Pull Alternative to Kanban"[120], and explored in depth in Hopp and Spearman's book *Factory Physics, Foundations of Factory Management*, CONWIP is a deceptively simple approach to using kanban in a discontinuous environment and in fact can be used to describe the buffering and pull mechanisms in any type of system. CONWIP is described by Rother and Shook in *Learning to See* as a *FIFO Kanban* and is similar to the WIP reduction aspect of Simplified Market Pull (SMP) described earlier in this chapter.

CONWIP can be explained by Little's law, which states that there are two factors that control the elapsed time required for inventory to flow through a system: the average rate of throughput and the average amount of WIP within the system:

Average WIP/Average Throughput = Average Flow Time

For example:

100 pounds WIP/10 pounds per hour = 10 hours average flow time

According to the authors, *push systems schedule throughput and measure WIP, while pull systems set the WIP levels and measure throughput.* WIP is highly visible and easily controlled, whereas throughput must be estimated, planned, and scheduled. This is a particularly difficult challenge in a discontinuous environment for reasons we've already explored: fixed lead times in the planning calculation, product variability, routing variability, congestion, and Murphy's law, to name a few. Pull systems simply reduce throughput time by limiting the amount of WIP in the system; this is accomplished by controlling the number and capacity of kanban containers.

Consider a bathtub: There is a faucet to let water in and a drain to let water out. If the rate of inflow and outflow are relatively equal (Average Throughput), the amount of water in the tub remains constant, and the Average Flow Time required for a molecule of water (a job) to enter and exit the bathtub depends on how much water (Average WIP) is in the tub. When you allow very little water to accumulate in the tub, the average flow time is very short. For example, if the average throughput rate is one gallon per minute, and if there are twenty gallons in the tub, the average flow time is twenty minutes. If there are five gallons in the tub, average flow time is five minutes—less WIP means less lead time. This bathtub analogy is useful in describing the influence of inventory reduction on lead time in any Lean environment.

Now consider a job shop as this bathtub: We don't worry about routings, setup, and run times, or other complicated scheduling variables, we simply determine the desired amount of WIP to leave in the system at any time. CONWIP controls the number of kanbans allowed on the shop floor, where each card or container has a roughly standard size (equal capacity by job is not a practical assumption in many job shops; we'll return to this issue in a moment). Available kanban cards signal for a new job to enter the shop and become available again only when the job exits the shop. Soon after a card exits the shop it moves to the front again, allowing a new job to enter. Expediters are not allowed to force the start of new work without a card present, even if the gateway workcenter is idle.

In a job shop implementation, CONWIP is typically a Generic Kanban system because part numbers are assigned from the order backlog at the start of production, although there is no reason it cannot be used in a Product-Specific or hybrid manner. The backlog may be prioritized by due date sequence, first come first served, critical ratio, or other methods. As a job enters the shop it is assigned the appropriate part number from the order backlog. If the discontinuous process builds to stock or to a level schedule (such as when feeding a final assembly supermarket) the next part number may be assigned by a mechanism similar to a heijunka box regulated by a priority

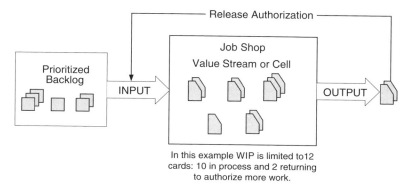

Figure 5-36. CONWIP

kanban; however, because the process is discontinuous it may not have a standard takt time and pitch.

Hopp, Spearman, and Woodruff suggest that CONWIP may be considered to be *input-output control* carried to its logical extreme. Traditionally input-output control is used to control the loading of a particular workcenter, but with CONWIP it is applied to a collection of shop floor resources (a cell, common routing pathways, or an entire value stream) where inputs are strictly controlled by the rate of output *from the entire system*. They suggest that CONWIP is similar to a technique used in air traffic control. On days with heavy air traffic, a departing plane will sometimes be held on the ground at the originating airport rather than be allowed to take off and remain in a holding pattern at the congested destination airport. Planes are held even if take-off runways are free at the originating airport, and the result is greater safety and lower fuel consumption with no added delay.

Hopp, Spearman, and Woodruff suggest that CONWIP cards should represent standard quantities, but a fixed kanban size is an impractical requirement in many job shops where each job may be different and the process batch size is controlled by the customer order quantity. If total WIP is regulated by the sum of all jobs in the system, and each job has a different capacity, then this requires a calculation upon job release to ensure that the total WIP threshold is not exceeded. This may require that when a small job leaves the shop another small job must be started, rather than a larger job that is ahead in the backlog sequence—waiting until the large job can be started may idle the bottleneck operation. If the large job is critically delayed by the advancement of several smaller jobs, then it may need to be issued to production without a free kanban card (with supervisor approval), temporarily overriding the WIP threshold.

As we explored earlier with SMP, as a result of inventory reduction CONWIP naturally causes bottlenecks to reveal themselves. As excess WIP is drained from the system, it finally accumulates only in front of the constrained operation. With highly variable parts and routings, however, transient bottle-necks may move around the shop according to the particular mix on the shop floor at the time. In either case, stationary or transient bottlenecks, if there is little congestion on the floor then workers may quickly recognize the situation and rebalance capacity to relieve the constraint.

Earlier it was said that CONWIP is deceptively simple, but it's important not to make it *too* simple. We must not interpret CONWIP as a big black box for the whole factory. At a macro level, perhaps CONWIP can measure the WIP in the entire plant, but that can also be done from the accountant's balance sheet. The problem with such an aggregate measure is that you risk starving individual bottlenecks. If there are several significant flow paths within a plant, each has a constraint that may be managed separately. Therefore, for CONWIP to be effective we must define common pathways within the plant, consisting of a collection of workcenters, resource groups, and cells. We then establish a WIP target for each pathway that optimizes its characteristics and constraints. Tom Knight of Invistics, a long-time proponent of CONWIP, describes the design of *Flow Paths*:

> A Flow Path represents a group of products that visit similar work centers. Flow Paths are typically defined to facilitate the logical division of the plant into multiple flows, each of which can be considered a "focused factory", independent of the others. Multiple Flow Paths can be defined for a plant, and a product can belong to only one Flow Path. Within a high-mix plant, high volume Flow Paths might best utilize traditional Kanban, while low volume Flow Paths might best utilize the more generic CONWIP approach to pull scheduling.[121]

In summary, CONWIP offers substantial benefits despite the simplicity of implementation—although the degree of simplicity depends on the nature of the environment and how many distinct Flow Paths are involved. CONWIP not only identifies bottlenecks but enhances process reliability, aids in the early detection of quality problems, results in less clutter, and reduces lead time. CONWIP offers a relatively simple approach to improvement by draining the swamp without paying particular attention to each and every operation step, and therefore it may be a valuable first improvement step in a congested job shop, leading to further improvement initiatives such as SMP, TOC, and POLCA.

SMP—Simplified Market Pull

SMP was described earlier as a technique to control a discontinuous operation when there is excess capacity, using familiar Theory of Constraints scheduling logic. SMP is essentially Drum-Buffer-Rope (DBR) with market demand, rather than a bottleneck production operation, as the constraint. DBR buffer management is therefore applied to the shipping buffer (finished goods supermarket) in a Make to Stock environment and to the order backlog in a Make to Order environment. If a shop using SMP finds itself suddenly capacity constrained, it may shift to traditional TOC methods, managing the CCR buffer.

In a Make to Order situation, new jobs are *authorized for release** to the plant at the due date plus the buffer time, which includes the expected production time plus the allowance for protection against disruptions. Note that, in TOC fashion, buffers are described by time and not quantity. Priorities are assigned to each order and are adjusted as conditions change based on the amount of buffer penetration—calculated as the percentage of the buffer used to date. Thus, if an order has 10 days of buffer and it is now three days since its release, its priority is 30 (percent). Another order with five days of buffer three days after release has a priority of 60. When a worker in the plant finishes the job he's been working and must choose between these two, he should work on the priority 60 order next. This prioritization rule applies at any workcenter, whether at the gateway or downstream.

For Make to Stock replenishment orders, SMP may trigger the release of an order when the quantity available in the shipping buffer (finished goods supermarket) drops below a predefined reorder point—a Product-Specific Kanban replenishment signal may be used here. The order will be given a priority that reflects the inventory status. If the desired level is 100 units and the order is triggered when the inventory drops to 80, the job will be released for 20 units with a priority of 20: 20% of the item's buffer (quantity in this case) has been consumed. Let's say that there is another shipment of 40 units from inventory brings the level to 40 (100 − 20 − 40 = 40 remaining). If the previous job in the backlog has not been started yet, its quantity will be changed from 20 to 60 and its priority will also be changed to 60. If the job has already been released to the shop floor, then a job will be created for quantity 40, priority 40 and the earlier order's priority will be changed to 60 to reflect that it is more critical to finish it and get those units into stock. The second order's priority reflects the expectation that the earlier order will arrive into inventory sooner and will cover some of the demand before this quantity is ready. These are dynamic priorities, meaning that they will change during the life of the work order in response to changing conditions. These priorities control the release of jobs to the shop floor and the movement between workcenters once the job has been released.

The two priority systems, using time buffers for Make to Order demand and quantity buffers for Make to Stock replenishments, work well together. Because of the dynamic nature of the priorities, software may be required to calculate SMP prioritization, and to communicate these changing priorities as jobs move through the shop.

SMP prioritization and CONWIP kanban techniques are complementary: CONWIP provides the pull and WIP control mechanisms to ensure the shop

---

* When using SMP in conjunction with CONWIP, for example, SMP calculates the release time authorization and priority, but the job is not issued to the shop floor until a kanban card is available.

is not overloaded, and SMP calculates priority of release and flow. It is important to note that SMP's only impact on the shop floor is the prioritization of jobs at each operation; SMP priorities may be familiar to many, because they resemble the popular critical ratio method. If the release of work to the shop is tightly regulated by CONWIP then congestion is minimized. If that is the case there should be few jobs queued up at each workcenter at any time, and lead time is reduced, minimizing the burden of reprioritizing jobs once they are on the shop floor.

A discontinuous operation may wish to implement CONWIP first, draining the swamp to relieve congestion while establishing a simple pull mechanism. Then depending on the nature of the operation (Make to Stock, Make to Order, or a mix of both), the quantity of jobs in the backlog and on the shop floor, and the frequency of changes to their priorities, the enterprise should determine whether an SMP prioritization system is appropriate. The shop should also be prepared to switch to DBR when demand exceeds capacity.

POLCA—Paired Overlapping Loops of Cards with Authorization

Introduced by Rajan Suri in his 1998 book *Quick Response Manufacturing, A Companywide Approach to Reducing Lead Times*,[122] POLCA attempts to optimize the flow of work within a discontinuous environment characterized by complex and variable products and routings. This is accomplished by using APS scheduling software to plan the release and prioritization of jobs, then controlling the flow within the shop with an elaborate kanban system.

CONWIP controls total WIP contained within each Flow Path, which may include a collection of workcenters, resource groups, and cells. Prioritization of job release to the Flow Path may be performed with FCFS, critical ratio, SMP/DBR, or other techniques, but once the job is released within the Flow Path, CONWIP does not intervene until the job exits and prepares to enter a subsequent Flow Path. By contrast, POLCA is a sophisticated prioritization *and* shop floor pull technique, regulating the movement of work *between operations and cells*. Although POLCA requires APS scheduling software to optimize the sequencing and releases in advance, once the job is released to the shop floor, POLCA kanban pull mechanisms aid the flow without further software intervention. This is consistent with a key principle of kanban, which is to control the flow of work with simple pull signals, not allowing work to start until there is available downstream capacity. According to the APICS Lean Manufacturing Workshop Series, although POLCA requires APS scheduling:

> There is frequently a problem of adherence to schedule when operators tend to second-guess the system by either doing easy jobs first or by taking the apparently logical step of reducing changeovers by doing two similar jobs consecutively instead of rigidly following the schedule. The operators may be right or they may be wrong. They may make considerable gains, particularly when the situation in the shop has changed since the schedule was calculated. Or this may

make the situation worse when following [downstream] workstations become disrupted. POLCA, like kanban, tries to remedy this by a continually updated and visible system.[123]

POLCA begins by analyzing and rationalizing the flow of products and processes, grouping flows into common cells whenever practical. The layout of the plant is then diagrammed, identifying the physical relationship of each cell (but not each individual workcenter within the cells). Each cell is then identified with a simple name, such as A, B, C. Common flows of work among cells are identified, such as A–B, B–C, A–C, and so on. Each cell pair is assigned a pool of kanban cards (POLCA cards) that controls the flow of work between the pair of cells. POLCA does not attempt to control the flow of material within each cell; rather, it controls the flow of material between cell pairs. Each BOM and routing is defined in terms of cells, not individual workcenters. This serves to simplify and flatten the BOM and routing, thereby simplifying the planning and scheduling process performed by the software. Suri calls this High-Level MRP (HL/MRP). The scheduling software then calculates the range of time that the job *may* start (not *will* or *should*), called *authorization times*.

The number of POLCA cards provided to each cell pair is controlled to minimize WIP. Suri suggests an application of Little's law to calculate the appropriate number of cards in each loop (total WIP threshold) based on a forecast.* Forecasting a unique or variable product mix in an Engineer to Order or Make to Order job shop can be difficult, so it's necessary to forecast at a high level using some logical grouping such as product families, so we can estimate the load on each cell pair. The HL/MRP system calculates total demand based on a planning horizon of one or several months, using firm orders on hand, as well as a forecast of products using aggregate families or product groupings. The planning department periodically adds or removes POLCA cards depending on the anticipated workload. If demand suddenly shifts, or congestion appears, then the value stream can be rebalanced by adding or removing cards for certain cell pairs.

It is important to note that forecasts are used only for planning and periodically adding/removing POLCA cards to control WIP on the shop floor based on anticipated workload. Jobs are not issued to the shop floor until 1) an actual customer order exists in the backlog, 2) the job has reached its authorization time window, and 3) there is a POLCA card waiting at the gateway workcenter of the cell. Jobs that arrive at a cell are sequenced by authorization time on a First Come First Served basis. If the first job does not have a POLCA card available, the team picks the next job and looks for its POLCA card. If none of the authorized jobs has a POLCA card available, then the cell

---

* Where LT(A) and LT(B) are the estimated average lead times of each cell over the planning horizon, and NUM(A,B) equal the total number of jobs that go from A to B during this time, and D is the number of working days, and assuming a standard job size, then # of A/B cards = [LT(A) + LT(B)] × NUM(A,B)/D.

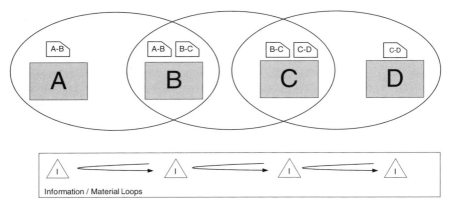

Figure 5-37. Paired Overlapping Loops of Cells with Authorization (POLCA)

remains idle until a new job arrives, a new POLCA card arrives, or a premature job changes to an authorized job. Note that POLCA cards can be either
physical cards or electronic signals. If there are a large number of jobs and
routings, an electronic POLCA dispatch system may be used to identify the
next job to start based on these three rules. Also note that because release of
the job is calculated by HL/*MRP*, the system is checking for material availability of the entire job before it is released.

When a job is released to the gateway operation of the first cell, the system
produces a routing sheet (work ticket, traveler) that follows the job, describing the sequence of operations, materials required, and work instructions. This
information can be distributed on paper or electronically.

Once the job is released, the Paired Overlapping Loops of Cards regulate
the flow of the job among the cells until the job is complete, as shown in Figure
5-37. The POLCA card for a cell pair stays with the job through completion
of *both* cells in the pair before looping back to the previous cell in the pair. In
this example the job moves through cells A, B, C, and D. This involves three
pairs of cells: A–B, B–C, and C–D, and thus three POLCA cards. When the job
moves into cell A it consumes an available A–B card. When the job moves
into cell B it picks up the B–C card as well. The job now controls two overlapping cards: A–B and B–C. In fact, a job will have a single POLCA card only
when at the first and last cells of the entire routing (A–B and C–D).

If the job finishes cell A and moves to cell B and there is no B–C card
waiting, then the job stops and waits for it to become available. The A–B card
is not returned to the front of the A cell until the job is complete at cell B,
which requires the B–C card to be issued. This is the principle of paired overlapping loops, which requires the availability of the next downstream cell pair
for the current cell-pair to complete. This cascading cell capacity signal
attempts to release jobs only when the sequential downstream cells are ready
for it. Once the job finishes cell B, the A–B card is returned to the front of cell
A, signaling availability. A roaming material handler can return this card,

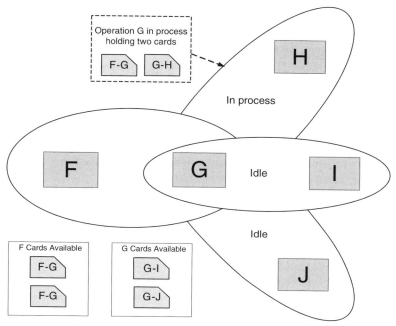

Figure 5-38. POLCA with downstream branching

although space and time permitting Suri prefers that the cell team returns the card as the final step to complete the job at their cell.

Of course even POLCA can cause limited congestion, depending on how tightly cards are controlled—there may never be a *perfect* solution for a discontinuous job shop. Consider this example where cell loop F–G feeds three alternate downstream cell loops G–H, G–I, and G–J, as shown in Figure 5-38. Let's say there is a job underway in G–H, but G–I and G–J are idle and their cards are available. There are a total of three F–G POLCA cards in circulation, two of which are available at the gateway workcenter of cell F. Now another F–G–H job comes along and finds an F–G card available so it is released, only to queue up at G–H, which is still working the previous F–G–H job. With three F–G cards available, there is nothing to prevent a second job from queuing up at G–H. This, of course, can be prevented by reducing the number of F–G cards from three to two, or even one—but this would starve an F–G–I or F–G–J job that has downstream capacity available in cells I or J. This slight but controlled congestion may be acceptable if limiting the quantity of F–G cards or changing the routings would add waste, or potentially create or starve a bottleneck.

Although Suri does not specifically emphasize the use of TOC concepts in POLCA, they are simple to apply. The kanban quantities leading to a constrained workcenter, or to several workcenters that are potential constraints depending on the shifting workload, may be increased, effectively creating a

CCR buffer. And because an APS system is used for POLCA planning, the system can backward schedule and prioritize to ensure that the CCR buffers are fed properly, and so plan ahead to alert the shop floor if the current backlog will overload a CCR, or create a new one.

Although this clever combination of APS and kanban does not appear to be widely used at this time, POLCA deserves consideration in many environments. In his book Suri attempts to coin a new approach to manufacturing called *Quick Response Manufacturing* (QRM), contrasting with Lean Manufacturing in that QRM applies to discontinuous operations, emphasizing lead time reduction as more significant than the general elimination of waste. His comparison of QRM and Lean Manufacturing is summarized in Figure 5-39.[124]

JIT or Lean Manufacturing	QRM
Systematic elimination of waste leads to continuous improvement	Relentless reduction of lead time results in continuous improvement and elimination of waste
Create "flow" by designing production lines so orders can proceed continuously without any backflows or stoppages. One piece flow is the goal.	Create cells based on families of products with similar operations. However, they need not have linear flow; products can go through cells in various sequences. One piece flow is not necessary; small batches may be necessary as a consequence of the customized nature of products.
Support flow using takt time and level scheduling. Use detailed analysis of tasks and standardization of work to achieve the balanced takt times throughout the production facility.	Support the ability to meet demand for widely differing products through organizational flexibility and techniques such as time-slicing, and by exploiting the understanding of system dynamics. Also, use different combinations of cells to create varying end items.
Suppliers meet flow requirements via pull signals and flex fences.	Suppliers support quick response by changing their operations and via redefining their interactions with the customer.
Pull signals material replenishment: "sell one, buy one" or "ship one, make one" implies there is a product ready in stock to sell, or there is a product in finished goods to ship. Hence inventory needs to be kept at each point in the supply chain. Too much inventory when there are a large number of end items. Doesn't work for custom-engineered products.	Tailored to companies that have very high product variety or engineered products. Goal of QRM is not to start a job until there is an order for it. Uses a combination of push for material planning and release, and a modified pull approach to prevent congestion.
Best suited for providing custom combination of predefined options for a baseline product.	Strongest when used for custom-engineered products.
Requires relatively stable demand, and largely for replacement products.	Use to forge new market niches such as emerging segments with unpredictable and rapidly changing demand, or where products must be tailored to individual customers.
Emphasizes on-time delivery as a primary performance measure.	Primary measure is lead time reduction. On-time performance is achieved as a by-product of the strategy.

Figure 5-39. JIT and Lean vs. QRM

As you can see in Figure 5-39, Suri's definition of Lean is relatively narrow, emphasizing repetitive production using a level schedule, Product-Specific Kanban, and one-piece flow as the ideal state. He suggests that the low-volume and high-mix agility of QRM may be leveraged to create new market niches. However, when this strategy is successful, a niche may become a mainstream market and the company may then wish to segment their product family into Runners, Repeaters, and Strangers, shifting some products toward the repetitive end of the product/process diagonal to capture market share. Emphasizing QRM as incompatible with Lean rather than simply representing different points along a single continuum may foster intellectual resistance to the transformation of a job shop into a mixed-mode operation. In this book I take the position that the basic principles of Lean apply across all types of operations using a variety of methods. In the end this is all just wordplay, of course, and all of these methods are useful in the appropriate circumstances, regardless of what they are called.

SEARCHING FOR THE RIGHT SCHEDULING SOFTWARE?

The Scheduling Software Market

It would be folly to attempt a thorough analysis of the scheduling software market in this chapter, for it would be out of date before this book itself was scheduled for publication. With that said, we can explore the four basic categories of scheduling software, offering some guidance for those considering an investment.

1. **ERP Scheduling Tools**—Integrated within the core application of an ERP system, these offer the tightest integration with MRP II planning and production control capabilities. Although in the early days many of these Lean scheduling and execution tools were primitive (such as a kanban system clumsily pasted over a traditional work order dispatch system) the quality and sophistication of embedded Lean tools have improved steadily through development and acquisition.

2. **MES Scheduling Tools**—Manufacturing execution systems emphasize tight integration with the shop floor, and sophisticated scheduling and (often real time) process control capabilities are commonly their strengths.

3. **SCM Scheduling Tools**—Supply chain management planning and scheduling tools orchestrate the demand and supply within the enterprise and among trading partners; these tools are especially useful when multiple stocking and production locations require schedule coordination. Figure 5-40 shows that production scheduling applications accounted for 17% of $1.36B spent on SCM software in 2003.[125]

4. **Third-Party Lean Tools**—These planning, scheduling, and execution tools often represent the latest advances in Lean theory and practice. Many are offered by smaller software companies with limited resources,

Application Segment	2003 Revenue	
Supply Chain Planning	$678M	50%
APS Production Scheduling	$225M	17%
APS Manufacturing and Distribution Planning	$204M	15%
APS Demand Planning and Forecasting	$197M	15%
APS Supply Chain Network Design	$52M	4%

Figure 5-40. SCM spending by application segment

and the choice of such a system should be considered both a product purchase and an important collaborative relationship with the publisher and its consulting team. Although these systems naturally require a significant investment in integration with your ERP system, as with the maturation of other best of breed software categories, you should expect that these publishers will develop sophisticated interfaces to popular ERP systems. If you want to fine-tune and push the envelope with Lean planning, scheduling, and execution, these advanced solutions deserve careful consideration.

Any enterprise seeking to acquire a Lean scheduling solution should choose on the basis of more than just functionality and cost; they must also find the best balance of complexity, integration, service quality, and vendor viability.

The Mosaic Approach to Scheduling

Carol Ptak is a past APICS President and former Vice President of Manufacturing Strategy for PeopleSoft. She is in an ideal position to observe global IT and operations market trends and emphasizes that many enterprises need a blend of Lean tools and techniques. In a 2004 interview with *Manufacturing Business Technology* she affirmed that more manufacturers are taking a *mosaic* approach by mixing techniques such as Lean and bottleneck management.[126]

Although TOC, SMP, CONWIP, and POLCA techniques may add less value in a repetitive environment, they can simplify the scheduling and increase the throughput of a discontinuous process, substantially reducing (but not eliminating) the requirement for sophisticated scheduling software. This is clearly the reason why TOC has become so popular in recent years, and why we've seen many constraint-based scheduling software choices appear in the market. The past decade has seen the refinement of many mixed push/pull scheduling and execution techniques, supporting the mixed-mode Lean Enterprise with a mosaic of visual, manual, and software-based scheduling solutions.

Here is *the bottom line* when designing a mosaic approach to any environment: Focus and simplicity are paramount. Whether managing a constraint or a pacemaker, it is essential to schedule a process at only one point. Each process must use a simple, logical, and preferably visual mechanism to signal demand pull. And it's important to carefully consider the placement of each inventory buffer, whether a kanban container or supermarket—its size, content, and specific value within the overall value stream. Lean software tools add value only when they are focused in this manner.

A Lean Constraint Management Success Story

Harry's Fresh Foods is a manufacturer of fresh (never frozen) foods (www.harrysfresh.com). Harry's was founded in 1978 and entered a period of rapid expansion in the mid-1990s as consumers turned to convenient and fresh foods. Harry's serves large and small grocery and food service organizations across a wide geography, with a variety of demand patterns. The frozen entree market is extremely competitive, with large suppliers and well-entrenched distribution channels.

The key to Harry's success and rapid growth are freshness and variety. They produce everything in small kettles Just-In-Time to satisfy customer demand, managing short shelf lives with little perishability waste. Because their processes are flexible and their test kitchen so creative, Harry's can bring new and creative products to market much faster than their competition, quickly satisfying changing and regional food preferences while rapidly developing specialized products on request for their larger customers.

In 2002 Harry's built a state-of-the-art processing plant designed for the smooth flow of materials from fresh, dry, and wet ingredients, through preparation, staging, cooking, packaging, and out the door to their customers. Soon after this plant was put into operation, they began the search for a new ERP system to help manage their significant growth.

As we investigated the requirements for this new system, we discovered that this brand-new plant was operating below expectations for total throughput. We quickly discovered a constraint. Although their prep and cook lines were optimized for small batch production, the more traditional packaging lines—with multiple package sizes, flavor and allergy sequencing, changeover, and cleanout issues—limited their sequencing flexibility and throughput.

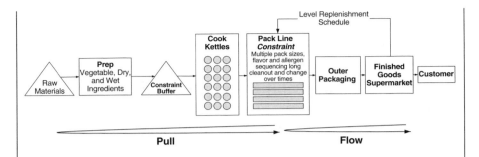

Harry's elevated the importance of the constraint management and scheduling portion of their ERP selection criteria and implemented constraint-based scheduling (CBS) in phase 1, going live with the entire system in just 120 days. Just like their production, the project team at Harry's was committed to small batch software implementation. According to Brad Paris, Vice President of Supply Chain for Harry's Fresh Foods, and project manager,

> We have two machines that fill one type of container, these machines are the bottleneck. One machine is twice as fast as the other. Traditionally, we thought the most effective way to manufacture was to use both machines simultaneously until we obtained the volume we needed. This created a relatively short burst of production that generated large throughput numbers. Intuitively, we thought large throughput generated the most efficiency and least cost.

> However, when we ran CBS in this scenario, it suggested that we spread the production out over a longer period, utilizing only the more efficient machine. At first glance, this was in complete contrast to our philosophy of small batches and maximizing total throughput. However, after studying the results, we discovered that CBS was suggesting that we utilize all the capacity of the most efficient machine (the constraint) before using the slower machine. Once we understood this philosophy (and studied the upstream and downstream processes to ensure there would be no disruption in the production flow) we decided to give it a try in production. The results were immediate. Our expectation over time is to get the same output with 15% fewer man hours, and we continue our setup time reduction efforts to increase our flexibility.

> CBS has also helped us to balance our production scheduling. We have seven fill lines that feed three chill lines. So there is always a balancing that needs to occur to maximize the chill lines by switching production effectively among the fill lines. This coordination requires a significant effort since we regularly run over 50 products per day. By using CBS, we are able to establish load limits for each fill line that creates the most effective use of our chill lines. Now, the system will not allow our schedule to become unbalanced, which will ultimately result in an increase in throughput of over 20%.

Finally, CBS helped us sequence our production based on certain characteristics that we have assigned to our products. For example, some of our products contain allergens which require a time-consuming cleanout of machinery before producing the next item. By assigning these changeover times to the different combinations of characteristics, we can use CBS to sequence all of our production in a manner that minimizes overall change over time. To date, we utilize different combinations of 7 characteristics per finished good to create the optimal sequence of production. This process resulted in a 12% reduction in overall downtime.

Since we clearly identified our throughput constraint and made this a priority for our ERP implementation, in less than 120 days we began to see the enormous benefits of constraint based scheduling tools. We have introduced minimal scheduling and execution complexity, while creating substantial production efficiencies and cost reductions.

During the research and development of this chapter, the staunchest proponent of the visual factory, and accordingly the avoidance of IT whenever practical, has been Jeffrey Liker at the University of Michigan. I am grateful for his determination and hold him in esteem as a Lean advocate of the highest order. During the past two decades Jeffrey has worked closely with Toyota, experiencing time and again the mastery of overwhelming complexity with relentless waste reduction, simplification, and the help of thousands of laminated kanban cards. I believe that his arguments against the introduction of IT onto the Lean shop floor do not stem from an inherent bias against IT, but from a recognition that IT can be mistakenly perceived as the easy way out, masking the underlying issues and inhibiting continuous improvement. Once introduced, IT can be very difficult to remove—a technology monument.

However, when you leave the domain of top-tier manufacturing companies like Toyota and Dell and consider the vast number of medium and small manufacturing enterprises, few may reach this advanced state of maturity any time soon. Furthermore, each industry has a different set of characteristics that constrain their planning and control requirements. Not every manufacturer can plan demand and a level schedule several months into the future. Not every manufacturer can cluster its suppliers in a campus setting. Not every manufacturer can control the configuration of its parts through carefully designed optioning programs supported by sophisticated sales and marketing strategies, and buffered by a global distribution network. And there are many manufacturers that have essentially discontinuous operations, who by design will never be able to accomplish repetitive flow. In fact, most small and medium-sized manufacturers must deal with a greater degree of uncertainty, and there are few that can claim to be as mature on the path to Lean as Toyota. Yet there is something for everyone with Lean, and all should aspire to move toward the ideals that Lean encompasses.

When IT helps a company along this path, then it is a useful tool. But the moment IT becomes an impediment, a crutch, a distraction, it should be eliminated as a cause of waste. According to a white paper entitled *Advanced Planning Systems as an Enabler of Lean Manufacturing* cowritten by Jeffrey Liker, this requires a new way of thinking and interaction between Lean and IT practitioners:

> Most manufacturing plants, even in progressive companies, are in a transition to Lean Manufacturing. For example, they may have plans for continuous flow assembly cells, but they are not fully implemented. They may have plans for pulling material from feeder manufacturing processes to the assembly cells, but they are still at the stage of scheduling production. There may be efforts to improve preventative maintenance, but downtime is still a problem. In these cases, APS software cannot be set up solely to support the future Lean system—this would waste the capability of the system. Instead, it should be used to improve the scheduling needed during the transition to Lean. For example, the model can reveal where the bottlenecks are to achieving flow and therefore help prioritize the Lean initiatives. It may show clearly that reducing setup time at a certain machine is a prerequisite for sending level pull signals back to raw material suppliers, as well as quantify the benefits of doing this. This information can be a powerful tool both for prioritizing Lean initiatives, as well as for providing the ammunition to persuade skeptics who are challenging the Lean systems.
>
> Many manufacturing operations are quite complex and careful planning is needed to determine the best mix of information technology and manual processes appropriate for a particular operation. In many ways the most promising application of APS is for mixed models that have combinations of scheduled operations and pull systems. It generally takes years to convert a plant of any size and complexity so the plant can get years of benefit from using APS to schedule at first and then to be used for planning purposes to support pull systems. This requires a whole new way of thinking. Traditional scheduling experts simply assumed the entire plant would be scheduled using the computer system. Lean thinkers often just assume the entire plant will use pull systems and scheduling is not needed, except to set up a leveled schedule for final assembly. Those with APS expertise have to work as a team with the Lean Manufacturing experts to design, implement and improve upon the planning and execution systems in concert.[127]

THE TRANSITION TO LEAN

Several years ago Toyota began designing the new Motomachi plant, Toyota's largest industrial complex. An IT specialist brought Mikio Kitano, who was to oversee the development of the new plant, a typical information system design flowchart describing the proposed IT infrastructure for the new operation. This diagram contained all the usual IT symbols—information flowing from computer to computer, storage devices, input devices, output devices, and the like.

Kitano returned the diagram to the IT specialist saying, "At Toyota we do not make information systems. We make cars. Show me the process of making cars and how the information supports that." The specialist returned with a new diagram, showing the body, paint and assembly lines representing how Toyota builds cars. The bottom of the diagram showed various information technologies and the way in which they would support the production of cars. As far as Kitano was concerned, the process flow diagram showed IT in its appropriate place.[128]

This story underscores the fact that IT exists to facilitate the value streams of a manufacturing enterprise. To ensure a valid contribution, each and every IT investment must be *critically* examined to ensure that it adds value, not waste, distraction, and confusion. This can only be done by teams getting close to the process, going to Gemba and observing the process firsthand, value stream mapping the current state, and identifying the desired future state and the specific capabilities that IT must provide to get there. Once IT investments are in place, they too should be subjected to continuous improvement efforts. If a scheduling system is an intermediate solution, helping the plant take another step toward Lean, then it is a worthy investment. But the moment the scheduling system becomes entrenched, hindering the further development of flow and visual mechanisms, it becomes an impediment to be eliminated. This suggests that the transition to Lean is made in evolutionary stages, as illustrated in Figure 5-41.[129] As you examine this table it is important to note that companies will rarely reside at either extreme of pure optimal scheduling or pure Lean, but will be in a state of flux among the intermediate stages— moving toward the ideal (and arguably unattainable, because improvement is continuous and never-ending) state of *pure* Lean.

Although most companies may never attain an ideal state of "pure" Lean as suggested in Figure 5-41, many can realize substantial improvements with carefully designed systems that plan and manage intermediate buffers, while synchronizing push scheduling and pull execution of production. It is important to note that a company may shift this push/pull point up or down the product/process diagonal *depending on the lead time requirements of the customer*. The guideline used to determine the optimal push/pull positioning of a particular value stream within the overall supply chain is called *Decision Point Analysis*, illustrated in Figure 5-42.

At the far upper left corner of the diagonal is Engineer to Order, where each order is unique. In most cases demand cannot be planned in advance, so most materials must be ordered from suppliers and all production is push scheduled, usually in a discontinuous job shop or on-site assembly project environment. Down the diagonal is Make to Order, where similar products are fabricated to customer order. In this case some basic inventories may be planned in advance, and some processes may be grouped (product flow analysis, group technology) to achieve some degree of flow and demand pull. Further down the diagonal is Assemble to Order, where core assemblies may be purchased and produced in advance, stored in a final assembly

Model	What is it?	Benefits	Limitations	Where appropriate?	Role of APS
Pure Optimal Scheduling	Optimally plan and schedule all work centers and operations. Each operation builds to the schedule.	Demand-driven system will develop globally optimal plans.	Assumes all operations and plants in the supply chain execute the schedule. Lack of a response to unplanned deviations might cause inventory buildups. However, APS can act as a fast decision aid in case of such events.	Non-level customer demand (>10% fluctuation from planned schedule). Many products that require shared resources. Optimized sequence of special, lengthy setups required. Low product yield (e.g., complex paint colors)	Develop global optimal schedule and dynamically size buffers. Provide advanced warning and visibility to problems. Provide direct link to procurement outside 4 walls of the plant. Provide customer and release due date visibility throughout the process.
Hybrid Pull and Optimal Scheduling Environment	Combine pull and scheduling as appropriate.	The best method for the particular circumstances.	Operations leadership might not pursue the most aggressive and comprehensive path for a full lean transformation.	Some products or phases of the manufacturing process have the characteristics appropriate for optimal scheduling (see above) and others have the characteristics that support lean manufacturing.	Mixture of roles in scheduling and lean. "What-if" scenario development for continuous improvement initiatives.
Transition to Lean	Lean phased in to product lines. Scheduled operations that have not yet transitioned to pull systems. Modeling to help prioritize lean initiatives.	Use the best scheduling methods for each part of the plant based on their evolution.	If not managed right the scheduled systems might discourage the movement to lean, i.e., become dependent on the schedule.	Any plant that has not yet fully implemented continuous flow and pull systems but is heading in that direction.	Schedule operations not yet changed. Prioritize lean initiatives. "What-if" scenario development for lean transition initiatives. Support parts of the plant where lean is implemented as described below.
Pure Lean	Leveled schedule for final assembly and then pull from all upstream operations with supermarket buffers.	Operations.are coupled with their downstream "customers." Highly visual and can lead to controlled inventory with continual reduction of inventory. Helps build to real customer demand.	If assumptions for pull do not hold (e.g., unstable demand or processes), the system can fail. Leveling not always feasible, e.g., continued emergence of the build-to-order customer model.	Relatively predictable demand. Stable customer schedule. Moderate product variety. Moderate customization. Stable process. Feasible to arrange equipment into product lines without major new capital equipment purchases.	Calculate kanban quantities. Calculate min/max levels. Develop leveled schedule. Early warning if unexpected events will overload system. Decision aid in case of unexpected events. Overall master planning.

Figure 5-41. The Lean transition

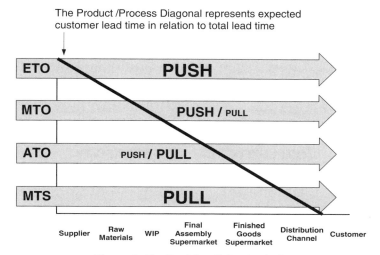

Figure 5-42. Decision Point Analysis

supermarket, and pulled by the customer order (or a level schedule where each unit is configured separately) into a rapid final assembly process. At the far lower right corner of the diagonal is Make to Stock, where production is repetitive, perhaps according to a level schedule or continuous flow, and all customer demand is pulled immediately from a finished goods supermarket. In each shift down the diagonal, the lead time to customer delivery becomes shorter until it is instantaneous when the customer pulls from finished goods at some point in a distribution channel.

As we stressed early in Chapter 4, positioning on this product/process diagonal must be a strategic decision, and an enterprise may offer products that occupy several positions. For example, a job shop may perform a product/process rationalization analysis and discover that 25% of their total throughput belongs to a single product family, which may be transformed to a cellular design. They may then define a strategy to invest in design, engineering, and production core competencies relevant to this product family, leading to greater product variety, increased throughput, and reduced lead time. This strategy may lead to increased market share in this particular segment, opening the door for expansion into related markets and technologies. This may ultimately lead to market recognition as a preferred supplier based on a carefully nurtured set of core competencies, products, processes, and customer relationships.

In this way, a job shop can strategically migrate along the product/process diagonal, evolving into a mixed-mode manufacturer with a finely orchestrated mix of push, pull, repetitive, and discontinuous operations. But even if a job shop finds its product mix so variable and its volume so low that it does not justify reorganization of portions of its job or project shop into cells, it can still

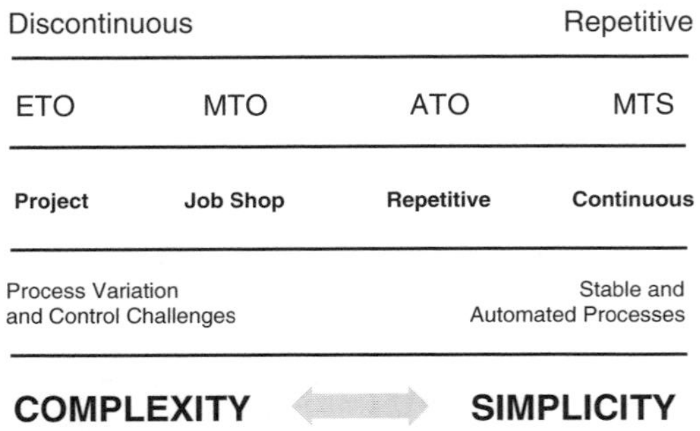

Discontinuous			Repetitive
ETO	MTO	ATO	MTS
Project	**Job Shop**	**Repetitive**	**Continuous**
Process Variation and Control Challenges			Stable and Automated Processes

COMPLEXITY ◁▦▷ **SIMPLICITY**

Figure 5-43. Complexity vs. simplicity and the role of IT

take many steps toward Lean performance improvement described earlier in this chapter.

As a manufacturing enterprise positions its products and processes along the diagonal, it must also determine the appropriate IT capabilities that add the most value. Generally speaking, as an operation tends toward the discontinuous end of the spectrum, complexity increases, requiring additional decision support and process control tools as suggested in Figure 5-43.

So naturally the degree of IT support required for planning, scheduling, and execution depends on the inherent complexity of the environment. We may therefore suggest the appropriate level of IT involvement for each particular process along the continuum, as illustrated in Figure 5-44.

There are many product and process transformations that must occur along the journey to Lean. The transformation process usually begins with simple initiatives requiring little IT intervention: basic education, 5S, standardization, value stream mapping, and so on. In this early phase of Lean Manufacturing development it may be appropriate to cautiously discard IT monuments erected by previous generations, after teams have determined that they do not serve a legitimate purpose elsewhere in the organization. As Lean evolves, however, at some point the value streams may reach a degree of complexity where IT becomes necessary.

So many transformations must occur before an effective pull system or level scheduling is practical, that teams may become discouraged with what lies ahead for them. We encourage them all that although the journey to Lean will take many years of sustained effort and continuous improvement, each step matters! The only fatal setback is to not take the next step.

How to begin this journey? When asked that simple question during our workshops and on-site assessments, our answer is always the same:

1. Start with basic education combined with individual and team exercises, emphasizing 5S housekeeping activities to reinforce the concepts.

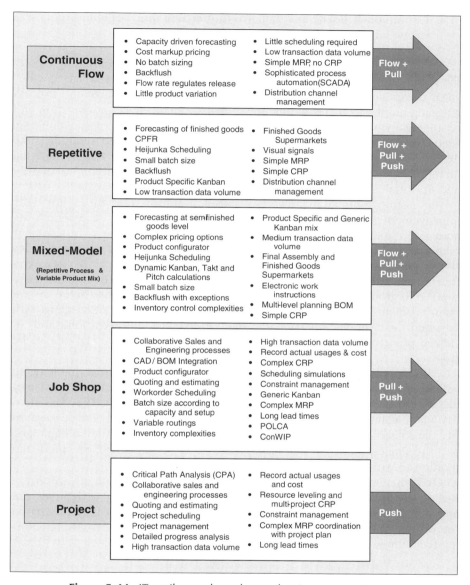

Figure 5-44. IT attributes along the product/process continuum

2. Move on to high-level entity and process mapping, helping cross-functional teams to develop a holistic understanding of the organization structure and the overall value streams.

3. Perform value stream mapping to develop a clearly defined current state and identify waste.

4. Continuously define and enhance future-state targets, guiding and measuring improvement initiatives as you go.

Education and practice develop common skills and understanding among the team members, helping everyone to speak the same vocabulary. 5S cultivates discipline, standardization, awareness, and self-empowerment. Process mapping builds strong cross-functional teams, enhancing communication and problem-solving skills, while chipping away at organizational silo thinking. Value stream mapping of the current state documents and quantifies the detailed flows of material and information, causing waste and simplifying factors to stand out from the background complexity and helping to identify and prioritize areas for future-state improvement.

But nothing ultimately matters without a future state target to guide improvement initiatives. According to Rother and Shook, "A current state map, and the effort required to create it, are *pure muda* unless you use your map to quickly create and implement a future state map that eliminates sources of waste and increases value for the customer."[130] Until there is a future-state value stream map developed by the cross-functional team, there is no solid ground for sustained progress. Without a future-state value stream map, problem-solving often degrades into cloud-sculpting, abounding with unstated assumptions, questionable data, individual interpretations, and localized objectives and measures, creating disagreements that may never be resolved.

With mapping a team develops a quantified, fact-based model that leads to valuation and prioritization of improvement alternatives. The team can clearly distinguish the flows, gaps, and interruptions in key processes. They can identify where value is added for the customer, and where it is not. They can quantify the constraints and pacemakers, the appropriate buffers and pull linkages, and the single point by which each process may be scheduled, executed, and monitored with the least complexity.

Once the development of these maps is under way, the team may return to the concepts explored in this chapter and apply whatever techniques are appropriate to their environment. No matter what positions they occupy on the product/process continuum, every enterprise will find Lean Manufacturing techniques that add value.

Part Two

Building Blocks of Information Systems

Chapter **6**

Charting the Enterprise Software Universe

In this chapter we'll explore the three primary enterprise software systems: ERP, CRM, and PLM. We'll examine the components and interrelationships among these multifaceted systems and explain how they may enhance Lean performance.

ENTERPRISE RESOURCE PLANNING (ERP)

Virtually every event that occurs in an enterprise—every plan, order, production activity, and shipment—trickles down through the system and onto the balance sheet and income statement as a source or use of assets or liabilities. All value streams converge on finance. So it should not surprise you that financial and managerial accounting are at the core of an ERP system.

The ERP database is the storehouse of the audit detail that may be used to prove that the company is managing its assets and operations wisely. This is why ERP is called the *backbone* of a business information system and the *system of record* for all financial events. By definition, an ERP system contains both financial and operational modules, and there is an ERP system for practically every type of industry. Although the rules of accounting are similar among all industries, the operational components can be quite specialized. For example, the operational ERP components of a professional services firm are designed to automate time and expense capture, billing, and project costing. By contrast, the operational components (MRP II) of a manufactur-

Lean Enterprise Systems: Using IT for Continuous Improvement, by Steve Bell
Copyright © 2006 by John Wiley & Sons, Inc.

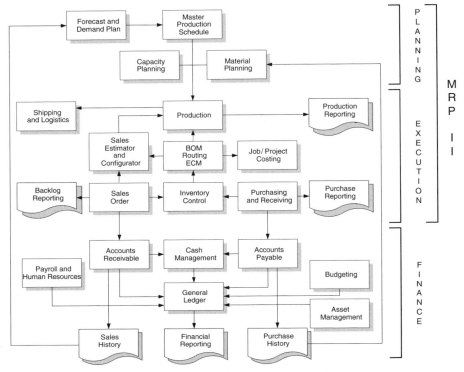

Figure 6-01. ERP manufacturing system schematic

ing ERP system contain the planning and execution functions illustrated in Figure 6-01.

It is unlikely that a single company would use all the modules offered by an ERP supplier; even when they do, it is most likely that these modules are implemented in phases and not all at once. Whatever ERP modules are used by the company, they are all integrated within a logical data model. This means that a transaction is introduced to the system at the beginning of the transaction life cycle, continuing through subsequent processes without duplication of data entry, creating a thread of facts. For example, a quote becomes a sales order, which initiates production, leading to shipment, invoicing, and collection of payment. The final result is revenue and cash appearing on the financial statements.

Although the operational modules have a greater impact on the productive performance of the organization, at the end of a transaction life cycle the final posting is made to the financial system—hence the ERP financial core is the termination of most transactions. High-level reports leading to detailed drill-downs on various performance measurements may be produced, because there is a transactional thread that naturally *relates* all events of the life cycle together into a logical process. Herein lies the business significance of a relational database.

ERP Extensions

Although an ERP system offers many capabilities to manage a complex business, it may not *completely* satisfy the particular requirements of an enterprise. The enterprise then faces a choice: either to make the best of these built-in ERP capabilities, perhaps enhancing them through customization, or to purchase a subsystem and pay the costs of integration with the core ERP system.

For example, a company may need specific CRM capabilities not offered by their ERP software vendor. As a result they purchase and implement a separate CRM system, then identify the necessary interfaces between their CRM and ERP systems, and develop a method for sharing information between them. This approach, where multiple applications are selected based on their individual merits, is called a *best of breed strategy* and usually carries a significant cost of integration and maintenance.

Here are examples where a core ERP system may offer limited capabilities, requiring integration with a separate subsystem:

- **Customer Relationship Management (CRM)**—ERP includes customer records and transactions and other supporting information such as salesperson, territory, pricing agreements, and payment terms. Some ERP systems offer tracking of customer interactions but do not provide the depth and process orientation that CRM offers.

- **Product Life Cycle Management (PLM)**—ERP manages product structures through the Bill of Material and Routing records and may automate the Engineering Change Management and Material Review Board processes. Some are able to import CAD drawings, which are associated with BOM records, and link to engineering specifications and work instructions during production work order creation. An ERP system does not offer the depth and process orientation of a dedicated PLM system.

- **Product Configurator**—Although many ERP systems offer basic product configuration capabilities within the order processing system, many do not offer the sophisticated, programmable rules-based capabilities of standalone configurator applications.

- **Forecasting and Demand Management**—Many ERP systems are capable of producing forecasts based on sales history and trends, importing this information into the Master Production Schedule. A separate forecasting and planning system may be required to extract and manipulate information from multiple sources with sophisticated forecasting techniques.

- **Advanced Planning and Scheduling (APS)**—ERP automates the Master Scheduling and Material and Capacity Requirements planning processes, usually providing basic infinite- and finite-capacity scheduling capabilities. More advanced APS systems may be incorporated within ERP or attached as third-party systems, planning production and distribution requirements for multiple inventory locations, managing constraints, optimizing schedules, and collaborating with complex supply chain planning environments.

- **Warehouse Management Systems (WMS)**—ERP usually manages multiple inventory locations for receiving, production, and distribution tasks. In some situations more sophisticated WMS systems may be required for bin and shelf locators, optimized picking and putaway instructions, integration with material handling systems, with bar code, radio frequency (RF), and radio frequency identification (RFID) support for a wide variety of material handling activities. A dedicated WMS is often needed to optimize large, high-volume, multilocation distribution operations.

- **Manufacturing Execution Systems (MES)**—ERP usually provides basic tracking of production activity and work in process on the shop floor. MES systems offer real-time monitoring of activity through automated and human data capture systems for sophisticated planning and control of the shop floor. MES systems are often used in specialized environments such as automotive, chemical, electronics, food, pharmaceuticals, and textile production.

- **Enterprise Asset Management (EAM)**—ERP systems usually offer basic asset management capabilities such as asset accounting, property tax reporting, depreciation, and basic asset location tracking. Advanced EAM systems provide preventative maintenance, scheduling, usage and performance monitoring. EAM can also help to manage asset hierarchies, components, version histories, serialization, replacements, warranty tracking, remanufacturing, field service, and support.

- **Shop Floor Control and Data Acquisition (SCADA)**—ERP may automate the capture of material movement and production activity through bar coding and other data capture interfaces. Specialized SCADA systems are usually required to monitor and control machine operations through the use of Programmable Logic Controllers (PLC), Computer Numerical Control (CNC), and other machine and device interfaces.

- **Quality and Compliance Process and Document Management**—By their nature most ERP systems codify basic business processes but do not offer the extensive process, workflow, digital signature, Statistical Process/Quality Control (SPC/SQC), and documentation controls required by ISO 9000, ISO 14000, QS 9000, FDA GMP, and other national and international quality and regulatory organizations.

- **Project Costing and Management**—ERP usually offers the basic capabilities to classify expenditures by project codes in addition to general ledger account numbers, providing simple project-based reporting. Advanced project costing subsystems track and report more complex project work such as new product development, marketing programs, installation and service projects, and other events. Some ERP systems specialize in project-based production. Project costing systems must be distinguished from (and often integrate with) project management systems that offer planning, scheduling, and resource management capabilities.

- **Logistics and Transportation Management**—ERP may store carriers, route, and rate tables and perform load planning, staging, and shipment scheduling. Advanced logistics and transportation management systems are often needed for interactive planning, load and route optimization, scheduling, and tracking capabilities.
- **Supply Chain Management (SCM)**—The category of SCM software is very difficult to define, extending the core capabilities of ERP to encompass virtually all forms of electronic communications, collaboration, integration, and transactions that occur between trading partners. From its early days as EDI (which is still going strong) to the latest forms of Business to Consumer (B2C) and Business to Business (B2B), eCommerce is enabled by the Internet and limited only by our ingenuity.

Larger ERP systems offer most of the capabilities listed above as integrated components. As the enterprise software market continues to mature and consolidate, and as more companies focus on delivering low-cost but full-featured solutions to the midmarket, smaller ERP publishers will offer these additional subsystems either as part of their internally developed product line, through acquisition of externally developed applications, or through strategic third-party alliances with other software vendors. Because an acquired or third-party subsystem was designed separately from the main ERP system, the integration may not always be complete or well-designed. On the positive side, the ERP publisher has absorbed most of the cost and risk of bolting the applications together. Furthermore, the ERP publisher is generally committed to the ongoing integration of these applications as the systems are upgraded. A note of caution, however: In the past some third-party integration agreements have been short-lived, leaving the customers to maintain the interfaces.

The Challenge of ERP

The greatest challenge is to think of ERP as an essential tool for process automation and continuous improvement, rather than just a comprehensive suite of software. Let's face it, ERP can be extremely complex, expensive, and risky. There are many well-publicized ERP project failures, and more than a few examples where a failed ERP project destroyed the company. ERP is so difficult because it automates and institutionalizes the core value streams and underlying processes of the enterprise—could there be a more difficult challenge than this? Perhaps the greatest mistake that a company can make is to implement an ERP system without first setting in motion the continuous improvement of its value streams. By overlaying a new ERP system on top of old business processes, you may discover that you've simply institutionalized poor behavior, or as Hammer and Champy put it, "paved over cowpaths."

When implemented properly, however, ERP becomes the foundation for organization-wide continuous improvement. Not only is the ERP system the primary source for feedback to support the operations-level continuous improvement initiatives, the enterprise software systems themselves must be continuously improved or they will surely lose their value and become obsolete.

CUSTOMER RELATIONSHIP MANAGEMENT (CRM)

CRM began as contact management software used primarily to track people and addresses, record conversations, and provide event reminders. To this day, many individuals and companies rely on the Contacts folder within Microsoft Outlook as a poor man's CRM system. Contact management software stores relatively static information. On the other hand, CRM manages not only the data, but also the processes and performance measures underlying the customer interactions.

There are three basic roles that may be supported by a CRM system: marketing, sales, and customer service. Each of these roles, which are often performed by a different functional organization within an enterprise, deals with customer interaction during the transaction life cycle. A macroview of this life cycle and its interaction with ERP is illustrated in Figure 6-02.

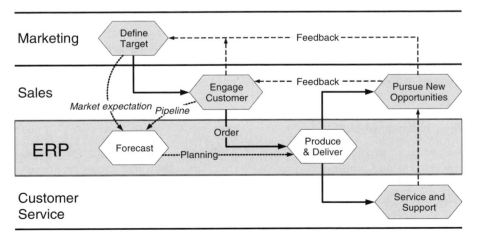

Figure 6-02. CRM transaction life cycle

CRM supports revenue growth through three basic strategies:

1. **Acquiring New Customers** through effective marketing and sales programs.
2. **Retaining Customers** by serving them well, addressing each relationship individually whenever possible, while carefully monitoring sales, delivery and service performance, and customer satisfaction.
3. **Increasing Customer Profitability** through revenue growth, pricing, rebates, branding, bundling, cross- and up-selling. More significantly, by becoming an integral part of the customer's strategic and product planning processes, CRM helps to add more value in ways that are meaningful to each customer.

Observe in this illustration that marketing defines the target market and may perform various activities to attract the attention of potential customers. Once a potential customer engages with the sales organization, they enter the *pipeline* phase and become a *prospect*. Depending on the nature of the sales process, there may be specific and measurable milestones during its progression that are used to rate the strength and timing of the pipeline. Once the customer commits to a transaction, the ERP system registers a sales order and delivers the product. Customer service then follows through to manage the postsale interaction with the customer. During the entire transaction life cycle, marketing may gather performance measurements on the customer, product, and sales process, in order to refine their tactics.

Although this diagram presents an oversimplified view of the roles of marketing, sales, and customer service, it illustrates the customer-centric process that CRM supports through a unified process and data model. Common components of this process and data model are shown in Figure 6-03.

A marketing and sales organization often begins its journey to CRM with each employee storing their own contact and interaction information on their local hard drive, Personal Digital Assistant (PDA), or hard copy records. This creates not only data fragmentation, duplication, errors, and process inefficiencies but also a serious security risk. Customer records, communications, and transaction history are valuable assets of the company and should be centrally managed and secured.

A challenge with a CRM system arises from worker mobility. Individuals responsible for business development often travel, working on airplanes, hotel rooms, and at the customer site. They may carry portions of the CRM database on their laptop or PDA, retrieving and entering data as they travel. This may include the generation of a quote or lookup of product information, which requires a subset of the customer, product, and pricing information stored in the ERP system to be contained in the local CRM database. When the traveling user dials in to the corporate network, connects through the Internet, or returns to the office, he *synchronizes* his local database with the central CRM

Marketing	Sales Automation
Campaign planning and management	Account management
Competitor tracking	Call center management
List management	Contact management
Partner and channel management	Correspondence and mail-merge
Trade show and event management	Customer loyalty programs
Transaction/process analysis	Estimating and quoting
Customer satisfaction measurement	Interaction history
Customer segmentation	Lead, Prospect and Opportunity management
Customer/channel profitability and potential analysis	Mobile sales support
Decision support	Pipeline/Forecast management
Lead and prospect management	Preferences tracking
	Pricing and Promotion management
Customer Service	Product catalog management
Automated routing, queuing and escalation	Product Configuration (sometimes contained in CRM)
Case management, incident response	Product management
Contract and service performance measurement	Quote/proposal generation and tracking
Customer self-service portal	Sales Force Automation
Email auto-response	Sales literature management
Field service and dispatch	Sales performance management and reporting
Help-desk management	Sales process management
Issue resolution measurement	Territory management
Return Merchandise Authorization (RMA) processing	Web commerce
Sales history lookup	
Service contract management	
Support knowledgebase	
Warranty service administration	

Figure 6-03. Common CRM components

system, uploading and downloading changes. However, with the increasing availability of the Internet and wireless communications, many traveling CRM users work *online* with a Web interface, eliminating the need for offline synchronization.

CRM and Forecasting

CRM can feed the demand management, forecasting, and S&OP processes. There are two primary approaches to forecasting: projection based on past sales history, and estimation based on customer interactions and pipeline activity, which was often not practical before the emergence of CRM.

1. **Forecasting Based on Sales History**—uses various statistical techniques applied to historical sales data to anticipate future demand. The sales

history data are often extracted directly from the ERP system, requiring no interaction with CRM. Historical forecasting may be used when the demand for a product is relatively stable over time, and may also account for predictable demand variations such as seasonality.

2. **Forecasting Based on Customer Interaction**—uses current market intelligence, customer interaction and pipeline information captured in CRM (also the ERP quote, blanket order, and EDI systems) during prospect and customer interactions. This approach may be useful when limited historical demand information is available, during new product introduction planning, when product life cycles are short, and when the market is volatile. Whereas the historical forecasting approach is very scientific (although not necessarily any more accurate), pipeline forecasting can be very subjective, dealing with less structured data on customer expectations and purchasing plans.

When the sales process is well defined, with weighting factors assigned to each milestone, then it is possible to have the CRM system calculate a forecast, for example:

Sales Milestone	Probability of Sale
1. Identification	5%
2. Qualification	10%
3. Demonstration	25%
4. Proposal	50%
5. Application for financing	85%

Each of these five steps in the sales process is a measurable milestone, and as a prospect moves through each of these stages, the likelihood of the sale increases. If the salesperson enters the expected amount and date of the sale, the CRM system may then use these weighting factors to calculate the anticipated total value of the pipeline for each future period. Of course, each of these three factors, estimated value, expected close date, and the probability of close, are subjective estimates. For this reason, many companies include forecast precision as part of their sales compensation plan to encourage thoughtful and accurate forecasting.

An enterprise may combine these two methods of forecasting (shown in Fig. 6-04) or perhaps use each method separately for particular markets, product lines, or customers. Looking back to the Sales and Operations Planning (S&OP) process, to plan production properly a company should attempt to understand its demand patterns and develop a technique to correlate the appropriate information from the ERP and CRM systems. As you can imagine, this process is difficult to standardize and often requires a degree of customization of the data integration to suit a particular company's needs.

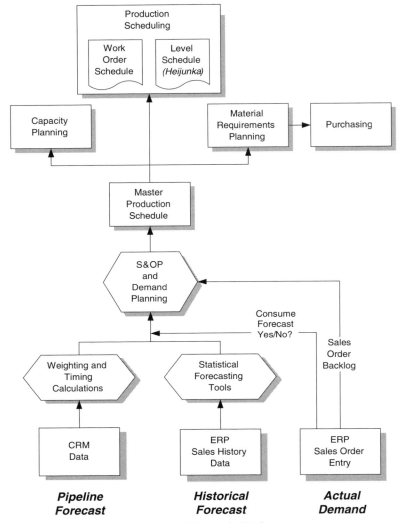

Figure 6-04. Combined CRM and ERP forecasting process

The Challenge of CRM

The greatest challenge of CRM is the natural but unnecessary conflict between structure and creativity. It is really quite simple to implement CRM in *three easy steps*:

1. Modify all the attitudes and behaviors of people within your company, to focus on the customers' needs and desires. Develop catchy slogans like "The customer is #1."

2. Reengineer all your business processes, organization structure, product/service development process, pricing, compensation models, marketing and sales strategies to reflect step 1.
3. Select and properly implement the right software and underlying technologies. Train your staff, as well as your extended sales and service force that are outside the walls of your company. Maintain and continuously improve the processes and software.

Well, perhaps implementing CRM is not *that* easy! But seriously, look carefully at the order suggested. Change the attitudes, change the behaviors and supporting business processes, then choose the right information technology to enable these changes. Although the rules of accounting and production operations management (although complex) are relatively well-defined and structured, the processes of marketing and sales are highly variable and creative, and often unique to each industry, company, customer, and individual. As a result, CRM software tends to be very fluid and configurable, requiring considerable process analysis and system design to be effective.

And then there's the challenge of getting people to use CRM. Let's be honest: Marketing and sales people tend to be right-brain creative thinkers with a tendency to resist standardization. In fact, the top sales performer within an organization is often a renegade, known for out-of-the-box thinking to serve the customer and close the deal. Although we don't want to discourage these key producers, we cannot reliably grow a company on them simply because we cannot replicate their behavior, and their creative deal-making often complicates the delivery of the product or service.

To grow a company beyond the idiosyncrasies of the individual sales stars and their occasionally eccentric methods, it is important to standardize the sales and delivery process so it is consistent and repeatable. The transition from a highly personalized sales process to a more structured approach often marks the difficult stage of growth from an entrepreneurial to a professionally managed organization. Companies that fail to make this leap may feel they have a glass ceiling above their heads, because too many one-off business transactions create an administrative and operational burden that naturally limits their growth. In our practice, whenever we see a company with an abundance of highly creative and difficult-to-administer customer pricing agreements, delivery methods, and sales compensation plans, we suspect that the organization is approaching this stage of development. It is this transition (among other challenges) that can make the CRM implementation process very difficult.

I am not preaching unnecessary rigidity in the sales process; there aren't many practices that can sour a customer relationship faster than the impersonal behavior that results from thoughtlessly scripting every interaction with the customer. However, a pillar of continuous improvement is standardization, because stable and repeatable processes are required for consistent performance. Herein lies an important distinction: Although the customer wants con-

sistent performance, which results from the standardization of sales and delivery processes, at the same time she wants to be treated as an individual. In
his book *Why CRM Doesn't Work*, marketing consultant Frederick Newell
stresses that "Customers have shown they don't want to be hunted like
prey. It's time to recast the discipline of CRM as one of greater customer
empowerment."[131]

CRM should help sales and marketing to establish clear and consistent customer expectations, built upon processes that produce repeatable performance
and quality, with the ability to forecast sales with reasonable accuracy, to make
a business more manageable and scalable. At some point every organization
faces the challenge of institutionalizing its processes in such a way that it does
not stifle creativity, competitive advantage, or the customer relationship. This
is all about people, culture, communications, process, and change management;
it's not a software problem.

CRM and Lean

A CRM system enables an enterprise to eliminate wasteful activity through
the improvement and automation of internal marketing, sales, and customer
service processes, creating what may be called the *Lean front office*. Of greater
concern to us at the moment, however, is how CRM enables various aspects
of the enterprise value streams to skillfully interact with and *add value to the
customer*.

Improved Communications and Quality. By definition, the customer is the
primary focus of every value stream. CRM can help to structure customer relationships, establishing consistent methods of communication and collaboration. For many customers, especially where product differentiation is limited,
the perception of value may be influenced more by the quality and consistency
of the relationship than by the product itself.

When evaluating a supplier relationship, many customers attempt to value
the total cost of the products they purchase, including administrative burden,
quality costs, delivery accuracy, and the ability to interface seamlessly with their
internal procurement systems. By automating the sales and service processes,
adding value throughout the customer transaction life cycle, a company may
lower the total cost of doing business while maintaining product margins.

Extending CRM into the supply chain using various forms of Internet- and
Web-enabled eCommerce and portals, a company can develop the infrastructure to address a larger market, enabling customer *self*-service while reducing
internal administrative, sales, and customer service costs. As many have
learned with their EDI systems, coupling your business processes and information systems with your customer may create a durable relationship.

Improved Demand Management. One of the greatest challenges in cultivat-
ing a Lean production environment is the capability to anticipate the demand

over the necessary forecast horizon. Although forecasting and demand management will never be perfect, the marketing segmentation and sales force automation capabilities of a CRM system may considerably improve your ability to plan for demand, while at the same time helping you to better focus your products and customers based on their contribution to your company's strategy and profitability.

A proactive CRM event notification and data mining system may be able to identify new behaviors and emerging trends in customers, products, and markets. If customer call volume, the number of new quotations, or the percentage of wins suddenly changes, a CRM system can provide a notification.

Customer Service Management. Effective customer service processes, which may include incident response and escalation, customer self-help through the publication of a knowledge base and other useful troubleshooting tools, field service, warranty management, and RMA processing make for a professional response when the inevitable problems occur.

In addition, the customer service system can provide nearly instantaneous feedback to design, production, and distribution, alerting the entire organization to problems in the field. This is an essential element of the closed-loop feedback system that is often managed haphazardly by companies that have not automated or integrated their postsale support activities.

Finally, a CRM system may be used to consistently measure customer satisfaction by a variety of internal and external measures. Because the value chain is defined by how the customer perceives value, this measurement and feedback system is valuable to the strategic direction of a Lean Enterprise.

PRODUCT LIFE CYCLE MANAGEMENT (PLM)

PLM tools enable concurrent and collaborative design, engineering, and support. By bringing together customers, suppliers, marketing, sales, customer service, design, and manufacturing engineers early in the design phase, supported by the relevant information workflows, products may be brought to market more quickly and with greater variation, shorter purchasing and production lead times, lower cost, and higher quality.

With increasing competitive pressures, many manufacturers have focused on low cost to develop competitive advantage, or simply as a means for survival. That may not be a viable long-term strategy, according to Dave Caruso, director of research at AMR Research:

> The low-cost obsession of the past ten years comes at a price. For one, R&D spending has drifted lower. The Product Development Marketing Association* says it sees a marked drop between 1995 and 2003 in introductions of innovative

* www.pdma.org.

Correct Identification Of Customer Needs	60.8%
Remaining Competitive	50.0%
Increasing Product Innovation	32.2%
Reducing Time-To-Market	30.4%
Managing Overall Costs	29.5%
Proper Allocation of Project Resources	28.6%

Figure 6-05. Ranking challenges for improving product development

products, as opposed to incremental enhancements. Even more troubling is the shift toward a fast-follower strategy. In 2003, 37 percent of companies said they were fast-followers—up from 27 percent in 1995. With all these followers, who's leading? Innovation on the product side has a powerful impact on the bottom line. Work done in 2001 by product development guru Robert G. Cooper shows that for most companies, the percent of sales from new products averaged 33 percent, while the best companies derived 49 percent of their revenues from new products. The payback is astounding as well: ROI for successful new products averaged greater than 96 percent.[132]

In the 2003 Value Chain Survey, cosponsored by Industry Week magazine and IBM Corporation, respondents indicated the challenges with improving product development shown in Figure 6-05.

Respondents suggested the Key Performance Indicators correlated to company performance shown in Figure 6-06.

In an increasingly global business environment, the challenges of building an effective and aggressive product development effort require a new level of sophistication. International product development consulting firm CIMdata says that global enterprises must:

- Make effective use of a widely distributed worldwide organization, creating a virtual value chain with no time, distance, or organizational boundaries.
- Ensure that corporate acquisitions and mergers work together.
- Create and enable virtual product teams composed of people who are spread around the world.
- Leverage the intellectual assets in these dispersed teams and organizations.
- Enable 24 × 7 development and product support using global teams.[133]

	Bottom 25%	Median	Top 25%
Percentage of sales from products launched in the previous year	10%	15%	25%
Time to market today, days	258	150	60
Products launched on budget	50%	75%	90%
Products launched on time	30%	60%	86%

Figure 6-06. Ranking challenges for improving product development

Components of PLM

PLM is appropriately described as a strategy supported by a collection of tools and techniques, rather than a single integrated application. Many of the information technology components of PLM have existed independently for years and have been gathered as integrated suites of PLM tools:

- **Authoring**—CAD (computer-aided design), CAM (computer-aided manufacturing), CAE (computer-aided engineering), and 3D Visualization tools
- **Requirements Management**—provides input to design and engineering processes and may be used in conjunction with techniques such as Quality Function Deployment (QFD) throughout the conceptualization, creation, manufacturing, and distribution of products. These tools keep track of marketing and customer issues, functional and technical requirements, quality, safety, usability, serviceability, manufacturability, and cost factors while helping to manage the flow of information, and evaluate constraints and trade-offs caused by design decisions.
- **Product Data Management (PDM)**—creates a unified record of the design, specifications, characteristics, production, and distribution of products. This includes structured data stored in various relational databases and unstructured data contained in a wide variety of electronic document formats. Product data include detailed specifications on the items, Bills of Materials, routings, delayed change effectivity information, work instructions, sourcing, and compliance information.
- **Engineering Change Management/Control (ECM/ECC)**—routes change order, notification, and approval information through various pathways, managing multiple document versions, revision audit control, new part signoffs, compliance validation, and quality management.

- **Configuration Management**—provides change control and tracking for as-designed, as-manufactured, and as-serviced product information. For example, if a part has to be replaced in a product years later, the system can locate the original version, configuration, design, and specific materials that are required for service.
- **Sourcing Management**—provides supplier management tools including specifications management and history, supplier performance management, certification, and testing.
- **Collaboration and Knowledge Management**—includes such diverse software tools as scheduling, communications, collaboration, groupware, visualization, documentation, version control, exceptions management, storage, search, retrieval, reporting, data analysis and mining, security, and administrative tools to enable a geographically distributed product development team over the duration of a lengthy product life cycle. Such a process-oriented, cross-boundary system exposes vital and confidential company knowledge to outside parties, requiring a strong system of security and document control.
- **Quality Assurance, Regulatory Compliance, Environmental Health and Safety Management**—offers capabilities that vary widely by environment: food, pharmaceuticals, chemicals, hazardous materials, consumer goods, automotive and transportation, aerospace, defense, government contracting, etc.
- **Program and Project Management**—provides control over scope, time, cost, risk, schedules, and resource requirements during the design and engineering process. When program and project management are integrated within PLM, they provide the capability to link and communicate changes in design to the overall project change management process, controlling overall scope, quality, cost, and risk.

PLM is most useful where products have a high knowledge content combined with a short life cycle. For example, it is not uncommon in electronics manufacturing to simultaneously manage the information flows supporting the production of several distinct versions of the same product, to suit the specific requirements of particular customers and service channels. This creates a significant version control and traceability challenge, following the transaction thread from the detailed production and shipment events all the way back to the product definition.

When the product life cycle is short, end-of-life planning, marketing, new product transition, long-term support, and sourcing issues must be considered when designing each new product. "In today's market, a product's end of life must be analyzed before the product is designed," says Tom Maurer of UGS, a leading PLM software provider. "For example, if a product's life cycle is short, but its components have long lead times, a manufacturer may need to purchase in advance all the components that would be used for the entire time the product is made. The resulting inventory carrying cost for the components may eat into profitability such that the decision is made not to develop the product."[134]

PLM is useful not only in discrete operations (component assembly) but also in process (blending) industries such as foods, pharmaceuticals, and chemicals. Process industries are often highly regulated, with strict formula/recipe management, process standards and documentation, safety, version control, laboratory testing, health, environmental, and other regulatory compliance requirements where PLM adds value. PLM is also useful when the product is a structured service, such as banking, financial services, or insurance. Although the product is less tangible than manufacturing, the essential requirements of product data and life cycle management are very similar. Thus PLM may be useful for a manufacturing company that delivers services in combination with, or in addition to, their manufactured products.

Examples of sophisticated PLM systems used by larger manufacturers for collaborative development are not hard to find. For example, the *Industry Week* article "Factories of the Future" describes fascinating practices in the automotive industry:

> Automotive-interior supplier Johnson Controls, Inc. is supplying the entire cockpit for the 2002 Jeep Liberty, integrating 11 major components from 35 suppliers. To further enhance collaboration at the product-development stage of future programs, JCI will manage design sessions within the company and with supply-chain members via a Web-based design collaboration system. With this tool, 3-D solid models of parts and components from different suppliers are imported into the same online, virtual design space to see if they fit and function together. Already General Motors is crash-testing vehicles in a computer, simulating vehicle system performance, and doing factory layout in a virtual environment.[135]

Although these stories of information technology wizardry are seductive, for many smaller organizations it is more important to develop the basic competencies for organizing the capture, storage, retrieval, and security of information, enabling effective team collaboration. When a large trading partner invites a small one to participate in its collaborative design and development program, it is helpful for the small partner to have its processes standardized and documented (and perhaps automated) to make for an easier coupling.

Taken separately, PLM tools can assist with the various tasks required during the design and development process. When these tools are integrated, not only are the design and engineering processes performed more quickly, and in the appropriate sequence, but the entire team collaborates in real time within a logical data model, ensuring a higher integrity and quality of information.

The Challenge of PLM

In addition to the contents of its relational database, PLM must manage a substantial volume of *unstructured* knowledge and intellectual property in the form of communications, spreadsheets, drawings, diagrams, still photos, video and audio clips, and other document types that are not stored in a transactional database. In a concurrent development environment, the information flow between customers, suppliers, outsourcers, marketing, sales, service, design engineering, and manufacturing engineering can be fast and furious. Engineering documents may be in a constant flux, routing through multiple creation and approval pathways, with individuals editing multiple versions during the document's life cycle. Concerns abound of security, ownership, version control, approvals, search, retrieval, reporting, and cross-referencing of information.

Manufacturing enterprises have historically developed separate systems for managing the product design process and manufacturing operations. These systems evolved to optimally address the differing needs of engineering and manufacturing.[136] Design engineering, manufacturing engineering, production planning, and technical sales all have a rightful claim to ownership of product data. In a fluid environment they must each concurrently review, edit, and approve the product data, without violating the integrity of the information or disrupting the smooth and rapid flow of the development and production processes.

There should be coordination among all of the suborganizations and entities that collaborate on a particular project. In their book *Developing Products in Half the Time*, authors Preston Smith and Donald Reinertsen explain that:

> Upper management must provide a team setting in which the team takes full responsibility for the project. There are two parts to this issue. One is to weaken the linkages between the team and the remainder of the organization so that the team can in fact move with some freedom. The other is to create a motivational structure where the team must indeed complete the project on schedule.[137]

We must remember that PLM is a strategy supported by a collection of tools and techniques, rather than a single, integrated system. Effective PLM implementation strategy also requires a change in organizational culture, extending the boundaries of confidential collaboration, workflow, and communication to customers and partners. If done skillfully, concurrent engineering practices supported by a collaborative PLM technology infrastructure may nurture strategic partnerships that add value far beyond the traditional supplier-customer relationship.

PLM and Lean

A PLM technology infrastructure can help to reduce waste and add value to a Lean organization in many ways.

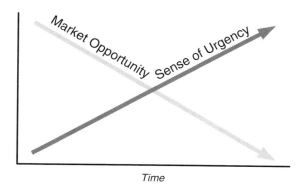

Time

Figure 6-07. The paradox of lead time and market opportunity

Lead Time Reduction. Project managers talk about the "fuzzy front end" of a project, where valuable time is exhausted answering basic questions before the project scope and particular issues and constraints are clearly identified. This is particularly true in an Engineer to Order manufacturing environment, where there is substantial knowledge and engineering content in each product. Authors Smith and Reinertsen explain the significance of this early phase:

> Three critical factors combine to make the Fuzzy Front End an area of extra-ordinary opportunity. First, it is a lengthy stage for most development projects. Second, it is full of cheap time-compression opportunities. Finally, it is an area in which the performance of individual companies varies dramatically. The market clock measures the time it takes us to respond to opportunities in the market-place. We should treat a week spent at the front end of a project with the same care that we would treat a week consumed at the very end.[138]

The authors suggest an interesting relationship between time and urgency, which they call the *urgency paradox,* shown in Figure 6-07. There is a natural tendency to squander time on the front end of any project, even though it is of the same value as time at the end, because issues, tasks, and priorities are not yet clearly defined. In fact, we may argue that time on the front end of the project is *more* valuable, because the opportunities for differentiation and the creation of competitive advantage are highest in the early stages of market opportunity.

Through the development of standard processes and information flows, aided by knowledge management and collaboration tools, PLM reduces lead time during the early design stage, accelerating time to market.

Manufacturability. As we learned in the Lean Job Shop discussion in Chapter 5, most production costs are designed into the product and process long before the job is sent to the shop floor. When manufacturing engineering cooperates with design engineering in the early stages of product design, they can influence decisions on manufacturability, cost, and quality for which they will later be held accountable. PLM facilitates effective collaboration among these and other participants.

Standardization. PLM becomes the system of record for product definition data, eliminating redundant and potentially conflicting versions of information that often proliferate within an organization. In addition to the central management of product information, the establishment of design, development, and manufacturing standards leads to greater standardization and reusability of designs, tools, components, and processes. This in turn may enable group technology, cellular production, reduction of inventory, reduced purchasing and manufacturing lead time, improved quality and serviceability.

Marketing and Sales Support. One of the often overlooked contributions of a PLM system is the value of the information to marketing and sales activities. By publishing product information through a public Internet or secure Extranet Portal, customers may be able to help themselves, leading to better decisions and a faster sales cycle. In fact, good Web-based product information may be an essential sales and customer service tool for a manufacturer that delivers high knowledge content within its products. According to John Schneiter in his *Industry Week* article "Taming the 5,000 Pound Gorilla," "A manufacturer of technical products whose Website fails to deliver the technical content that the prospective buyer needs or wants is like a retailer who spends serious money on a great storefront, but once the shopper steps inside, the sales staff is totally clueless."[139]

With that said, however, in some environments there are legitimate reasons to restrict the amount of technical content that is offered freely without qualified interaction with a human being, because there may be a risk that the information may be misunderstood or misused. In that case, a PLM system may provide the appropriate information to a sales team, who then skillfully manage the customer relationship, often through the combination of PLM and CRM systems.

Communication and Collaboration. PLM offers a variety of tools for communication and collaboration. By making vital product design standards and information available to the team, all stakeholders may be united in a streamlined process. Development projects often involve co-location, where project personnel representing the various stakeholder companies are located together, enabling more rapid and frequent decision-making and exchange of ideas. On the other hand, collaboration, knowledge management, and security tools can help development teams electronically co-locate anywhere in the world. When a team works around the world they also work around the clock, further accelerating the pace of development. Regardless of the physical arrangement of the team, when a company uses PLM and concurrent engineering effectively, it creates a framework for a strong working relationship among the parties, which may nurture a lasting competitive advantage.

Intellectual Property Management. Vital intellectual property is often stored in people's heads. With the approaching mass retirement of the baby boomer generation, this poses a significant risk to companies that have significant intellectual property valuation bound to their aging workforce. In addition to retirement, there are many causes for knowledge to be irretrievably lost: hiring by a competitor, disability, death, relocation, or role change. If a manufacturing enterprise does not capture and institutionalize its knowledge and processes, it may be guilty of not protecting vital company assets. The positive side of this argument is that by documenting this knowledge it may be preserved and extended throughout the organization.

Reduction of Administrative Waste. Without PLM tools, much energy is devoted to manual document management. As physical documents propagate across an enterprise, more time and effort are required to complete any task or process. When vital product and process data are not available electronically, and made accessible through intuitive search tools, then a higher-skilled individual must invest valuable time to retrieve and interpret the information to address low-value questions.

When multiple versions of key documents are circulating, the risk of serious errors, omissions, oversight, and miscommunication is great. And the use of manual documentation creates a security risk, while naturally limiting an organization's ability to extend its collaborative development efforts across suppliers, customers, locations, languages, and time zones.

Aid to Continuous Improvement. PLM tools add value to continuous improvement efforts by codifying design and development, production, and customer service practices to ensure standardization of work. Automotive and aerospace supplier Lord Corporation describes the widespread benefits of the concurrent development process. It is useful to note that although PLM software is not specifically mentioned, its components enabled the project to thrive. This underscores the principle that value is delivered through process improvement, which is enabled by IT:

> Our overall objective has been to shorten the time to introduction of new products by 75%. We began the initiative with our most important resource— our people. We formed cross-functional teams that included members from engineering, analysis, and manufacturing. Each team was responsible for quotation, design, development, testing, customer interface, and launch of full-rate production.
>
> It turns out the simplest changes led to some of the most immediate and profound improvements. For example, daily team meetings, co-location of team members, removal of wall partitions, as well as an emphasis on project planning,

risk management, and problem resolution, all helped focus team efforts on satisfying customers.

After just six months into the program we have seen measurable improvements. For example, one effort yielded eight new part numbers—design, production molds, process development, testing, and shipment—in 5.9 weeks as compared to the typical 26 to 30 weeks. Applying Lean principles to design work also raised the team's cross-training level from 12% to 39%. Moreover, the number of tasks involving two or more proficient team members jumped from 9% to 61%. This made teams less dependent on individual specialists, which cut response time and boosted overall effectiveness. Average on-time performance improved 40% in just the first year. Equally important are the intangible gains in focus, communication, sense of ownership, and job satisfaction.[140]

PRODUCT CONFIGURATOR

A *product configurator* is what we once called an *expert system*, a software application that enables the rapid configuration and pricing of a product by defining the relationships among product options, materials, and manufacturing processes. By integrating the configurator interface and logic into the sales order processing system, it ensures that a viable design is created every time, with proper costing and pricing information, while requiring less involvement of a product engineer or customer service representative. A product configurator encourages design and component standardization and grouping, which facilitates mass customization, and is therefore considered an essential front-end tool for many Assemble and Configure to Order manufacturers.

Occupying a unique niche entwined amid ERP, CRM, and PLM systems, the product configurator is an extremely challenging subsystem to design and implement, not only because of its inherent complexity but also because the configurator may orchestrate the flow of design/build information simultaneously with ERP, CRM, and PLM (Fig. 6-08). Fortunately, some ERP vendors provide built-in product configurator software. However, so do some CRM and PLM vendors. And there are several powerful product configurators that run stand alone, with integration adapters to common ERP, CRM, and PLM systems. With so many choices, which approach should a particular enterprise select? Once the choice is made, how should the product configurator integrate with the other subsystems? And which department (engineering, sales, planning, or production) should be *primarily* responsible for its design and maintenance? We can't begin to answer those questions here, but we'll briefly define the scope of the challenge.

How a Configurator Works

A product configurator begins with a base product specification (configurable Bill of Materials) with a modular design to which various options are added.

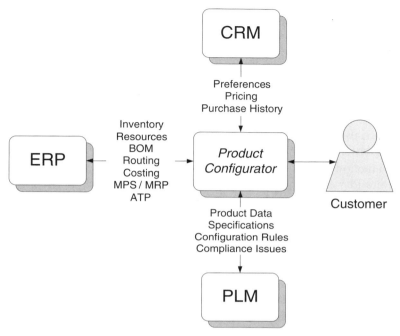

Figure 6-08. Three-way product configurator integration

Let's take the example of a computer, where the base product may be a particular model. Options may include disk drives, memory, keyboards, printers, pointing devices, and hundreds of other accessories. There will be practical and physical constraints that limit the combination of certain components: physical space of the unit, total capacity of the power supply, compatibility of certain components, and so on.

An entry-level product configurator may work with straightforward include/exclude and multiplier rules. For example, the choice of a particular disk drive may require a specific connector cable, or a certain case can only accept a limited quantity of internal cards. These relationships can be maintained by the software with data entry grid forms or drag-and-drop lists. Although definition of all these rules may take considerable time, there is a *finite* set of possible combinations that may be defined. However, with just a handful of options and variables, that finite number can quickly approach the ridiculous. For example if there are six options with four choices each, that results in 4096 distinct configurations to potentially identify ($4 \times 4 \times 4 \times 4 \times 4 \times 4 = 4096$). At some point a different approach may be needed.

Rules-based, also called *parametric*, product configurators allow for fully programmable rules that govern variables and interactions such as weight, internal and external dimensions, organization of components, total power consumption or other mathematical or physical constraints, color, chemical, and physical property interactions, regulatory requirements, and so on. With a rules-based configurator, there is potentially an infinite variety of end items that may result, so the underlying engineering rules require a great deal of effort to design, build, and maintain.

A product configurator is not limited to the engineering design of the product and may also help to determine:

- Material requirements of the finished item, using a configurable BOM to calculate the gross quantities required for a particular assembly
- Production routing and instructions, based on the selected options, required workcenters or cells, operations, equipment, tooling, resource time, and detailed work instructions
- Estimated cost roll-ups
- Appropriate pricing based on the cost roll-up, separate options pricing, or a combination of both
- Additional pricing rules applied to the final configuration including promotions, quantity discounts, special configuration or kitting discounts, and customer contract pricing agreements
- Available to Promise (ATP) date for delivery of the configured item
- Appropriate part number when a new item is created, mapping design parameters and characteristics to specific digit positions and segments when using a smart part number
- If the identical configuration has already been built, displaying the existing part number, sales history, and production details

The Value of a Product Configurator in a Lean Enterprise

Because a product configurator crosses so many organizational, functional, and application boundaries, it can be difficult to clearly identify and prioritize the appropriate requirements to select the right system. According to Gene Thomas in his article "From Make-Sell to Sell-Make":

> Product configurators are like part-numbering schemes—everyone has an opinion and wants to get in the act to make sure their requirements are met. Marketing often gets impressed with flashy graphical front ends; production management is concerned about integration with its legacy MRP/ERP shop order and scheduling system; and engineering is consumed with product structuring and parametric drawing systems [. . .] flattening their BOM's, and CAD/CNC interface potential.

> The requirements of all functional areas need to be taken into account to avoid single function commitments that may turn out to be inappropriate or shortsighted. [The author suggests] the relative time and effort required of each functional area within a company when product configuration is to be implemented:

- Marketing: 20% to support sales force automation, quoting, and visualization
- Sales: 10% for support of sales order processing integration
- Preproduction: 40% for the rules-based support to generate BOMs, routings, instructions, tooling, and attributes to interface with CAD graphics
- Planning and Execution: 30% for support of broadcasted shop order instructions, constraint scheduling and sequencing, and computer numerical control (CNC)[141]

This last point is quite important to the implementation of a product configurator to genuinely support Lean production. If we wish to reap the full benefits of a product configurator, we must consider the other operational aspects of mixed-model production including takt time, cellular flow, and heijunka scheduling. This suggests a dynamic interface between the product configurator order entry system and the near-real-time planning and scheduling functions that direct the shop floor. The configurator should not only be able to verify material availability through ATP, but when the job is released it may then be queued up for insertion into the heijunka production sequence. This is reminiscent of the Toyota manager's comment in Chapter 4 that they are a *change to order* operation, modifying the level schedule as specific customer-configured orders are received and inserted into the production sequence. Regarding this scheduling interface, Gene Thomas emphasizes that:

> For product families that are built-to-order in focused factories, simplification of scheduling and controlling paperwork now become major objectives. Shop floor parts lists, manufacturing instructions, and rules-based operation sequences are broadcast directly to the focused factory cells from the configurator. Ideally, not more than several hours of paperwork are committed in any cell, a tactic that facilitates unprecedented rescheduling flexibility. With only a few hours of released operations committed to the floor, customer to order tracking and rescheduling can truly become accommodating from the front through the back end.[142]

Although a product configurator enables the rapid design of products without consuming engineering time for each customer interaction, this benefit is partly offset by the effort required to design and maintain the configuration database. It also requires a logical grouping of the products and processes, according to Richard Bourke in his article "Product Configurators: Key Enablers for Mass Customization":

> To gain the full benefits of product configurators, companies must also consider the following activities within the scope of product design and development:
>
> - Modular design—the concept of building smaller subsystems, designed independently, and able to function properly when assembled and tested as an end item.

- Parts standardization—reducing duplicate parts to increase flexibility, reduce costs and encourage design reuse of preferred parts.

For some companies, an additional process, product line rationalization, may be mandatory. This process is critical when a company has problems such as bloated and unprofitable product lines and a mish-mash of features and options. Loading such information into a product configurator may only lead to disappointing results.[143]

Reading this, you may recall the product/process rationalization and transformation approaches discussed in Chapter 5. The final conclusion you may draw is that a product configurator can be a very useful front-end tool when supported by a PLM strategy and used in conjunction with other Lean transformation techniques. But attempting the implementation of a product configurator before these fundamental design and process standardization efforts are under way is likely to yield disappointing results. In his article "The Burden of Choice", G. Berton Latamore stresses that:

Companies fail to recognize that they don't have to do mass customization for every customer, product, and process to be successful. Before you start redesigning [or buying product configurator software] for mass customization, you need to rationalize your products. The first question is this: Should we continue to provide these products in these volumes to these marketplaces or would we be better off with a higher focus on core businesses?[144]

It is this self-inquiry process that may lead a job shop to focus on particular product families and group technologies, moving down the product/process diagonal toward more repetitive production.

ERP, CRM, AND PLM WORKING TOGETHER

Value Stream Integration

Now it is time to understand how the core enterprise systems work together to enable a Lean Enterprise:

- ERP manages company assets.
- CRM manages relationships.
- PLM manages intellectual property.

To illustrate these roles it is helpful to define a value stream and map the contributions of each system within it. For purposes of this illustration (shown in Fig. 6-09[145]) the value stream encompasses product conceptualization through postdelivery customer service, with five milestones:

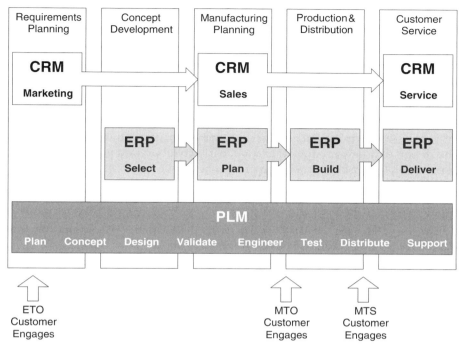

Figure 6-09. ERP, CRM, and PLM value stream interaction

- **Requirements Planning**—market research, strategic analysis of competitive factors, economic assessment of market potential, preliminary concept parameters
- **Concept Development**—conceptualization and design stage: when a viable result is achieved, this usually leads to prototyping
- **Manufacturing Planning**—defining product structures including BOM and routings and planning variables such as replenishment logic, kanban sizing, lead times, constraints, sourcing and quality issues, production planning and scheduling processes
- **Production and Distribution**—all activities from the release of work until the product is received by the customer
- **Customer Service**—continuing opportunities to serve the customer through warranty service, maintenance, repair, general customer care, and support

This illustration encompasses the three essential elements of the value stream described by Womack and Jones: 1) problem-solving that leads from concept through design and engineering to product launch, 2) physical transformation from raw materials through production and delivery to the customer, and 3) information management of the entire process.

Note that early in the value stream, the marketing research function of CRM explores the potential for new products and markets. At the same time, the PLM system facilitates the conceptualization and planning of new products and improvements to existing products and families. When and how the customer becomes engaged in the value stream influences the nature of how these two systems and their constituencies work together.

In an ETO environment, it is the customer's specific requirement that drives the design of a new product; therefore, early collaboration with the customer is essential. In an MTO or ATO environment, market research, as well as interaction with potential customers, may be used to develop concepts for a broader market. Thus there is a balance between general market research aided by CRM, and specific customer design collaboration through PLM. The sales order is entered into the ERP system when the customer places an order for the delivery of a configured final assembly.

In an MTS environment, the product is designed and built to a forecast and delivered to the market through a distribution channel. Depending on the nature of the product, opportunities may exist for customer or channel partner collaboration during the design and engineering stages. Cobranded or store labeled products may require collaboration with the design, packaging, and quality standards groups within the branding retailer. The marketing department may collaborate in a variety of design concept research techniques, including focus groups comprised of existing customers. But the consumer engages only when he sees the advertisement, approaches the shelf, or opens the catalog.

Although these are generalized examples, they illustrate the varying interactions among ERP, CRM, PLM, and customer demand depending on the nature of the product and process. Observe that CRM is involved during three phases: marketing, sales, and service. ERP is first involved once the product is conceived and specified, when initial supplier and material decisions are required. PLM, however, is involved during the entire value stream, from conceptualization of a new product through management of its end of life, and the service and support of that product long after production is discontinued. The intellectual property managed by a PLM system is omnipresent, touching every step of the value stream—this is why PLM describes a vast process rather than a specific information technology tool.

THE COPERNICAN VIEW

In the customary view of enterprise software, all subsystems revolve around the ERP financial core. The MRP II operational functions within ERP, as well as CRM and PLM, are often visualized as subcomponents of, or extensions to, the core financial system. There is a simple explanation for this ERP-centric view. Every financially substantive event that occurs throughout the entire organization eventually creates a transaction in the General Ledger, affecting the financial value of the company. Traditional performance measurement systems have been driven primarily by financial factors—for example, shareholders pay close attention to periodic financial reporting and management compensation is often linked to financial results.

In many cases the activities that manage customer relationships and product development are only *indirectly* related to financial performance. For example, what is the measurable financial value of an improved customer relationship, faster time to market, or reduced lead time? There is no doubt that these activities contribute to the value of the enterprise, but because they are difficult to measure, they do not fit cleanly into many traditional financial measurement models. And if we attempt to force-fit these operational measures, or worse yet ignore them entirely when making decisions, we may stifle important Lean initiatives.

Beyond this finance-centric viewpoint, there is an even more significant reason why ERP is considered to be the core of all other enterprise applications. Over time, as companies integrate their functions, they may extend the core ERP capabilities by adding software components such as CRM, PLM, APS, MES, WMS, and so on. To integrate and work together, these disparate systems must all integrate in a logical and nonredundant manner. What results is a unified transaction flow model in which all operational applications push transactions toward the ERP core, where the primary data entities are managed: customers, suppliers, inventory, resources, and assets. As an enterprise continues to invest in extending the ERP core outward with additional software components, integrating additional capabilities within a logical data model, there is a point at which the complexity may become overwhelming. Dave Caruso writes about this significant trend in his January 2003 *Manufacturing Business Technology* article "ERP as Infrastructure?":

> The nature of the [enterprise integration] questions have moved to strategies to manage the complex array of systems underlying the fabric of today's corporations, and away from the selection of an appropriate ERP system. [This is due to] the widespread success of strategic add-on applications such as global sourcing, supply chain management, product lifecycle management, and now, enterprise performance management (EPM). Unfortunately, the resulting system often has much redundancy, overlapping functionality, and usually requires complex integration to successfully execute some of the more unique business processes.

> ERP is no longer just an application, but has become the very infrastructure that allows a company to rapidly provide information to the strategic business processes [value streams]—and very likely—the new strategic applications that are powering the restructuring of industry, and are the source of substantial cost reductions. For this reason, information technology planners must embrace ERP as the core of their infrastructure, and work with their ERP providers to begin building an information architecture that anticipates the interconnectivity and services needs that will come at an ever-increasing pace.[146]

Focus on the Value Streams

As we have learned, ERP, and particularly its financial core, is the hub to which all other systems naturally integrate; it also provides the mechanism for financial controls and performance measurements. But this focus does not necessarily correlate with *where the value is created*. As we learned in Chapter 3, the value stream is defined by Womack and Jones in *Lean Thinking* by what adds value from the customer's perspective. They describe the three essential components of the value stream as the *problem-solving task*, the *physical transformation task*, and the *information management task*.

Once an enterprise identifies the value it intends to create, then it should define a set of strategies to achieve it. In their book *The Discipline of Market Leaders*, authors Michael Treacy and Daniel Wiersma describe three distinct strategies for value creation:[147]

1. **Operational Excellence**—streamlined operations and optimal cost of production and delivery for a large audience. The authors suggest McDonalds, Wal-Mart, Dell Computer, and General Electric appliance division as examples.

2. **Product Leadership**—focus on invention, innovation, rapid product development, and market exploitation. The authors suggest Johnson & Johnson, Nike, Sony, Hewlett-Packard, and Intel as examples.

3. **Customer Intimacy**—Organizations focus on relationships, not transactions, and are geared toward carefully selected and nurtured clients, creating a culture that embraces customer-specific rather than general market solutions. They become a trusted and influential advisor to the customer. The authors suggest IBM, Nordstrom, and Airborne Express as examples.

There is a surprising correlation between these two definitions of value creation and the design of enterprise software, shown in Figure 6-10.

Lean Thinking Value Stream Womack and Jones	Physical Transformation	Problem Solving	Customer Value
Discipline of Market Leaders Treacy and Wiersma	Operational Excellence	Product Leadership	Customer Intimacy
Enterprise Software Component	MRP II	PLM	CRM

Figure 6-10. Two perspectives on value creation

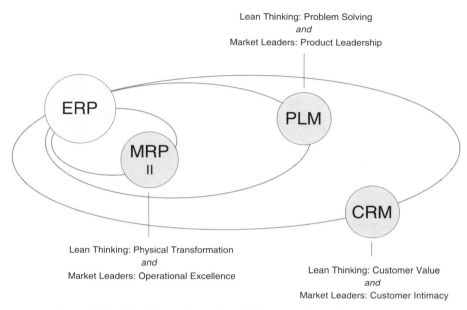

Figure 6-11. The Copernican view of the enterprise software universe

This correlation clearly illustrates that although the financial component may be at the core of the ERP integration model, it contributes to value creation by acting as a financial controller and scorekeeper for the operational activities. The three operational applications, on the other hand, *directly* enable the essential value creation processes—they are the IT backbone of a Lean Enterprise. Thus we can chart what I call the Copernican view of the enterprise software universe, shown in Figure 6-11. We will return to this comprehensive model later in this book.

Chapter 7

Integrating Value Streams

Virtually every enterprise must cope with information system integration challenges. Enterprises are usually organized by department or function, and although enterprise software is typically designed within the framework of these silos, business processes and supporting information freely flow across these boundaries (Fig. 7-01).

In *Reengineering the Corporation*, Michael Hammer and James Champy write that although traditional corporations still follow a hierarchical management structure and a rigid, repetitive task orientation, disruptive information technologies have enabled them to fundamentally change the way they organize and behave, for example:

Old Rule—Managers make all decisions.

Disruptive Technology—Decision support tools (desktop computers, database report writing, business intelligence systems)

New Rule—Decision-making is part of *everyone's* job.[148]

Hammer and Champy argue that information flow and accessibility, enabled by modern information technologies, can break down the barriers of a highly structured organization, empowering employees to engage in dynamic and responsive business processes. This is similar to principles of Lean, suggesting many enhancements in the way that application software may interface with the individuals and teams performing the work. For example, information

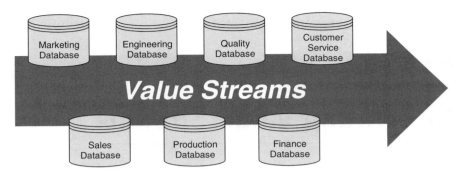

Figure 7-01. Information silos and the value stream

should be captured only once. Skillful design of the software user interface, with search and help screens, instant validation of the information, and interaction with the user, facilitates and error-proofs data processing—electronic poka-yoke.*

Hammer and Champy point out that fundamental reengineering was not possible before the emergence of sophisticated information systems. However, they insist that the first order of business is to improve the process itself, and IT is simply an enabler of the reengineering and process improvement efforts. For a business to run efficiently, information systems must therefore facilitate the flow of business processes across application software and organizational boundaries. ERP software is important because it provides a prepackaged suite of software incorporating many, though usually not all, core business processes. ERP software developers have acquired or formed alliances with publishers of complementary systems including CRM, PLM, WMS, MES, SCM, and others. Even these prepackaged integrations rarely satisfy 100% of their unique requirements, so most enterprises invest in assembling, integrating, and maintaining a patchwork quilt of supplemental applications surrounding their enterprise software core.

An enterprise often resorts to time-consuming manual interfaces as information moves between application software boundaries, and there is a proliferation of physical and electronic documents, spreadsheets, and small databases filling the gaps and cracks as information flows across disconnected systems. There are many disadvantages to these manual integration workarounds:

- Unnecessary throughput time is added to the process.
- Administrative cost is added to the process by entering, maintaining, and reconciling multiple sets of data.

---

* The APICS Dictionary defines poka-yoke as methods of performing operations so that actions that are incorrect cannot be completed. For example, a part without holes in the proper place cannot be removed from a jig, or a computer system will reject invalid numbers or require double entry of transaction quantities outside the normal range.

- Process flexibility and adaptability is limited, requiring new or modified interfaces each time the process changes.
- Unnecessary complexity and potential data integrity errors are caused.
- Fragmented and unsecure data jeopardize intellectual property.
- Costly errors result from reliance on bad data.
- Fragmentation of data inhibits measurement, reporting, and timely feedback to the people and process.
- Dependence on the people that know how to make these special interfaces work.
- Coherent audit trail of the transaction workflow, and the capability for historical reporting and trend analysis on the information thread, is lost.
- Labor-intensive manual procedures limit the ability to scale transaction volume without a disproportionate increase of administrative cost.

Although quick-fix integration workarounds may solve an immediate problem, at some point an enterprise may face a decision: continue applying reactive interfaces or develop a comprehensive architectural approach to integration, investing in the infrastructure and skills to proactively manage the complexity and reduce the NVA activity.

This chapter is intended to provide the ordinary person (not a technical wizard) a basic understanding of integration concepts, because most businesspeople during the course of their work will be confronted with information system integration challenges, or at least they will be responsible for mopping up behind a misguided integration attempt. It's therefore important to understand the basic issues and approaches of integration, which require some effort to explain and understand.

This chapter will explore several technical concepts that you are not required to master. If, however, your enterprise chooses to embark on system integration at any of the levels explained in this chapter, you're advised to have good technical talent on staff or close at hand. And it should be clear that this technical staff must also be capable of *understanding the business processes* that are to be integrated and orchestrated. This is why these technical resources are often joined by an internal team that is responsible for these processes—a continuous improvement team with a technical competency.

Mechanics of Enterprise Integration

To understand several important considerations when integrating events, applications, and data, let's begin with a simple integration scenario using software applications A and B, shown in Figure 7-02.

For example, application A could represent a Web order entry system, with B representing an ERP system. Suppose that a business process originates with a customer entering an order on the website, creating data in application A, and an instant (or a day) later there is a handoff to application B for fulfill-

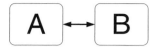

Figure 7-02. Point-to-point integration of two software applications

ment. Perhaps as certain events take place in B, such as confirmation of the order, processing of payment, and shipment of the goods, feedback data is sent back to A to support Web customer inquiry. This sounds simple, just moving some data back and forth between two systems. But what if there are thousands of transactions a day, with occasional variations, errors, and exceptions?

Basic Integration Considerations

When designing even a simple point-to-point integration scenario there are many potentially problematic issues that must be thoroughly considered; thirteen are described here.

1. *Online/Offline*. Systems can integrate either in a *tightly coupled* fashion (*online, real time, synchronous*) or in a *loosely coupled* fashion with a buffering time delay (*batch mode, offline, asynchronous*). Systems must often be integrated offline, because of either physical or logical constraints. For example, if the communication link between tightly coupled applications is lost for any reason, then the entire system cannot work properly. Online integrations can also create unnecessary performance overhead in a high-transaction-volume environment, and they are generally more difficult to design, operate, and maintain. Offline systems generally require elaborate transaction messaging management to work properly, but can be more forgiving in many situations.

2. *Transport*. The physical movement of data from one application to another can be sophisticated or as simple as *sneakernet* or *swivel chair* manual data integration. Network and telecommunications choices are rapidly evolving, including wireless technologies that offer tremendous flexibility for plant-level integration, and bandwidth becomes less expensive each year. The approach to routing electronic messages between applications depends on many considerations, including cost, timing, and security. In a mission-critical environment it is important to have an automated failover option in case the primary communication link is lost.

3. *Synchronicity*. Redundant information is typically stored at each node of an application integration scenario. For example, a CRM database will usually hold customer, prospect, and vendor records, whereas an ERP database will hold these same customer records in accounts receivable and vendor records in accounts payable. When a customer record is added, edited, or deleted, how many times must the information be manually updated within various systems throughout the organization? Ideally, just once, and the update should automatically flow to all integrated systems. This usually requires a hierarchal sequence for file updates, where one file is the *parent* and all others are *children* (this is also called a master and slave relationship, but I like to avoid those terms). Edits are made to the parent and replicated to the children, and ideally system controls should be in place to avoid edits directly to child records. Alternatively, multiple systems may be updated independently and synchronize data among themselves on a peer basis, but this many-to-many synchronization can be difficult to design and manage.

4. *Timing.* In an environment where information ages quickly, and if the integrations are loosely coupled, batch updates of shared information must be performed often. The Lean principle of batch size reduction can be applied to the logic of data integration, where the batch size and lead time window between updates can be quite small. On the other hand, frequent updating of small batches may create a significant processing overhead cost, so it's important to consider the timing requirements when designing an integration scenario. Inappropriate integration timing can create significant problems, take for example an order placed on a consumer products website. When the customer places her order, the website may confirm the availability of the item, providing an order confirmation with an expected ship date. But wait—what if the inventory values stored on the website database are not current? The customer who believes that she will soon be receiving her product will instead receive an e-mailed back order notice. What has happened? The inventory values stored on the Web server system were not regularly updated by Available To Promise information managed by the ERP system.

5. *Business Process Workflow Rules.* When attempting to diagram a business process you often encounter branching, looping, and exception-handling rules. For example, when an order is processed, if the credit check succeeds, the order is routed to shipment; but if the credit check fails, the order is suspended and a message is sent to the sales department for corrective action. Exceptions and variable routings are a fact of life, and if an ordinary business process has just three or four branches in its life cycle, the number of resulting unique transaction flows can number in the dozens. When flow control is needed, a workflow management tool may supervise the progression of each transaction amongst one or several applications, routing the transaction according to predefined rules and notifying a human being if an approval or intervention is required.

6. *Data Mapping and Transformation.* Mapping identifies the logical changes required for each data element to move from one application database to another. Transformation is the process of changing the data as it moves. Let's say that both software applications have a file that stores customer records: The source application stores the customer number as a six-digit numeric value, and the target application has a ten-digit code with a four-digit alpha prefix and a six-digit numeric suffix. Mapping can be simple, where customer number cross-references are stored in a lookup table, or it can be complex, involving programmatic calculations and combined field transformations.

7. Data Integrity. When data is ready to be uploaded to the target software application, it must be tested for validity. For example, does each field contain the proper number of characters, with the right mask,* and a valid date? Does the transaction violate any processing rules? Usually the role of data validation is left to an *Application Programming Interface* (API) provided by the target application, a programming toolkit that helps a developer connect to the program. When designed for a specific purpose (such as the receipt of a sales or purchase order from an external source) an API may be called an *Application Adapter*, which is designed to feed specific transactions in an acceptable manner prescribed by the logic of the target application software.

8. Transaction Integrity. When application A sends a transaction to application B, but communications fail, how does A know the transmission did not succeed? Or, if the message is received by B but fails to process and update B's database, how does A know it must roll back the transaction and try again? Imagine receiving a deposit slip from your neighborhood ATM machine, only to discover a month later when you reconcile your checkbook (you do, don't you?) that the deposit never updated your bank account in the central computer?

This sort of control (called a *two-phase commit*) has existed within the database world for decades. However, it is considerably more difficult to verify a transaction when it is committed across multiple databases, contained in several different applications, owned by several different companies, and transported over potentially unreliable lines of communication. Obviously without this type of assurance that all application databases update properly, many types of mission-critical eCommerce cannot work. *Transaction boundaries* must be defined that encompass all applications involved in a particular transaction, and these rules may be enforced by a central transaction monitoring system. This ensures that the collection of applications maintains unified transactional integrity, responding appropriately when this integrity is violated by any type of processing or communications error.

9. Audit Trail. Enterprise integration scenarios often involve transactions of significant value or risk to the organization, its business partners, and stake-

---

* A field mask defines the required structure of the field; for example, a three-digit alpha and three-digit numeric customer number would have the mask ZZZ999 and a phone number would mask as (999) 999-9999.

holders, and must therefore be auditable. If something goes wrong, we need to figure out what happened and why, correct the errors, and take preventative measures. Like central transaction integrity monitors, there may be a separate service that maintains an unbroken audit trail of all integration events. We cannot rely on the individual application database logs to reconstruct the complete audit trail, because errors may occur outside the domain of a particular application.

10. Performance. Enterprise integration scenarios often involve periods of dormancy followed by peak load spikes. If an integration engine cannot keep up with the peak load, this creates a bottleneck that will delay the business process. Integrations should therefore have sufficient hardware, software, and communications capacity.

11. Upgrade Management. Software applications are perpetually maintained, updated, upgraded, and occasionally replaced. When they are tightly coupled with other applications, then the entire integration framework must be shut down to change or replace a single component. When applications are loosely coupled, however, the environment is more forgiving, and a single application may be temporarily uncoupled to make changes.

12. Security and Privacy. An enterprise must protect its vital assets, and among these intangible information is perhaps the most difficult to secure. Threats to information security and privacy come from within and outside the enterprise. Some threats are deliberate and well-planned, whereas others are random but no less malicious. There are so many ways to slice, dice, store, move, and secure data that security management has become a full-time job for many IT staff. In fact, many larger enterprises have named Chief Security or Chief Privacy Officers. Security and privacy rules must be extremely well-designed and carefully monitored, if they are to effectively secure the hands-off integration of enterprise applications.

13. Ownership. The issues of synchronicity, business process workflow rules, transaction integrity, audit trail, security, and privacy all point to organizational questions: Who owns the data, and who defines the processing rules? As an enterprise coordinates and integrates its information systems and value streams, this will lead to questions of ownership and control. Take the product configurator, for example: Decisions on the design, use, and maintenance of the product configurator system may be equally claimed by engineering, sales, planning, and production. According to Tony Baer, industry analyst and technology writer for *Manufacturing Business Technology*:

> For all things data, process, or infrastructure, there is somebody who owns that piece of it. In one case somebody may own the data model, in the next case somebody may just own the data; that person may or may not own the process

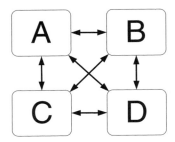

Figure 7-03. Point-to-point integration of four software applications

for ensuring data integrity. So whenever any decision is made on any of these elements, the owners must be in the loop; better yet, they should drive it. The bottom line is that you'll end up with a federation that owns and manages all of this.[149]

Middleware

Now that we have explored the basic considerations for integration, let's consider what happens when multiple integration points are involved.

Observe in Figure 7-03 that six distinct interfaces exist among four applications; the number of individual point-interfaces increases exponentially with the number of applications. An integration scenario involving just a few applications, transactions, and exceptions can become extremely difficult to manage, which means the performance of the value streams depending on the flow of this information are jeopardized.

Let's consider an example of a multipoint integration scenario, where a company employs an ERP system with separate applications feeding customer sales order transactions, as shown in Figure 7-04.

In this scenario, five independent sources of sales orders feed ERP, each with different operators and business rules:

1. **Order Entry**—This is a native component of the ERP system, with sophisticated configuration and pricing capabilities, staffed by trained customer service operators.
2. **EDI System**—This separate application receives a large quantity of electronic orders with standardized product, price, and delivery conditions. A separate EDI mapping and transformation system is required to interface to the ERP software.
3. **Web Order Processing Interface**—This separate application offers a limited number of standard products sold via the Internet direct to customers and through various affiliate organizations. These are customers, not highly trained data entry operators, using the order entry system, so the choices must be limited. Although the product catalog and pricing

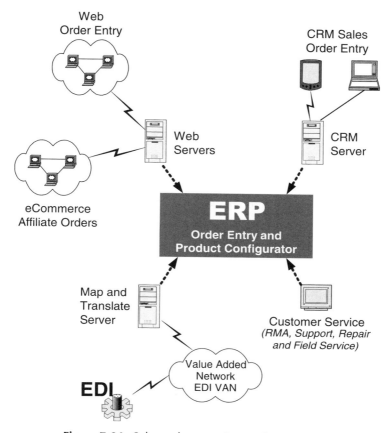

Figure 7-04. Sales order entry integration scenario

logic must be kept simple in this environment, back-end integration issues related to synchronization, timing, error prevention, cash fulfillment, promise date, order acknowledgement, and shipping/backorder confirmation can be challenging. This website runs on a separate server using Web commerce tools, periodically transferring transactions to the ERP system and receiving updates in return. And this application is updated frequently with new features, business rules, pricing, and promotions, thus creating a moving target for ERP integration.

4. **Roaming Sales Force**—This is not just a single application running on a separate system, but a collection of laptop computers and Personal Digital Assistants, with wireless and Web interfaces running CRM or route automation software with separate databases containing customers, products, pricing, and orders to be synchronized.

5. **Customer Service**—This system enables customer interactions in a variety of ways, looking up sales history and configuration information,

processing return, exchange, repair, and field service orders. This system may also include field service personnel using various offline or wireless handheld devices.

Obviously this order processing scenario presents many integration challenges, including the management of redundant stores of data. It also involves critical timing of transactions such as order and inventory commitments through a central Available To Promise (ATP) processor, because multiple offline order processes may be competing for limited inventory and productive capacity. This scenario is in fact quite common, and to manage a complex environment like this an enterprise should consider an *integration hub* approach.

From an engineering point of view, integration hubs have been around for a long time. For example, airplane control systems are a large collection of complex instrumentation, produced by different manufacturers, sending different signals, all coming together in a dizzying array of indicators in the cockpit. Needless to say, each of these systems is vital and must perform independently, with redundancy, and be replaceable without affecting any of the other systems. An integration hub is a backbone communications framework whereby multiple applications can be independently connected to (coupled) and disconnected from (uncoupled) without affecting the operation of the overall system. Enterprise integration transactions may be orchestrated by an integration hub called *middleware*; other names for the same concept are *Message-Oriented Middleware* (MOM) and *Object/Message Brokering*. Whatever the name, a middleware integration scenario looks like Figure 7-05.

Middleware tools communicate with two techniques: *store-and-forward* and *publish/subscribe*. A store-and-forward design is like e-mail, where data packages are sent to a specific address. These messages are then either automatically forwarded to the target application or stored for the target application to check in later. In a publish/subscribe model an application can send (publish) a single data package to the message broker, where other applications can then independently subscribe to particular types of message without

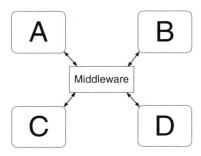

Figure 7-05. Middleware integration

the source application knowing who is receiving the data or what they are doing with it. In a store-and-forward model the same data package may be sent several times to multiple target systems, creating significant traffic across the network; whereas a publish/subscribe model creates the message only once, routing to the appropriate subscribers.

The middleware approach to message handling has many advantages over tightly coupled integrations. Loosely coupled applications can be modified, upgraded, or even replaced, with limited effect on other applications. And the flow of integration messages passing through the middleware can be monitored, audited, and fine-tuned for performance and fault tolerance. However, there is a cost to a middleware system, and it requires significant technical expertise. Depending on the number of applications and processes and their complexity, there is a point at which the cost of complexity outweighs the cost of middleware solution. A small or medium-sized company should look very carefully at the cost/benefit equation of a middleware system compared to point-integrations.

Middleware abstracts the integration process, taking the responsibility for flow and control of the data out of the hands of the source and target applications. It's important to understand that middleware *does not reduce the inherent complexity* of an integration scenario. Nothing can simplify an integration scenario except simpler design of the business processes and underlying systems! Nevertheless, middleware can help to make a complex integration scenario *manageable* by providing capabilities such as:

- Flexibility for coupling and uncoupling of applications, providing application independence for simpler design, maintenance, and replacement
- Fault tolerance of the entire system against the failure of any component
- Messaging to allow disconnected systems to communicate over large distances and with substantial time delay
- Centrally managed data mapping, transformation, and validation services among multiple systems
- Centrally managed rules (such as pricing and promotions) shared among multiple systems
- Mechanisms to ensure data and transaction integrity and auditability
- Scalable design for high performance

Data and Method Integration

After determining the appropriate architecture for enterprise integration, the underlying logic must be determined. There are two basic technical approaches: data and method integration.

Data Integration. Data-element integration involves the packaging of a set of data representing one or more transactions into an electronic message that is

```
                         ASCII Record

  "John Smith", "123 Central Avenue", "New York", "NY", "98023"

                          XML Record

  HEAD><TITLE>Customer Record</TITLE></HEAD>
  <BODY>
  <Name>John Smith</Name>
  <Address1>123 Central Avenue</ Address1>
  <City>New York</City>
  <State>NY</State>
  <Zip>98023</Zip>
```

Figure 7-06. Comparison of ASCII and XML data formats

then sent to a receiving application for processing. Modern data structures usually break a message into *header* and *payload* (*detail* or *body*) elements. The header element specifies content, type, routing and destination, security, audit, and validation information. The payload element contains the transactional data.

Although data are originally structured according to the design of the source application database, they must be encapsulated in a universal format that can be read by a variety of applications. Historically, this format has usually been ASCII (American Standard Code for Information Interchange), but XML (eXtensible Markup Language) is now the standard for the exchange of data. Contrasted with ASCII, which simply contains raw data, XML utilizes *tags* (shown in Fig. 7-06) that identify the structure and content of each data field and is thus a self-describing data format.

XML <tags> describe the definition and meaning of the fields they delimit, unlike the predecessor HTML format, whose tags simply control page layout and formatting. The XML format also incorporates a *schema* (structure) or Document Type Definition (DTD). Thus individual fields within an XML document can be described with attributes such as field size, field type, field masks, and content validations; an XML file is not only self-describing, it is self-validating. XML is more content-rich than ASCII, but it carries a performance overhead cost—an XML file is considerably larger than its ASCII counterpart. But with bandwidth and storage costs continuing to decrease, and the importance of system integration increasing, the benefits of XML far outweigh the costs. In fact, most database and application software developers incorporate many powerful XML capabilities within their products. This is particularly true of Microsoft, who has embedded easy-to-use XML handling capabilities into all of its desktop Office products, including Word and Excel. With a $3 billion annual R&D budget, and desktop software on the majority of computers on this planet, Microsoft has made, and will continue to make, significant strides toward integration for the common user and the small to medium-sized company.

EDI is another popular form of data integration messaging, and it has been used effectively for decades. EDI documents traditionally use ASCII data organized into standard document types that prescribe in great detail not only the structure of the header and payload, but also the specific logical content of the data within certain transaction types* to make the data consistent and easily transferable between trading partner systems. The reason why this is no longer the golden age of EDI is that EDI is expensive and complex, depends on highly standardized transaction types, and relies on proprietary and expensive Value-Added Networks (VANs).[150] There are substantial movements under way to blend the best characteristics of EDI and XML into open and flexible standards for B2B eCommerce across the public Internet.

It is common in the EDI world for a single transmission to contain hundreds or thousands of detailed transaction records. It is also common in other eCommerce environments to process thousands of transactions, each contained in a separate message. It does not require much imagination to realize the importance of proper controls, fault tolerance, data and transaction integrity, system performance, and audit trail under these conditions.

Method Integration. The second basic type of enterprise integration involves methods, where applications talk to each other directly. Like data integration, an encapsulated message is involved, but this message contains more than just data—it also includes program instructions.

For example, let's say that during customer order entry an ERP application needs to call another program that calculates shipping cost. The ERP application sends an electronic message that starts the carrier rate application (stored in another program within the company network) with data parameters identifying the destination zone location, number of parcels, weight of the parcels, and desired date of delivery of the transaction. The carrier rate application then executes, using these parameters as inputs, and returns a message to the ERP application with the cost of the shipment so the order may be completed.

Rudimentary method integration has been used for many years; an early example called a Remote Procedure Call (RPC) dates back to the mainframe era. Another example of method integration many readers may recognize is Microsoft's Object Linking and Embedding (OLE) technology.

---

* For example, an invoice is described by the ANSI X12 EDI standard as document type 810, with voluminous information on how the hundreds of possible fields in the header and payload sections are to be used.

For example, an Excel spreadsheet can be copied and *embedded* into a Word document by using the *paste special* command. When the user double-clicks on the Excel object within the Word document, it launches Excel, allowing the user to edit the worksheet with Excel functions while remaining in the Word program.

Method integration opens up new frontiers for process automation. The emergence of large-scale enterprise systems with open architectures (interoperable programming tools and databases) enabled independent applications to interoperate with standardized methods far more sophisticated and scalable than the RPC. More importantly, however, it signaled the shift in corporate computing from centralized toward distributed computing. Later in the 1990s, robust integration architectures began coalescing around the evolving Internet with seemingly unlimited possibilities.

The Future of Enterprise Integration—Web Services

The future of enterprise integration, enabling the construction of widely distributed applications across any local, wide area, or public network with industry standard technologies, has been named *Web Services*. The struggle on this frontier may define the future of the enterprise software application and integration industry and the fortunes of its players, so the Oracle/Sun Microsystems and Microsoft camps have fought vigorously on this important battleground. The Oracle/Sun Microsystems Web Services initiative is based on Sun Microsystems' Java 2 Enterprise Edition (J2EE), whereas Microsoft has named its tools and initiatives collectively as .NET (pronounced "dot-net"). For a while each withheld compatibility from the other, defeating the value proposition of platform-independent integration. Then in early 2004, in the face of a shared threat from the Linux Open Source software community (free is difficult to compete with) Microsoft offered a settlement of $1.6 billion, Sun called off its antitrust allegations, and these two new business partners agreed to ensure the future compatibility of the two environments. Time will tell.

So What is a Web Service?

Let's build on the example of an ERP application talking to a carrier rate application. In the earlier example, both systems resided on the local network. But what if carriers such as UPS and Federal Express maintained their own programs, accessible through the Internet, for customer applications to automatically place an order or track delivery services? Or what if a sales tax software provider offered a software/service to users across the Internet, so a company no longer needed to maintain its own sales tax rate calculation program? The company does not have to determine how to interface from its ERP application to the Internet-based sales tax system, because the interfaces are standardized and published as Web Services. The possibilities are endless when standard application capabilities become "services" provided on a charge-per-usage basis (or free) rather than as a purchased product, available anytime, anywhere. And because the service is available with a standard XML interface, an application can access its capabilities without concern for what is going on behind the scenes (how the application works, how data are validated or exchanged, or are there any maintenance or upgrades needed) as the provider takes care of these responsibilities and provides a completely uncoupled service.

Many clever computing devices are emerging to take advantage of the wireless communications and integrated Web Services. Personal Digital Assistants, cellular phones with screens and keyboards, pagers, and countless dedicated computing devices for specific industrial purposes, all seem to be converging on the Internet for real-time interaction. One XML message can be presented in a variety of formats—from a full-size screen, to a tiny cell phone or PDA display—using a pseudo-alphabet for stylus data input; even voice activated interfaces are becoming commonplace in many environments. Web Services creates a vision of a wireless Internet bursting with universal applications that can communicate with each other in real time with an endless variety of friendly computing and communication devices.

However, more than an agreement on basic technical standards must exist for Web Services to simplify the integration of business processes among multiple applications, both within and outside the boundaries of an enterprise. EDI has been sharing data among ERP systems for years, but the rules for various transaction types are rigidly defined and do not penetrate within the process itself, but merely touch the surface. Even so, EDI interfaces allow for creativity, and as a result the creation of a single transaction map between two trading partners may take days or weeks to design and implement. Standard Web Services, if they simply create an external touchpoint among systems, may suffer the same fate. The hope is to some day define standardized business processes and interfaces that penetrate beneath the surface of any enterprise system, allowing for smooth workflow among applications without requiring substantial design, programming, and testing.

For flexible *process* integration, standard process definitions (such as how to automatically receive a purchase order and handle the myriad exceptions that often occur) are needed. Organizations like *ebXML* and *RosettaNet* support the development of open XML-based standards, and with Microsoft, IBM, and Oracle sponsoring the collaborative Business Process Execution Language (BPEL), we seem to move closer to universal plug-and-play process integration with each passing year—but we may never arrive. As with the development of any standard in the IT industry, application vendors will always see an advantage to keeping some proprietary logic and integration components within their control, because it serves to differentiate their offerings.

With that said, is it possible that the most powerful ERP vendor, SAP, with a worldwide market share of the largest companies near 50%, can establish universal business process standards? In the *Managing Automation* article "SAP to the Rescue?," Joshua Greenbaum suggests they might:

> The future according to SAP is about business processes [and] process repositories. As the owner of a considerable amount of process knowledge already, what is new is the clout that SAP can bring as the promoter of these worthy ideas—not to mention 24,450 of the world's largest and most successful companies as its customers. With SAP in the driver's seat, a process repository of recipes for how to get the job done can actually move it to the center of the functional universe. With the repository comes a massive, industry-wide decomposition process that will result in the componentization of everything we consider enterprise software. Once that's done, recomposition takes over based on business processes, not sequential lines of code and monolithic applications. How we work will be defined by what we want to accomplish, not by the limits of the programming arts. The best way to accomplish our daily tasks will be outlined in a process repository that will define the assembly of software components that get the job done; with no more technological expertise required of the user than what is needed to drive a car.[151]

Although this vision of process standardization may sound far-fetched or utopian, SAP may have the time (they plan to have a complete repository by 2012), the resources, and the market muscle to get it done. If such a vision is realized, will it be beneficial to all, or an ominous sign that the way worldwide business processes are performed may be controlled by a single company? Again, time will tell.

Beyond Web Services, the Service-Oriented Architecture

Finally we arrive at the concept of the Service-Oriented Architecture (SOA). As we discussed earlier, beyond the point-to-point integration among applications (enabled by Web Services) a framework for enterprise-wide integration suggests a centralized middleware application. Until now, middleware software engines have been proprietary applications offered by companies such as IBM, Microsoft, Tibco, BEA, WebMethods, and others. Shouldn't there be a standardized set of middleware capabilities upon which Web Services can rely? That is the purpose of an SOA. Until SOAs, middleware architectures were complex, expensive, and proprietary. According to Tony Baer:

> Software history has been larded with ill-fated attempts to standardize the way programs integrate. Integration continues to be the costliest part of any software project. More recently, middleware has become the most popular integration approach, with most tools using proprietary technology-based hubs to direct interactions between applications, messaging systems, and data sources.
>
> All of these approaches failed because of their rigidity. That's not surprising. Conventional software programs were monolithic—they combined lots of

functions in lockstep, running every transaction against preset targets. So if you changed the data or business logic of a given transaction, those changes had to cascade back to every process, message, and data source they touched. When Enterprise Application Integration (EAI) middleware emerged, it worked the same way.

Now a reinvention of an old idea, the Services-Oriented Architecture (SOA), threatens to replace hard-wired connections with a service request model that could prove far more pliable. Unlike conventional systems, SOAs simply specify information or service requests, they don't specify which system must respond. In the long run, that has huge implications for making integration easier.[152]

SOAs offer great promise to manufacturing enterprises, who often face significant integration challenges and opportunities. According to Byron Miller of Forrester Research:

Especially in manufacturing where there is a lot of outsourcing and contract manufacturing, SOAs support quicker integration between companies. You get a high level of flexibility and lower overall integration costs. It may take until the second half of the decade [2000] before most enterprise suites extensively leverage SOAs, but vendors will get there, pressured by users demanding lower integration costs, and a need to fuse acquired product sets around common components and a supporting architecture. Both users and vendors need SOAs, it gets the market close to the plug-and-play vision it has long been after.[153]

Cautions about Enterprise Integration

All this technical jargon can seem very abstract to the nontechnical, so let me try to summarize how the pieces fit together:

- *XML* **is a Data Format**—a tag-based language for representing data. Because of its flexibility and built-in intelligence it has become the preferred neutral format for moving data among systems, but other formats (ASCII, EBCDIC, binary, and others) may be used by Web Services and SOAs.
- *Web Services* **are Methods**—a language for systems to connect and work together, a set of standard techniques for requesting and providing services. Web Services usually use XML data, but they do not have to.
- *Service-Oriented Architecture* **is a Framework**—an infrastructure and a way of designing software to expose capabilities to the outside world. An SOA may use Web Services to communicate with other systems, and it may use XML data formats, but these are not required. As the term "architecture" suggests, this is a structural design decision made by the software developers to incorporate service-request capability within the framework of their software system.

Don't let this simple summarization fool you; these are complex matters, and integration standards emerge and fade away with surprising regularly. Enterprise integration projects can lead a company into dangerous and unexplored waters, so the small or medium-sized enterprise should consider the following precautions:

- In support of general business process improvement projects, enterprise integration can eliminate waste and enhance performance, important issues in a Lean environment.
- Enterprise integration can be complex, costly, and risky, often requiring considerable planning, ROI analysis, and a substantial ongoing investment. Even a small enterprise integration project can easily cost five or six figures (in US$).
- Don't fall for "it will be easy with the latest .NET, Web Services, and SOA technology." Nothing about integration is easy, because it involves the interaction of dynamic business processes and complex technologies. Carefully analyze, design, and justify every investment in integration. Sneakernet creates many forms of waste, but a failed or inflexible integration project may be far worse. Smaller companies may be advised to focus on optimizing the capabilities offered within their ERP system before automating extended business processes outside their boundaries.
- The future of enterprise integration, Web Services, and Service-Oriented Architectures offers many seductive possibilities for the creation of value and competitive advantage. Small and medium-sized enterprises should whenever possible wait for these technologies to mature, as standardized interfaces are introduced as packaged solutions by their ERP vendors. Of course, a small or medium-sized company may suddenly need to move ahead with an enterprise integration or eCommerce project if a large trading partner encourages them. In that case, they should leverage whatever knowledge, tools, and assistance the trading partner provides, controlling the project scope carefully, prototyping each phase as they go, to prove the feasibility and benefits of each step.
- Lean organizations should approach enterprise integration from a holistic perspective like any other kaizen project. The team should simplify the value stream first, eliminating unnecessary tasks and transactions, doing everything they can to streamline the process before applying information technology.

Return to Copernicus

We have used the term "enterprise integration" when discussing the interface of data between applications, both within the Lean Enterprise and through-

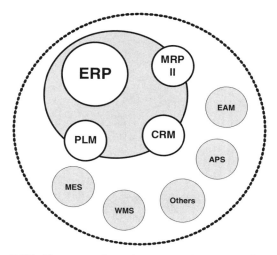

Figure 7-07. The enterprise value stream integration framework

out the Lean Network. The former is often called Enterprise Application Integration (EAI), whereas the latter may be called EDI, B2B eCommerce, or Supply Chain Management. Is there a technical difference between the tools and technologies used to integrate internal and external transactions? Not really.

Consider the design of a *value stream integration framework*, an architecture that enables multiple applications to work together in support of business processes. This framework may encompass the three primary enterprise applications, ERP, CRM, and PLM, plus other supporting applications. The value stream integration framework within the enterprise may appear diagrammatically as a bubble surrounding these applications, managing the integration of data and methods within. This framework architecture, illustrated as a dashed-line bubble in Figure 7-07, may be facilitated by sneakernet, point-to-point integrations, a middleware engine, Web Services, Service-Oriented Architectures, or any combination thereof.

Now extend the concept of the enterprise value stream integration framework to orchestrate the flow of data across enterprise boundaries, enabling the Lean Network. From a mechanical point of view, once an external trading partner transaction pierces the bubble of the value stream integration framework, moving inside the corporate security firewall, it's handled much the same as any internal integration transaction.

This integration framework facilitates the flow of information to support business processes and value streams within and among enterprises as illus-

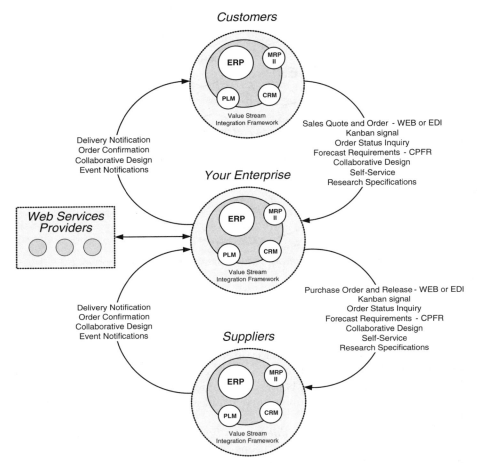

Figure 7-08. SCM, eCommerce, and the enterprise value stream integration framework

trated in Figure 7-08. The integration framework that helps multiple applications communicate, enabling the flow of information in support of value streams, also enables the reporting and analysis of that information in order to manage value stream performance; this is the focus of Chapter 8.

Chapter **8**

Managing Knowledge for Competitive Advantage

Alluring terms come and go with amazing speed in the software industry, and the term *Knowledge Management* came and went several years ago. It was an appealing concept really, that somehow information technology could help us manage "knowledge", similar to the abstract appeal that "artificial intelligence" and "expert systems" once held. But when the time came to actually define knowledge management, and to develop software that might enable it, the industry came up short. Software marketers replaced the hollow term with a collection of practical offerings: business intelligence, content management, portals, and collaboration. This chapter will explain what the concept of knowledge management ideally represents and examine the collection of tools and techniques required to achieve it.

The big problem isn't too little information but too much, and we must channel this deluge to our best advantage. Here's a short exercise—ask yourself: Where were you in 1990, one year before the World Wide Web was created? How were you employed? What were you doing with computers? Now imagine that someone had suggested that in just a few years a free service called "Google" would come along. Could you have imagined 24×7 real-time access to a global reference librarian? A system where you could ask almost any question, in many languages, toss it into the electronic ether, and instantly receive hundreds of responses that expanded your knowledge on virtually any subject matter? Would that have sounded like science fiction to you?

Now ask yourself this: How will the availability of knowledge change in the next five or ten years? How will it affect the competitive dynamics of your industry? What are you planning to do about it?

"Why is Knowledge Management such a difficult executive problem?" asks John Sviokla in his *CIO Magazine* article "Knowledge Pays":

> The primary issue is that knowledge itself cannot be directly measured, only its indirect effects can. That's because knowledge exists in the context of its use.
>
> How does this knowledge impact finances? It begins with the basic order to cash cycle. From 1989 to 1998, Dell's working capital moved from a positive 70 days to a negative 11 days. What enabled such a dramatic shift? Knowledge of every part of its value chain: configuration, customer demand, part availability, supplier quality. Most important, Dell shared this knowledge with its suppliers and customers.
>
> W.W. Grainger, the $4.5 billion industrial distributor, has this same kind of deep knowledge of the maintenance, repair and operations (MRO) space. If you want to find one of five million parts online, you can enlist a sophisticated search engine at Grainger's website. If this doesn't work, you can go to a service called FindMRO with specialized search tools, or you can submit a request online to ask Grainger to find it for you. Don Bielinski, group president at Grainger, says the typical online Grainger order averages $250. FindMRO orders average $1200, and 80 percent of the goods are shipped directly from the manufacturer to the end user, converting demand to cash much more quickly.
>
> Superior knowledge management frees companies to operate on fewer assets, collect their cash faster and have less volatility.[154]

A knowledge management strategy is critical for any enterprise striving to become Leaner. Effective knowledge management enables an enterprise to deliver products and services better, cheaper, and faster; it also strengthens relationships with business partners through efficient collaboration. In a world of increasing information overload, the personalization of knowledge addresses each interaction in the most effective manner. Personalization provides people with what's most relevant to them, whether it's a product or piece of information, a document they need to approve, or a portal with single-click access to the activities they perform most often. *Connecting people with relevant information creates value.*[155]

Structured and Unstructured Information

Any discussion about knowledge management must begin with a definition of structured and unstructured information, because this distinction guides how information is created, captured, stored, managed, secured, searched, retrieved, and presented. Knowledge exists in structured and centralized databases, as well as unstructured repositories of electronic documents scattered

Figure 8-01. Structured and unstructured data

throughout an organization. According to the Giga Information Group*, unstructured information outnumbers structured information by 4:1, and the vast majority of knowledge remains unrecorded and in people's heads (Fig. 8-01).[156]

In addition to the information in people's heads, volumes of information are stored in nonelectronic forms within an enterprise—physical documents stored in filing cabinets, buried in stacks of paper on desks, voice messages, and handwritten notes pinned to bulletin boards. How much vital intellectual property belonging to your enterprise is stored on Post-It Notes™?[157] It's evident that information technology can only go so far to manage the full spectrum of knowledge within an organization. According to Megan Santosus of the Knowledge Management Research Center,

> Peter Drucker was among the first management gurus to say that the key challenge for knowledge workers is creating a structure that will promote and support how those knowledge workers can be most effective. [In Jeff Nielsen's book *The Myth of Leadership* an organization that] eschews hierarchy and

* Giga Information Group was acquired by Forrester Research in February 2003.

rank-based leadership in favor of peer-based thinking represents the future of business. The most effective organizations will [develop] collective groups of employees who share everything they know and make company decisions accordingly.[158]

For a Lean Enterprise to thrive, it is essential to create a learning organization that channels this collective knowledge to improve performance. Continuous improvement emphasizes teams, education, communication, and collective problem-solving—much of the knowledge and wisdom required resides in people's heads and not inside a computer. We will return to these intangibles in Part 3 of this book, demonstrating how appropriate use of IT can invigorate and focus enterprise-wide continuous improvement. However, the emphasis of this chapter is on specific tools and techniques to manage structured and unstructured *electronic* data, information, and knowledge— leading to empowered workers and fact-based decision-making.

Structured Information

Structured information is managed by an application such as ERP or CRM (and within some aspects of PLM) and stored within a database comprised of tables, records, and fields. Data *tables* are organized according to the type of information they contain: customer table, parts table, invoice table, and so on. Each table contains *records*, which store a specific set of information on each type of entity or event. For example, a customer table may contain thousands of individual customer records. Each record contains *fields*, storing various elements of data related to each record. For example, a customer table contains fields such as name, address, and payment terms. Imagine a spreadsheet (which is a data table) containing a customer list, where each row represents a record for each customer and each column represents a field.

A *relational* database is used to manage data stored in multiple tables. Two kinds of tables exist in a relational database: *master* and *transaction* tables. A master table represents an *entity* such as a customer or inventory item. A transaction table records an *event* such as a sales order or shipment to a customer. A relational database associates one or more master records together, which relate to a particular transaction event, and includes a transaction identifier such as an invoice number. A simplified example of a sales history database relationship is depicted in Figure 8-02:

In this example a salesperson is assigned to each customer and the salesperson number is then stored on each transaction record. When a sales commission report is needed the primary source of the report would be the sales history table, sorting invoices by date range, grouping and totaling them by salesperson number. The calculation would then *join* to the salesperson table, using a *query** to find the commission percentage to be applied for each sales-

---

* Queries are questions posed to the database and are created with a *Structured Query Language* (SQL).

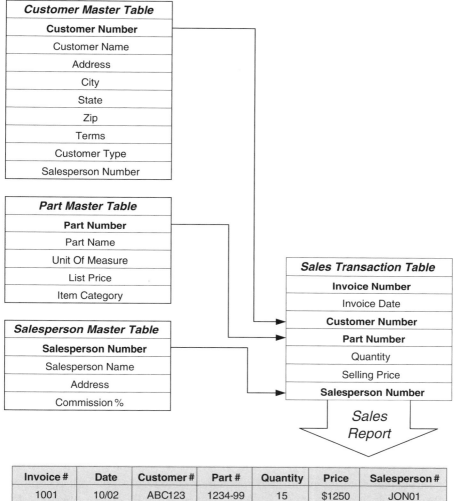

Figure 8-02. Relational database table structures

person and to find the salesperson's name and address to print a commission check; this information is not stored on each transaction record in order to speed processing and conserve storage space. It may help to study these table relationships for a moment to see whether this example makes sense to you, but if you don't care to learn the structure of databases and queries, then feel free to move on.

This example uses three master tables for the customer, salesperson, and part and generates only one transaction table containing the sales history for

*Two-database query
joined by common
customer records*

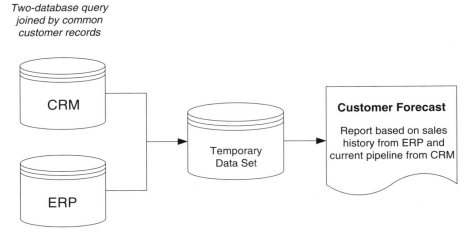

Figure 8-03. Multiple databases joined into a single query and report

the invoice. In reality a single transaction event in an ERP system may use several master records, while creating dozens of transaction records in several tables, each representing a specific slice of the data. A typical ERP system contains thousands of tables and millions of transaction records. Each record is a tiny grain of sand on a large beach, but collectively they represent the lifeblood of a business. This data is used to prepare management reports, control processes, and guide decisions on how to best manage the enterprise. Databases from separate software systems can be related together by common records, such as a customer number that is mapped across multiple databases. When such logical multiple-database relationships can be created, then a single query may be used to join these databases to produce a single report, as illustrated in Figure 8-03.

This capability to join multiple application databases to provide a comprehensive view of a transaction life cycle across a value stream underlies the enterprise value stream integration framework depicted in Chapter 7. When all enterprise value streams, applications, and their underlying data are logically interrelated into a coherent framework for reporting and analysis, we then have the necessary foundation for *structured* knowledge management suggested in Figure 8-04.

Unstructured Information

Unstructured information is a grab bag of electronic file types: word processing documents, spreadsheets, images, CAD drawings, video and audio files, e-mail messages, as well as financial and operating reports generated by ERP and other applications, all saved as data files in permanent storage. These files

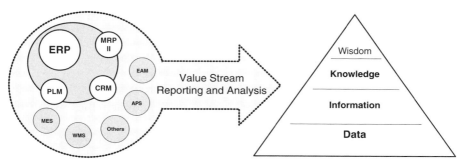

Figure 8-04. The value stream integration framework and structured knowledge management

may be stored in directory locations across countless local and networked hard drives and in other storage media.

Compared to the naturally tidy organization of a relational database, unstructured data must be carefully managed. If unstructured information is left unmanaged it can create virtual information anarchy. Attempting to organize unstructured data requires us to ask many difficult questions:

Organization

- Where is the data stored? In central repositories, local hard drives, archives, offline storage media, or most likely, a combination of these?
- How is the structure of hierarchical storage directories (taxonomy) organized within the filing system? By department, project, customer, product, or other categorization? How are search pathways and logic organized to provide the fastest return of relevant results?
- Who is responsible for creating and maintaining each branch of the hierarchy?
- How can we ensure that conflicting versions of the same document do not exist in several locations?

Validation

- How do we ensure that the contents are valid?
- What is the source?
- What happens when the information becomes out of date?
- How many chronological versions of the same information should be stored, and how are they cross-referenced?

Security

- Who has access to this information?
- Who can read, write, and delete each document?

- What sort of security model is required—by individual or role?
- Do individuals outside the company have access?

Compliance

- What regulations require us to document our processes?
- Must we be able to monitor user access and activities with this information?
- What health, safety, or legal liability issues would result if the information management system does not perform properly?

Search

- What criteria (content classifications) must we be able to search the information by?
- Should each file contain particular searchable information related to the contents (table of contents, index, or metadata*), or do we need to search the entire file each time?
- How will we search graphics and binary files that do not contain text?
- How much information is stored, and how long will it take to search?
- How many locations must be searched?
- What if a particular data source is offline (turned off or disconnected from the network)? Should the system cache metadata on the contents to let us know the data exists but is currently unavailable?
- Can searches be extended by the user to include public content on the Internet?

Change Management

- What is the life cycle of a document?
- How do we manage multiple versions of the document as it is created, updated, phased out, and deleted or archived?
- What controls are required to check in and check out documents for editing?
- How do we store a document that is being edited but not ready for public release?
- What workflow is needed to facilitate the routing, review, and approval of a document when it is ready for release?
- Can we automatically archive documents, making them available for an extended search but somehow identified as not being current?

---

* Metadata is "data about data", providing information on structure, content, context, intellectual property rights, and other characteristics. It can also refer to a usable subset of the data stored on a separate repository for performance or security reasons.

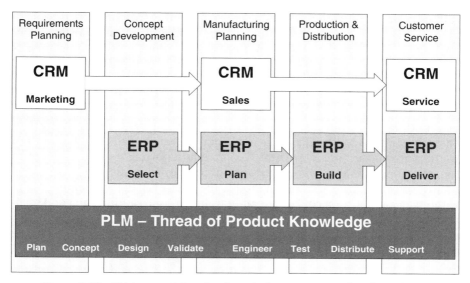

Figure 8-05. PLM—sustaining the thread of enterprise product knowledge

Many sophisticated tools exist for the management and reporting of structured data; powerful database technologies have been around since the early days of computers. On the other hand, the flood of unstructured information is a recent phenomenon that the software industry continues to wrestle with, and the term *content management* describes a variety of tools and techniques to manage it.

Product Life Cycle Management and Quality Management systems are specialized enterprise software applications that manage structured *and* unstructured data together. In the case of PLM, design and engineering documents (unstructured data) are interrelated to engineering and manufacturing processes (transactions in the ERP database) and the customers, suppliers, parts, and BOMs (entities in the ERP database). As we described in Chapter 7 and as shown by Figure 8-05, PLM manages the underlying thread of structured and unstructured information that binds together the entire life cycle of the product and customer.

The Components of a Business Intelligence System

Now that we have explored the distinctions between structured and unstructured information, let's consider the tools we may use to build a comprehensive knowledge management system. We'll begin with *Business Intelligence* (BI), which refers to a combination of reporting and analytical tools, based on data gathered from structured databases. Another popular term for this approach is *Decision Support System* (DSS). ERP and other business applications typically provide a library of preconfigured reports, as well as tools to modify existing reports and create new ones. Many of these reporting tools, although capable, may be less than user-friendly, which motivates companies to use off-the-shelf reporting tools. The most popular of these tools are Crystal Reports and Microsoft Access*—both offer user-friendly interfaces for the creation of database queries and the design of sophisticated and nicely designed reports. Crystal Reports is more flexible when it comes to the variety of reporting tools, whereas Microsoft Access is the Swiss Army Knife™ of desktop database tools, with a powerful database engine built in. One word of warning for those using Microsoft Access as a reporting tool: In addition to report writing, it is a software development toolkit. With Microsoft Access the user (given read-write access) can modify the data, even deleting entire tables, without any application controls or audit trail. Extreme caution and good security measures are advised or catastrophic data damage may result.

Banded Reports

The most common type of operational (as opposed to financial) report is a *banded* report, where records are sorted, grouped, subtotaled, and totaled by ranges of dates, transaction numbers, customers, and so on (Fig. 8-06). Each line (band) of the report represents a record of data, and each collection of lines represents a group of records sorted in a particular way, with subtotaling and totaling lines underneath. Most report writers have the capability to automatically produce banded reports: Crystal Reports and Microsoft Access both provide user-friendly *wizards*—helpful utilities that ask the user questions, then automatically construct the report format with the appropriate sorting, grouping, totaling, and layouts.

Crystal Reports continues to redefine the boundary between a report writer and a software development tool. With versions that support .NET and Java integration, application developers may incorporate Crystal functionality within the fabric of the software application. At some point the distinction of an external report writer is lost, and Crystal may simply become part of the application itself. This trend will most likely maintain Crystal's enviable position as the market leader, as its ability to seamlessly weave into the fabric of enterprise software applications, combined with its power, innovation, and ease of use continue to improve.

Financial Reports

Financial reports such as the balance sheet, income statement, and statement of cash flows are particularly challenging for a banded report writer such as

---

* Although some would prefer calling these simply "report writers" rather than business intelligence tools, saving the latter for more sophisticated analytical tools, the fact is that many people have done remarkable things with these relatively simple tools. Categorically, any useful form of analysis on information, even visual information on the shop floor, may arguably be called "business intelligence."

Sales History Report

Report Date : September 28, 2004		From 1/1/04 to 12/31/04		All customers		Region : West	

Date	Invoice #	Salesperson		Amount	Discount	Net	
Customer : ABC001		ABC Electronics	Region :	West			
2/3/04	121476	WAY001		$ 5,250.00	($ 150.00)	$ 5,100.00	
3/31/04	123753	WAY001		$ 4,500.00	($ 120.00)	$ 4,380.00	
8/12/04	138750	WAY001		$ 8,400.00	($ 275.00)	$ 8,125.00	
		Customer ABC001 Total		**$ 18,150.00**	**($ 545.00)**	**$17,605.00**	
Customer : DEN002		Denton Industries	Region :	West			
3/5/04	121899	SMI003		$ 1,200.00	($ 20.00)	$ 1,180.00	
4/28/04	125758	SMI003		$ 4,800.00	($ 110.00)	$ 4,690.00	
9/18/04	139370	SMI003		$ 3,450.00	($ 150.00)	$ 3,300.00	
		Customer DEN002 Total		**$ 9,450.00**	**($ 280.00)**	**$9,170.00**	
		Region West Total		**$ 27,600.00**	**($ 825.00)**	**$ 26,775.00**	

Page 1

Report Header

Banded Detail

Subtotal

Group Total

Report Footer

Figure 8-06. Banded report

Crystal Reports or Microsoft Access because they are formatted in an intersecting row and column style similar to a spreadsheet (Fig. 8-07).

Most ERP systems offer an embedded financial report writer that reads information directly from the general ledger transactional data. The financial report writer is often (but not always) a separate tool from the banded report writer, because it requires particular reporting capabilities. A typical financial statement uses rows to represent general ledger accounts (cash, accounts payable, revenue, etc.), as well as subtotals and totals. Columns may represent periods (month to date, year to date), budget figures, or organizational units (where each column represents a department or location). Columns may also be used for calculations, such as a budget variance column that subtracts the actual and the budget values in the preceding columns. More sophisticated financial reporting tools offer spreadsheet-like capabilities, allowing the user to define calculations at specific row and column intersections.

The financial reporting tool may offer dimensions representing the organizational structure, where the chart of accounts is segmented into departments, locations, or other organizational units—giving the appearance of a hierarchical structure that may be called a *reporting tree*. Individual branches or subtotals of this tree may be used to drive columns, or separate pages for each branch (or combination of branches) of the tree. This flexible design capability is especially useful in an organization with a complex reporting

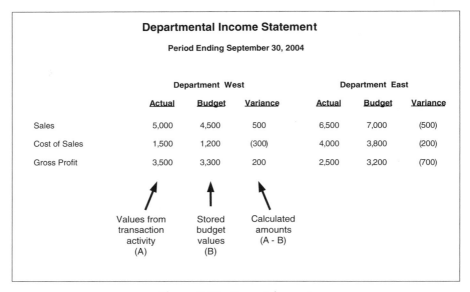

Figure 8-07. Financial report

structure, or where there are frequent reorganizations or shifting of financial responsibilities.

Drilldown Reports

A *drilldown* report allows the user to double-click on a value within an on-screen report, and the software automatically displays a subreport with the next level of detail. Drilldown reports may be launched from banded or financial reports and may even cross from a financial to a banded report, encompassing several layers of transactions within the database. For example, the user may drill down on the total sales figure on the income statement, which then presents a list of all the sales values that comprise the total. Drilling further may take the user deeper into the supporting details within the database, displaying the invoice and payment details behind a particular sales transaction and ultimately leading back to the original production work order, shipping documents, and sales order from which the invoice originated.

Drilldown reports are a logical result of a relational database design, and in many cases eliminate the need for digging through file cabinets for source documentation, quickly answering the universal time-wasting questions: "Where did this come from?" and "Why did this result occur?"

Pivot Tables and Data Cubes

Imagine a report designed like a Rubik's Cube™, where the user can pivot the data and view the results from a variety of *dimensions*. These dimensions may

Name	Invoice#	Price	Cost	Salesperson	Territory	Item
Smith	123	100	50	Jones	West	Chairs
Franklin	124	95	50	Baker	East	Tables
Martin	125	110	65	Jones	South	Chairs
Johnson	126	105	55	Baker	West	Chairs
Michaels	127	50	35	Yates	East	Cushions
Johnson	128	125	75	Jones	South	Chairs
Edwards	129	65	30	Jones	West	Cushions
Martin	130	145	80	Yates	East	Tables

Figure 8-08. Source data for a pivot table

Dimensions
Rows
Columns

Sum of Price		Item			
Territory	Salesperson	Chairs	Cushions	Tables	Grand Total
East	Baker			95	95
	Yates		50	145	195
East Total			50	240	290
South	Jones	235			235
South Total		235			235
West	Baker	105			105
	Jones	100	65		165
West Total		205	65		270
Grand Total		440	115	240	795

Figure 8-09. Pivot table

include customers, customer types, territories, salespeople, product types, date ranges, and so on. When analyzing a large volume of transactional data by pivoting dimensions, and presenting the results in a tabular or graphical format, hidden relationships and trends may show themselves.

Microsoft Excel offers a *pivot table wizard*, which guides the user through the development of a data cube that is then manipulated directly within a spreadsheet. Anyone that is not yet familiar with this powerful capability of Excel should invest an hour (or less) in learning it, so here is a quick lesson.

Start by creating several rows of fictitious data, like that shown in Figure 8-08. The first row of the table must contain the field names, because these are used to automatically name the dimensions. Highlight this range of data and from the menu bar and choose *Data, Pivot Table*—and let the wizard walk you through the rest of the process. Excel will create a pivot grid with empty columns, rows, and a body—you then click and drag each dimension into one of these areas, and you now have a pivot table to play with, as illustrated in Figure 8-09. As the user clicks and drags the dimensions around, the pivot table automatically recalculates. As you can see, a pivot table is a very simple yet powerful analytical tool that takes only minutes to learn and create.

In this simple example we created the pivot table using data contained within the spreadsheet itself. Using a query, Microsoft Excel can also point to an external database (such as ERP) for the source of the information to be used in the pivot table. Each time a pivot table is rearranged, the values at each intersection are dynamically recalculated. There is a practical limit to how much data can be handled in this way, depending on the number of records and dimensions used. A modest volume of sales history may contain thousands of transactions, using a handful of dimensions, and the total of all possible intersecting values in the pivot table can easily grow into the millions, far more than can be calculated dynamically each time the user pivots the table.

A *data cube* (also known as an OLAP cube, which stands for Online Analytical Processing) precalculates all possible intersections of these values periodically, storing them in a special multidimensional matrix database format. The slicing, dicing, and pivoting process is then just a matter of selecting the right values to display, because the burdensome calculations have already been done. Once the precalculated data exist in a cube format, Microsoft Excel may be used to view the data. Alternatively, more sophisticated reporting, visualization, and analysis tools may be used* to view and manipulate the data in a variety of tabular and graphical formats.

Data Warehousing and Datamarts

The moment we create a separate repository to store data for analysis, whether in a cube or transactional record format, we have created a data warehouse or datamart. A data warehouse stores a large volume of general data for later analysis with a variety of reporting tools (Fig. 8-10), whereas a datamart stores a more limited, topical set of data for a specific purpose, or for use by a specific department or workgroup. In either case, they maintain a store of data separate from the online transactional data source.

Offline data repositories are useful for a variety of reasons:

- Large queries and calculations against the live transactional database may create security, performance, and communication bandwidth challenges.
- The system may need to precalculate values (such as a data cube) and store them for rapid presentation and analysis.
- The system may need to combine data from multiple sources into a single database for various reasons:
  - The data sources are updated on different time schedules and must be synchronized.
  - The data sources may have limited availability or communications bandwidth.

* Available from Cognos, Business Objects, Hyperion, Information Builders, Qliktech, and other vendors.

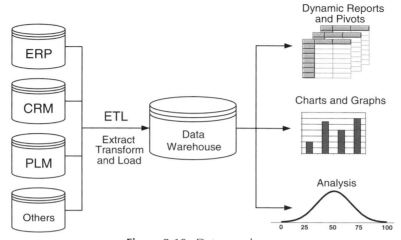

Figure 8-10. Data warehouse

- Some data sources may have stringent security controls.
- There may be considerable mapping, transformation, combination, validation, and summarization of the data from multiple sources into a single set of information.

Event Notification

With a powerful transactional system like ERP, the database can automatically issue alerts of exceptional events and conditions that deserve immediate attention. This is in stark contrast to common reporting practices where individuals waste valuable time and energy periodically scanning reports, searching for problems, while knowing that the time when these problems could have been prevented or mitigated is long past. Event notifications may be related to any condition or event within the database, such as a salesperson exceeding his quota, an unpaid invoice becoming overdue, or an A-level inventory item falling below the safety stock. These conditions and events must be clearly defined so a programmer can develop a *query* or *database trigger** that initiates some form of automated notification.

An event notification can execute in many ways—sending an e-mail or pager message, printing an exception report, changing default values in the system (i.e., placing a customer on credit hold), or launching a software program (i.e., running MRP to review material requirements of a particular

---

* A query is a question posed to the database that is run periodically and can launch an event when a predefined condition is found. A trigger is logic that is hard-wired into the database, firing the instant a prespecified condition occurs; triggers are instantaneous but create more processing overhead than queries.

item where there is a suspected shortage). When defining event notifications it is important to clearly understand the relationship between the business process and the underlying flow of information that represents that process; the end users and the data analysts must share ideas, to direct the system to react appropriately.

When experimenting with event notifications, a company should start slowly by selecting a few critical thresholds and exceptions to manage. Nothing will frustrate a good event notification system more quickly than when users find dozens of repetitive or otherwise unnecessary messages appearing in their inbox each day.

Executive Information Systems

An Executive Information System (EIS) is designed to portray Key Performance Indicators (KPIs) in a graphical format and may incorporate all of the business intelligence capabilities described so far, composed in a *dashboard* view (Fig. 8-11). The goal of an EIS is to present a holistic and intuitive view of enterprise performance across the many levels of the functional and team-based organization. An EIS should quickly call attention to exceptions, pro-

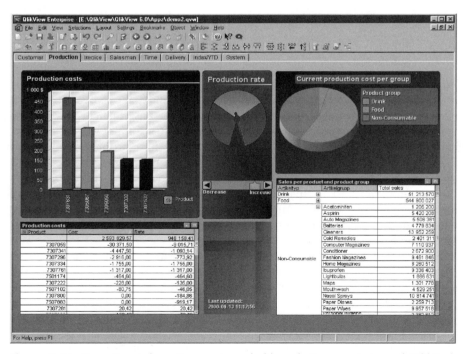

Figure 8-11. Executive information system dashboard. By permission of QlikTech International[159]

viding a mechanism for the individual to visualize a complex situation, then drill down from high-level KPIs into the transaction details when appropriate.

From an organizational point of view, the interrelated perspectives presented by a comprehensive EIS system can link strategic goals, tactical objectives, and team-based initiatives into a sound management framework. The KPIs may also be presented in a *scorecard* fashion, where various aspects of business performance are organized into particular contexts such as customer service or Lean performance. Scorecards may be designed to support a particular performance management methodology, such as the Lean Scorecard shown in Figure 8-12; we'll explore the Balanced Scorecard approach in Chapter 10.

EIS systems have long been the holy grail of IT, and in the early years many large enterprises spent millions of dollars developing their own proprietary EIS reporting systems, cobbling together a variety of back-end data sources. Contemporary EIS tools are now standardized, and many ERP applications offer preconfigured EIS systems. Even with these standardized tools, however, EIS systems can still be quite complicated to design and build, because the preconfigured views and reports seldom meet all the needs of a particular enterprise.

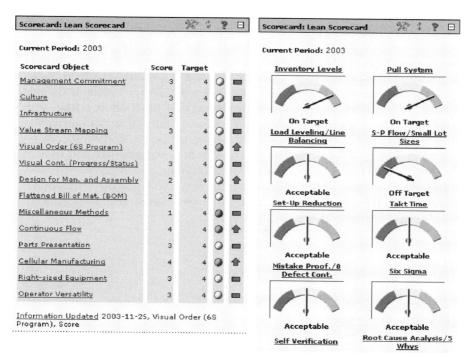

Figure 8-12. Lean performance scorecard. By permission of IFS North America[160]

A Business Intelligence Success Story

WARN Industries (www.warn.com) is a $150 million+ manufacturer of automotive off-road products sold to consumer markets and built into vehicles through OEM relationships with DaimlerChrysler, Ford, General Motors, Nissan, Toyota, and other automakers. WARN has been pursuing Lean Manufacturing initiatives for over a decade. WARN completed their ERP implementation in two years, with a total project budget of $3 million, replacing a variety of legacy systems. The business functions integrated into the new ERP system included order entry, EDI, planning, scheduling, manufacturing, shop floor data capture, advanced warehousing, logistics, and finance.

Although the ERP implementation was considered a success, providing a powerful transaction engine to run the business, it did not provide a satisfactory data analysis tool. According to Travis Pierce, IT Manager for WARN, "The ERP front end could give you what you wanted, but the time it took to realize those results was incredibly painful. For example, if we wanted to identify our leading customer for a specific part number in a specific state for the prior calendar year, it would take days or even weeks to have the report written and posted to the system for use."

WARN decided to implement a third-party Business Intelligence system and hired our firm to assist. We conducted interviews with key personnel in all departments to gather requirements, and to identify members for a cross-functional selection and implementation team. The team identified four vendor candidates; the final decision was based on a combination of cost, functionality, and ease of use. With the ERP system already in place, the team quickly began work on the BI implementation, completing the first phase in less than two weeks, delivering useful dashboards within four weeks. WARN employees are now able to make tactical decisions on a daily basis instead of having to wait weeks for new reports to be developed.

According to Pierce, "Today we can produce valuable ad hoc data in seconds, and a nice side effect is a significant reduction in our license fees. We had many users in the ERP system that required logons strictly for data inquiry. We have since moved those users into the BI system, freeing up expensive ERP licenses."

How to Begin with Business Intelligence

Considerable effort and investment are required to tailor a comprehensive EIS system to the goals and objectives of a particular enterprise. As small and medium-sized companies get started with an integrated information system, however, many would be happy just with timely and accurate reporting on their business, and they often struggle with developing the in-house talent to make use of the basic reporting capabilities included in their new system. Often our clients will ask: "How much skill and time is required to learn how to write reports, and who in our organization should we dedicate to the task?" With a few classes and a little practice, most people can learn the basics of a report writer. Given a predefined library of reports and queries, and a little practice, an individual can learn to create her own reports, formatting them as she likes, and running them whenever she wishes—this can be very empowering for someone who was previously starved for information.

Although it may be simple to format reports with the latest tools, creating the queries that gather the data underlying these reports is another matter entirely. If a report requires a set of data that does not currently exist in the report or query library, someone must identify the source of the information within the database, map it to the business process, and develop a query that returns the appropriate data to the report writer. Often the query must be parameter driven, asking the user to specify ranges of dates, customer numbers, and other relevant criteria, then sorting and grouping it properly each time the report is generated. Even aided by graphical query building tools this requires expertise that most users do not have. In fact, it would be counterproductive and potentially risky (from a data integrity and security point of view) to have too many people working directly within the database. Furthermore, some enterprise systems may contain thousands of tables, each with cryptic names such as "A100015", and with extremely complex and dynamic relationships among them. And, frankly, some databases are just designed more comprehensibly than others. In some cases it would seem that the developers competed with one another to create intricate data relationships and obtuse table and column names. Queries that cross multiple databases usually require additional skill, because mapping of key relationships and programmatic transformation of the detailed data are often required to achieve useful results. The development of a separate data repository, whether a data warehouse or datamart, is the domain of true software development, and the design and construction of even a small data warehouse can easily cost five or six figures (in US$).

Within a small or medium-sized company, at least one individual "power user" should receive extensive training in the database structure, query, and reporting tools, to develop and maintain a library of queries, and to support end user reporting requests. In a small company this is often the same individual who becomes the internal training and application support coordinator. It is important to make this a formal role, allocating sufficient time and focus if satisfactory results are desired. In larger organizations it is customary for this to be a dedicated role, comprised of one or more individuals belonging to the application support team.

Most applications come with a library of existing queries and reports that serve as design templates to create new reports. Many applications also offer database *views*: These are predefined queries that combine commonly used tables, fields, and filters, producing useful results with relatively little effort.

Finally, many software companies publish a *data dictionary* that describes in detail the content and logic of the database and Entity Relationship Diagrams (ERD) that display the table relationships and process flows within the database. Data dictionaries, ERDs, and views can demystify the database structure, expediting query and report development.

As an individual develops the skills to either create queries and reports for the users, or to design the queries and coach the users in developing their own reports, it is important to encourage discipline whenever a new report is desired. When a perplexing business issue is encountered, the first reaction is often to gather large quantities of data in an undisciplined manner, hoping a solution will magically appear. As a result, an organization may burden itself with the production of countless reports that are never eliminated, when in fact the information does not add value. This reckless reporting creates administrative waste and does not lead to better decision-making.

The individual responsible for supporting the end user reporting and analytical processes is the first line of defense against wasteful reporting practices. In our experience we have found it useful, when approached by a user asking for a new report, to take two precautionary steps. The first is to apply the Lean *five whys* technique, asking Why? as many times as it takes, going beyond symptoms and assumptions to root causes (we'll discuss five whys in Chapter 10).

If the request for a new report is appropriate, the second precautionary step is to ask the individual to design the desired report. This design must be very specific, and the company may consider a report request document containing a checklist of questions such as:

- What is the purpose of this report?
- Who will use it and when?
- What security restrictions are necessary?
- How is the report physically organized?
- What fields of information must appear on this report?
- What ranges of field values (dates, document numbers, etc.) should be selected and restricted each time the report is run?
- How is the information sorted, grouped, subtotaled, and totaled?
- What line and page breaks are needed?

Once the report request is approved, it becomes the *design specification*. There are several benefits to this approach. First, the user is required to think critically in terms of the data and process, defining exactly what is needed, and what value the information provides. Second, it discourages spurious report requests, because the individual must invest time and energy, rather than just tossing a fuzzy report request over the wall to the IT department. Finally, with a report specification in hand the developer may be able to identify an existing query or report that may be used as a basis for the new one, reducing work, increasing component reuse, and encouraging the development of a query and report library. As the inevitable changes to business processes and system design occur over time, by maintaining a central library of queries and reports, change management of all end user reports is simplified.

In our experience, many small and medium-sized companies treat query development and report writing as an afterthought of a system implementation, asking someone with little training and no formal allocation of time to accept the burden. Once an organization commits to the investment in an enterprise software system, it is necessary to invest the proper resources in reporting to produce satisfactory results.

Enterprise Portals

Now it's time to bring all of this structured and unstructured knowledge together, making it available to the right individual or team, at the right time, and in the right format, to help identify and solve problems, serve customers, and continuously improve. Enterprise portals have become an indispensable tool for many organizations; a portal is a doorway, a single user interface through which all knowledge and activities within an organization may be accessed.

An enterprise may now deliver information to anyone through a portal, as long as they have access to the Internet. A portal can be a central control panel and navigation system with access to commonly used applications and information sources. An individual may have the option to configure the home page of his portal to support how he works. Furthermore, a portal environment may provide the user or team with a library of hundreds of supplemental portal window frames, and a user may construct several portal windows to support different roles and activities. For example, there may be a separate portal window specifically designed to support the collaborative Sales and Operations Planning process, and for measuring daily KPIs on the shop floor, or for managing projects as shown in Figure 8-13. All of these portals may be accessed as links through the primary portal, which may then become the user's primary desktop and menu interface.

The possibilities for arranging portals to support an individual or collaborative team's workflows are practically endless. Along with this power come the obvious questions: *Who needs access to what information, in what format, when, and why?* At the highest level, we may categorize portal users into three constituencies: public information consumers, business partners, and employees, which are described below.

- **Public: The Internet**—the public domain of marketing, communications, and customer self-service:
  - Many companies have learned that it pays to empower the customer, providing her with an abundance of automated product information and customer service. By providing self-service capabilities, they have

Figure 8-13. Project member portal example. By permission of IFS North America[161]

found that a content-rich site not only attracts customers, but can also significantly reduce transaction costs.

- Limited and generalized public content is published here. There is no access security.
- Personalization may identify the user by name, role, geography, language, preferences entered by the user, or by observing the user's search and browsing patterns, altering the presentation and content appropriately.
- Transactions may be performed on the site, with public offerings that require limited pricing and configuration, fulfillment, shipping, and credit card payment (i.e., www.amazon.com).
- Company contact information including address, e-mail, phone, and fax numbers may be published here. Some provide access to a customer self-service knowledge base, whereas others offer e-mail or online chat with customer service representatives and online service parts ordering.
- A secure link may be created from this site to an Extranet site for partner access.
- **Business Partners: Extranet**—the domain of secure partner communications:

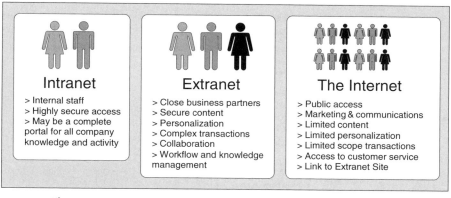

Figure 8-14. Comparison of Intranet, Extranet, and Internet portals

- Content is secure and requires a login.
- The site is highly personalized, with specific content for each user, including confidential documents, pricing, transaction history, order information, inventory, delivery, and account status, design, engineering, and configuration information.
- The site may automate complex transactions including quotations, pricing and availability lookups, interactive product configuration, sales orders, and returns.
- The site may enable sophisticated collaboration, communication, advanced search capabilities, and participation in knowledge capture and publication.
- **Employees: Intranet**—the domain of highly private and secure company communications:
  - Internal staff and highly privileged guests are allowed.
  - All secure company knowledge, activity, and communications may be delivered through this portal.

The architecture of the systems, storage, and security to provide such a multilayered environment may be highly complex; a conceptual overview of its design is illustrated in Figure 8-15. Observe in this diagram that the Intranet includes all private systems inside the primary firewall*, like a castle keep guarded by its internal defenses.

The structured and unstructured data are tightly secured, while a replicated set of less-sensitive data is often published outside the firewall onto the public Internet web server. Arms-length replication of metadata into a separate

---

* A firewall is a software and/or hardware device that restricts access from the outside world. There may be several layers of firewall security; in this narrative *primary firewall* describes the innermost secure layer.

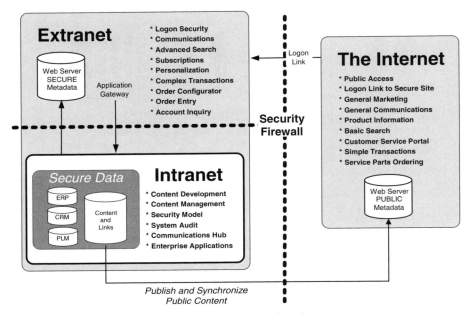

Figure 8-15. Enterprise portal architecture

repository outside the firewall ensures that under no circumstances may an intruder gain access to the internal data stores. This approach requires the internal systems to update information to this secondary metadata storage on a scheduled basis, or as documents are approved for publication.

Visitors to the Extranet site may be authorized to access the innermost systems, and the data to support these secure activities may exist either inside or outside the primary firewall. For example, a layer of metadata may be stored on a web server outside the primary firewall to support external transaction processing; these data then periodically update the enterprise applications within the primary firewall. Alternately, an *application gateway interface* may be provided on the Extranet site, communicating directly with the enterprise applications that reside within the firewall with a Web-enabled application user interface.

With so many secure enterprise systems, users are often overwhelmed with passwords. System administrators require each password to be changed regularly, while preventing the use of easy to remember passwords such as birthdays and children's names. As users collect a large number of passwords that they cannot possibly remember, they tend to store them in unsecured locations such as files on their hard drive (containing the word "password" so they can be easily found by an intruder), poorly hidden in drawers, and on laptops and PDAs. This natural yet irresponsible practice jeopardizes the entire security system.

One of the benefits of an enterprise portal is the potential for a consolidated security layer that provides each user with a single password, called *Single Sign-On*. Although this may be technically complex from a system design and security maintenance point of view, it should be transparent to the users. Once they are authenticated upon entering the portal environment, access rights accompany the users throughout the portal session, regardless of how many separate applications and interfaces they may access.

Most importantly, the underlying architecture of the portal should be completely transparent—once an individual logs on and identifies himself, the environment should be personalized so the user is provided with a friendly interface and access to three vital services: *Communication and Collaboration*, *Content Management*, and *Application Gateways*.

Communication and Collaboration. A portal's primary display area may provide a friendly welcome screen, displaying messages, links, and communications from various sources, as illustrated in Figure 8-16, including:

- A viewing area for e-mail, calendars, and task lists
- A posting area for general notices, similar to a bulletin board
- Exception notifications generated by enterprise applications

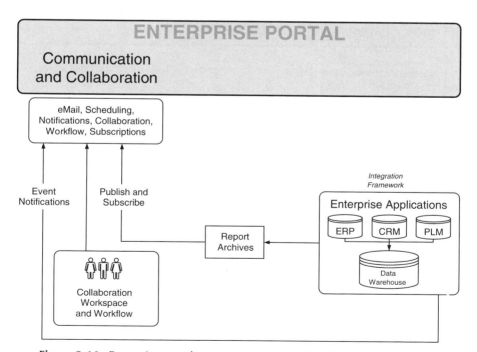

Figure 8-16. Enterprise portal—communication and collaboration interface

- Publication notifications of documents that are subscribed by the user
- Documents or transactions routed to the user by a workflow approval system
- Various collaboration workgroup, content sharing, and discussion group areas organized by task or project, with tasks and information requests organized by due date and priority
- A page for the user to create and arrange information according to personal preferences and workflows

Content Management. The second important function of the portal is to provide search services to locate *unstructured* data within the enterprise (Fig. 8-17). This search may also be extended to content outside the enterprise, including search of partners' secure sites with the appropriate security permissions automatically applied and public Internet search services such as Google and Yahoo. Note that the output from enterprise applications (structured data) becomes unstructured data the moment a static report is saved to permanent storage.

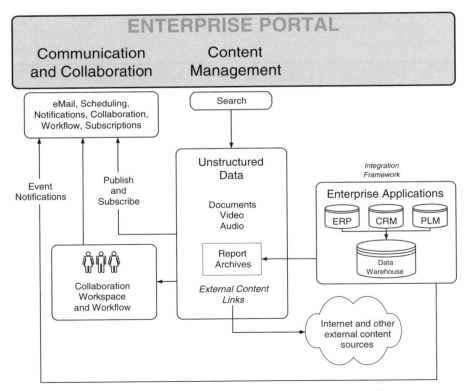

Figure 8-17. Enterprise portal—content management interface

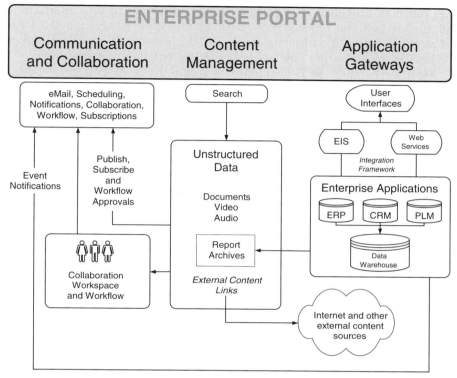

Figure 8-18. Enterprise portal—application gateways interface

Application Gateways. The third important function of the portal is to provide unified access to the *structured* data of the enterprise, through user interfaces to multiple applications and presented through various EIS and business intelligence tools (Fig. 8-18). This allows the user to sign on to the portal once, and then *navigate* through a variety of integrated and nonintegrated applications and EIS pages without having to negotiate a jumble of screens and application boundaries. In this regard a portal becomes a fluid menu system and control panel for each user. Although a completely seamless and dynamic interface may never be practical for many organizations, a well-designed portal can help to automate and simplify many burdensome tasks.

An enterprise portal can be a friendly and empowering tool. With a portal each user may cut through information anarchy, organizing disparate sources of information into relevant knowledge to make better decisions and serve the customer. Despite the relatively low cost and standardized tools available, the construction of a comprehensive portal is not a simple task. A small or medium-sized company with limited IT budget and staff should seek a reasonable balance of cost and benefit, prioritizing and simplifying access to key functions instead of attempting to deliver universal one-click interfaces to all applications and reports. Nevertheless, these tools have reached a stage of maturity where, if a business can justify the investment, such information technology scenarios are now not only possible but within practical reach.

Most importantly, to deliver optimal results from an enterprise portal, information flows must align with value streams. An enterprise may attempt to build elaborate interfaces, search engines, decision support systems, and portals to disguise the structural flaws within their content management, transactional systems, and business processes. Although these efforts may simplify user navigation and eliminate some administrative waste, they only perpetuate underlying structural wastes that should be continuously improved.

Part Three

Managing Change with IT

Chapter 9

The Event-Driven Lean Enterprise

In business, excess information must be suppressed. Toyota suppresses it by letting the products being produced carry the information.

Taiichi Ohno, *Toyota Production System*[162]

Material and information flow are two sides of the same coin. You must map both of them.

Mike Rother and John Shook, *Learning to See*[163]

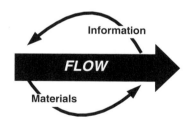

AUTOMATED DATA CAPTURE

The output of a production operation is only as good as the quality of the inputs; likewise, an information system is only as good as the input it receives from the outside world. Data capture describes the countless methods by which this input may be obtained, and we begin with three important questions to ask about data capture in any manufacturing environment:

1. How much data should we capture?
2. How do we capture this data?
3. Why should we capture this data?

Let's begin with the first question: How much data should we capture, particularly from the shop floor? Our answer: *As little as necessary.* If data does not add value then capture is wasteful. As you will learn, however, the question of data adding value may depend upon your perspective.

The dilemma of data capture is like the Heisenberg uncertainty principle of high-energy physics, which states that one cannot measure both the velocity and location of a particle, because doing so affects the state of the particle itself. Likewise, one cannot measure a process without affecting it.[164] Data capture not only consumes the time of shop floor staff, potentially reducing throughput, but it may cause confusion. What we measure communicates what is considered important, and by measuring useless or counterproductive activities we send the wrong signals; inappropriate measurements cause poor performance and misdirected workers.

From the shop floor perspective, data is captured to control and improve the process. From the financial and corporate governance perspective, the concerns of shareholders and management, data is captured to measure and audit the process to ensure that the company's financial objectives are met and resources are used wisely—another form of control. As an organization moves toward Lean it expects to gain considerable performance benefits, but it must learn to think differently about the nature of control. According to Brian Maskell and Bruce Baggaley, authors of *Practical Lean Accounting*:

> As we move to Lean Manufacturing the burden of data collection becomes worse. If we make smaller batches we have more work orders, which leads to more tracking, more labor reporting, more machine time reporting, and more waste. Many organizations "perfume the pig" by automating these transactions, but they are merely automating waste.
>
> So a Lean company needs to manage and control Lean in a way other than by creating paper or computer transactions every time material is moved or altered during the production process. That is why we say that transactions are to Lean Accounting as inventory is to Lean Manufacturing. Transactions are pure waste and for the most part are in place to bring control into a manufacturing process that is out of control.[165]

As Lean processes are simplified they become easier to control, and ideally self-controlling. Pull signals regulate the release of materials; less WIP in the plant means faster throughput and less need to count and track inventory. Automated displays (andons) on machines visually signal problems, work-center greaseboards and other visual tools provide instant feedback on takt time, throughput, and other vital statistics to help keep the process flowing smoothly. Mistake proofing (poka-yoke) is built into each task to ensure near-flawless performance, and to quickly identify and correct a problem when it surfaces. Operators may be empowered to halt production if a serious problem appears, launching an instant Kaizen Strike to assess and eliminate the root cause. These and many other physical and visual controls may be built into Lean Manufacturing processes, reducing the need for data capture *for control purposes*.

Although Lean simplifies the processes, the number of distinct transaction events can multiply exponentially, obstructing a Lean initiative. Data capture in a discontinuous operation can be particularly challenging with variable operations and routings. As a job shop moves toward smaller transfer batch sizes, the volume of individual transactions moving through the shop floor begins to grow. Orphaned parts belonging to large jobs spread across the entire plant; escorting each unit through the plant with routing information, work instructions, and job cost data capture forms would create a mountain of paperwork. However, judicious placement of data capture and control points is important in a discontinuous operation, because by their nature these processes will *never* be in control to the degree of a repetitive environment. Certainly a job shop can make layout and cellular improvements in many areas, but these improvements may create a substantial increase in transactions unless an earnest effort is made to eliminate them. Herein lies a warning for job shops making a half-hearted effort at Lean improvement.

Why Collect Data?

Several reasons may be cited to explain why ill-advised data capture projects are launched:

- Customer-mandated automation requests that are not accompanied by internal process improvement efforts
- Naive assumptions of simplicity, accompanied by a lack of discipline needed to identify exactly how the process will be automated
- Vague assumptions of costs and benefits
- Attempts to overcome the deficiencies of the core ERP system with ill-conceived data capture activities, which may simply add yet another layer of complexity and waste
- The assumption that with increased data collection, more information will be available, which automatically makes for better decision-making
- Envisioning automated data capture as a point solution, rather than asking how it contributes to improvement of the overall value stream

Later in this chapter we will explore Radio Frequency Identification (RFID), an emerging technology that will greatly expand the possibilities for data capture, while at the same time inviting opportunities for misguided use,

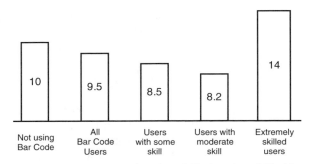

Figure 9-01. Inventory turns correlated to skilled and unskilled bar code use

unleashing a deluge of data. Initially though, we'll focus on bar coding, because we can draw conclusions based on many years of practical experience.

The National Association of Manufacturers sponsored a study that yielded very interesting results (shown in Fig. 9-01) by measuring the correlation of bar code use with inventory turns.[166]

The survey indicates that companies *not* using bar code reported better inventory turns than the *average of all* companies using bar code. The real wisdom of this study emerges when you differentiate bar code users by *some skill, moderate skill,* and *extremely skilled*. Observe that the inventory turns achieved by bar code users with some skill is less than the average and by users reporting moderate skill *is even less*! Not surprisingly however, the inventory turns of extremely skilled users are remarkably high.

You may draw your own conclusions from this research, but here are mine: Many companies assume they know what they're doing with automated data capture when actually they do not. Those who are less humble, who hold to the belief that they are more skilled than they actually are, can make the situation far worse than better. We have seen this scenario played out many times in the field.

According to the study:

> The use of bar codes enables the implementation of technologies such as JIT (Just In Time), SQC (Statistical Quality Control), CIM (Computer Integrated Manufacture), automated inspection, CAD (Computer Aided Design), CAM (Computer Aided Manufacturing) and many hard and soft technologies. Thus, if bar codes are not used with extreme skill, the use of several other technologies may suffer and inventory turns may not improve.

This study presents a compelling argument that automated data capture must be considered not as a point solution, but as an enabler for the continuous improvement of the overall value stream.

Feedback is essential. The "C" in PDCA and the "MA" in Six Sigma DMAIC (measure and analyze) are an important foundation of any continuous improvement effort. Lean performance *management* requires performance *measurement* based on relevant information. Note here the emphasis on *information* and not *data*. Data is the raw input to a decision process; raw data only adds value if it can be transformed into the right information to answer the specific question being asked at the time. Although this seems like an obvious notion, it implies that we must capture a vast breadth of data, so that we are prepared to answer all the questions we are likely to ask in the future; just-in-case data capture seems terribly wasteful.

On a Lean shop floor, information should be as simple, visual, and relevant as possible to facilitate real-time decision-making as work flows through the plant. But do other legitimate reasons exist to justify more invasive and thorough data capture methods? Accountants may insist on capturing detailed resource and material consumption at every step along the routing so they can manage financial reporting, costing, profitability and pricing analysis, make versus buy, and plant and equipment investment decisions. Material managers may wish to measure inventory quantity, movement, and consumption at several points on the shop floor to manage inventory levels and reduce stock-outs. Plant engineers may wish to monitor key processes for consistency, efficiency, throughput, preventive maintenance, and quality factors. But production managers may wish to *backflush* all material and labor activity in a single transaction when the job is complete, eliminating all data capture on the shop floor because it distracts workers and inhibits throughput—recall the comments of the production manager from the story in Chapter 1: "Just keep that @#$% ERP system away from my Lean shop floor!"

These arguments are often never settled conclusively because the various constituencies may harbor conflicting motives for capturing data. Although some of these motives may lose their potency as people learn to apply the principles of Lean Manufacturing to their decision-making processes, the underlying attitudes and policies may be deeply entrenched and won't vanish overnight.

Four Reasons for Automated Data Capture

To develop an effective and *interactive* information system we must understand how the system interfaces to users, events, and the outside world. We must design the appropriate interface to suit each need, working closely with the users to ensure that the system supports the process, not what the designers think they know about the process. We must identify what conditions create exceptions and then design the appropriate trigger and signal mechanism. Finally, the system must be accepted and used, rather than rejected or tolerated as a nuisance. To implement a useful system we must therefore understand the four fundamental reasons for data capture: Process Automation, Process Control, Performance Management, and Compliance.

Data Capture Reason #1: Process Automation. Process automation is an industrial engineering discipline supported by Supervisory Control and Data Acquisition (SCADA) software; examples include controllers for oven temperature, valves, scales, and countless other machine-level interfaces that automate or regulate a physical operation. Although these systems are under the watchful eye of technical specialists, they may provide useful information for managing production, especially when combined with data from the ERP system.

Detailed exploration of industrial engineering and process automation is beyond the scope of this book; however, it is important to note that although process automation's primary purpose is to improve engineering performance, detailed process data is a natural by-product, captured at little or no additional cost. It is also important to remember that process automation investments such as material handling, processing, and packaging lines are more prevalent in repetitive operations characterized by greater product/process standardization and volume. Discontinuous environments are more variable, and so generally require more human interaction and may be less suited to hands-off automation.

Data Capture Reason #2: Process Control. Process control is different from process automation. Although no automated supervisory control of equipment may be in place, *measurement* of process performance and quality can still occur, and that feedback may be used to:

1. Control the process in real time with visual or auditory feedback.
2. Evaluate the causal relationships to improve process design and quality over longer periods of time with data-based problem-solving tools.

In a visual plant the first stage of process control involves educated and alert workers and teams observing the process, empowered to quickly resolve problems as they arise. More complex problem-solving and process control challenges may require an empirical and rigorous approach such as Six Sigma, which relies heavily on gathering and analyzing process data. It is here we find many questions to ask, to determine whether data capture is required to support longer-term continuous improvement efforts:

- How much detail is needed to support a continuous improvement process, particularly if it involves Six Sigma analytics?
- Must we capture and store data if it is not currently needed but we suspect a need in the future, or can data capture be performed on a temporary or random sampling basis only when it is needed?
- What is the source of the information? Must it be captured by the operators on the shop floor? Can automated hands-off capture be employed? Is the data already available through machine interfaces? Can the super-

visors capture summary data with wireless handheld devices during their walkthroughs?

Of course, the answers to these general questions depend entirely on the nature of your environment. A paradox of continuous improvement is that with maturity your new challenges may become increasingly complex and subtle, often requiring more rigorous analysis using large volumes of data. The important point here is that a need may always exist for the capture and accumulation of detailed historical transaction data beyond the requirements of daily shop floor activities. The complete elimination of data capture that does not *immediately* add value to the process may therefore not be realistic in a Lean environment, but we can certainly strive to make data capture less invasive.

Data Capture Reason #3: Performance Management. When we use data for continuous improvement we're pursuing future-state improvements based on our assessment of the current state using feedback from testing new ideas. The feedback that guides our decisions results from measurement, and it's true that "you get what you measure." Proper design of measurement is essential to continuous improvement, just as it is to the scientific method in general. However, many companies confuse cause and effect when designing their measurements, holding individuals accountable to results without a clear understanding of the causes. According to Steve Geary and Kate Vitasek in their article "Cause and Effect":

> There are two types of metrics: process and results. Process metrics describe cause, and results metrics measure effect. By understanding the cause and effect relationship between the underlying process and resulting performance, the practitioner can design a system of process metrics that will yield the desired result.[167]

Result Measures. A result measure examines the output of a process, or of an entire value stream comprised of many interrelated processes. Two common examples of result measures are on-time delivery of customer orders and inventory record accuracy.

If a result measure is within our target range, then we can assume that *some* processes are performing reasonably well. That is not to say there is no room for improvement, or that a particularly successful process is not masking the poor performance of another. When a result measure falls outside our target range, however, it tells us something is wrong, but it will not identify the specific cause. Furthermore, the message arrives after the fact, so we have no opportunity to prevent the problem. We must turn to our process measures for root cause analysis, so that we may prevent the problem from happening again.

Some result measures may be more useful than others, depending on how clearly they help to identify the root cause. For example, an important difference exists between the effectiveness of periodic physical inventory and cycle counting methods to promote inventory record accuracy. A periodic (usually monthly, quarterly, or annual) physical inventory count is a result measure that tells us very little about the source of the problem. In fact, the typical plant-shutdown count-carnival imposes a significant throughput penalty, while often causing more inventory confusion than it solves. Plants that perform monthly inventories often record wild variances in one direction, followed by offsetting swings the following period. These gyrations may continue indefinitely with no explanation but much speculation and finger-pointing. Cycle counting is a far more effective result measure; because it is highly focused and frequent, cycle counting can quickly spot the activities that cause inventory variances so they may be corrected. Over several months an effective cycle counting program should eliminate wide swings in variances, leading to the consistent record accuracy necessary for Lean inventory planning and control. And as you reduce inventory, there is less to control and count. With rigorous cycle counting you may even be able to convince the statistically minded auditors that periodic physical counts are no longer necessary . . . but don't count on it.

Process Measures. A process measure reflects the performance of an event, such as the setup time of a particular machine. Process measures help to identify the source of variation leading to a substandard result measure. Consider the result measure of on-time delivery. Using an Ishikawa *Fishbone* cause and effect diagram as depicted in Figure 9-02, the Kaizen team may define several

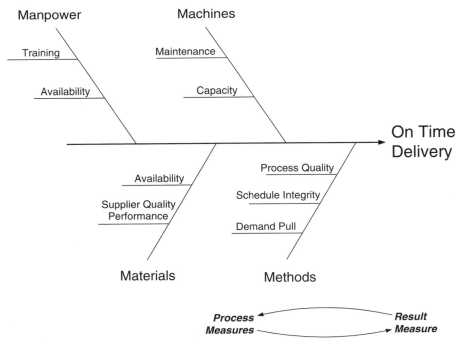

Figure 9-02. Ishikawa "Fishbone" cause and effect diagram

factors that influence on-time delivery. The team then prioritizes these influences, establishing process measures on the most significant probable causes. By proactively monitoring these process measures and adjusting the process based on the feedback received, the result may be controlled.

Data Capture Reason #4: Compliance. Why are periodic physical inventory counts still performed if they're not an effective result measure? Why are traditional cost allocations used to impute factory overhead and other indirect costs when these costs are not practically useful, and can actually be quite harmful to achieving Lean results? Is this information useful for enhancing performance of the organization? No, that's the unfortunate nature of most compliance measurements.

The ideal approach for compliance is to gather all of the data necessary for operations management, finance, and compliance purposes with a limited number of shared data collection points, storing all of the data in a single transactional database. The various consumers of this data should create their own "views" of the data with queries and report formats. Financial views, operational views, and compliance views may originate from the same database of facts. But problems arise when using the right views for the wrong purpose, for example, making operational decisions based on financial or compliance views.

Each nation maintains its own laws and regulations protecting the health, welfare, economic, and security interests of its citizens. Many international alliances and organizations also enforce regulations that affect a global manufacturing and distribution enterprise. With the continuing threat of terrorism, mounting global environmental crises, corporate misbehavior, and cyclical actions balancing free trade against protectionism, these regulatory requirements are surely not going away. Aside from the obvious societal value these regulations produce, they also create NNVA that hinder Lean performance.

One particularly troublesome issue for many companies is product traceability throughout the supply chain. Whether involving rigorous lot traceability for food and pharmaceutical products, or serial component traceability for medical devices and aircraft parts, traceability introduces a significant amount of data capture and manipulation at a *highly granular* level beginning on the shop floor and continuing through the entire value stream, creating an overhead cost that adds little value to the product. Skillful data capture and workflow methods are required to mitigate this cost and disruption.

How to Evaluate a Potential Data Capture Opportunity

Bar coding in particular, and automated data capture in general, are often perceived as a panacea that will solve a variety of problems, such as inventory control. There are many root causes for poor inventory record accuracy—lack of discipline, poor processes, inadequate storage facilities, etc.—that bar coding alone cannot fix. When you evaluate the potential application of any data capture project, you should start by asking a few questions:

- Have we mapped the process?
- Do we clearly understand the root causes of the problems we're trying to solve?
- Should we automate the data capture or simplify/eliminate the task?
- What benefits do we expect from the data capture inputs and outputs?
- Are these localized benefits, or will they affect the performance of the overall value stream?
- How will these benefits be measured and valued?
- Will these benefits add value to the customer?

If the data capture concept passes the preliminary justification test, then it should pass the test of physical practicality requiring us to study the application carefully, asking the following questions:

- Who prints the bar code (or RFID tag)?
- Where and when is the bar code printed?
- How many labels are printed at a time?
- How many labels are needed for each job? One per item? Per container? Per kanban?
- How much operator time is required to print and affix the label?
- How much training is required to prepare workers to use the bar code equipment properly, and to troubleshoot common problems that arise?
- How can we mistake-proof bar coding procedures?
- What preventative maintenance measures will be required to ensure bar code equipment is kept in good order and functioning properly?
- What controls are required to ensure the proper label is affixed?
- What type of label substrate material is the bar code printed on?
- How is the bar code affixed to the product? What is the adhesive? Does it damage the product?
- What environmental conditions does the product experience during the lifetime of the label? Will a readable label survive?
- When is the bar code scanned? How many times during the production and distribution life cycle?
- Does the product change form or configuration during the production process?
- Are lot or serial numbers consumed or created? Are there one-to-many or many-to-many relationships in the lineage? How does the labeling reflect this lineage?

- What physical orientation of the product is needed for the label to be accessible throughout the process without causing movement waste? Will this require unstacking or moving to access the label?

By asking a few simple but disciplined questions data capture proponents are encouraged to think critically and visualize a value-adding solution. In many cases asking these questions will initially lead to confusion or frustration, but don't let this cause you to abandon the idea—this investigation should lead to a thorough and objective evaluation of data capture across the entire value stream, not just as a point solution. This evaluation may begin with the formation of a team, a detailed walkthrough of the process from the end to the beginning (following demand-pull signals), then again from beginning to end, followed by value stream mapping to quantify the flows and constraints. You're assured of discovering value with this approach whether or not you decide to invest in automated data capture.

EVENT MANAGEMENT

We need no reminders that the pace and complexity of global commerce continue to increase. We can point to many recent technological advances in transportation, communications, and computerization as contributing factors to this acceleration. Ironically, we must often rely on information technologies to help us cope with this increasing speed and complexity.

With an overwhelming flow of information washing over it each day, an organization must become skilled at event management. Vivek Ranadive is CEO of TIBCO, a publisher of enterprise integration software; in his book *The Power of Now* he explains the importance of developing an event-driven culture:

> The event-driven company manages by exception, directing the vast majority of the company's human attention to the small minority of out-of-the-ordinary business situations that present both the most risk and the greatest opportunity. Though we seem to be drowning in information today, there will be orders of magnitude more information in play in the networked world [the RFID explosion had not begun when Ranadive wrote this] increasing the business necessity of systems that automate as many processes as possible and filter what is worth our attention from what is routine.

> Event-driven companies [. . .] define themselves as being, above all, customer-centric. They keep their sales and marketing ahead of the competition. They put the best information management tools in their employees' hands. They implement true knowledge management programs to leverage their valuable intellectual capital wisely. They update workers, customers, and partners instantly with crucial business information and events.

Putting the majority of your effort into the minority of tasks that hold the most promise and the most risk [constraint management] is threatening to those who crave the comforts of the familiar, but it is the only way to have a chance of shaking the rust and leading the competition.[168]

So now we understand the strategic significance of event-based management, but *what is an event*? In the real world an event is an activity, a transaction, a process of transformation from one state to another, a relationship of cause and effect. From an information system point of view, an event is a record of one step of the transformation process, stored in a database. At what *granularity* (or *atomicity* in the IT vernacular) must an event be recorded? How much detail is necessary for planning and control?

At one extreme, we might monitor a production process, capturing detail at every step, in increments of minutes and seconds. Every time inventory moves, or a human or a machine performs a task, we could record volumes of data for immediate process control and future process analysis and improvement. We may also capture the state of the product and process at each step, measuring many characteristics such as size, weight, temperature, and color for conformance to specifications of quality and compliance. At the other extreme, we might capture limited data only once at the end of the process, backflushing material and resource consumption at predefined standards.

How much data is enough to manage a process on an event-driven basis? At what point do we create waste? And if we gather this data with a non-invasive technique that does not impact the process, does it create waste? When a tree falls in the forest and there's no one there to hear. . . .

In addition to granularity, how immediate must the flow of data be? We often hear the term *real time*; what does this mean in the context of determining data capture requirements? Is real time measured in days, hours, minutes, seconds, or nanoseconds? The appropriate time horizon for a particular event is quite different if we're discussing a supply chain or semiconductor fabrication. From a business point of view, let's use a practical definition of real time: the period of time necessary to gather data on an event, enter it into an information system, interpret the information, and take preventative or corrective action. According to Lee Hudson, Manager of Manufacturing Information Technology at Becton Dickinson and Company, a $4 billion medical products manufacturer accustomed to rapid change in markets and technologies:

> History for our operating organization is anything that is older than a day. After "the day" it is all history and analytics.[169]

In his article "The Reality of Real-Time Intelligence", Tony Baer suggests a critical distinction in time measurement between the business office and the shop floor:

> When process engineers and operators hear buzzwords like "real-time" business activity monitoring, they probably chuckle. Real time to the front office is

nothing like the split second deterministic response times required for real-time control of industrial processes or machinery. Not surprisingly automation solutions for the front office rarely have penetrated the plant floor.

At best, plant-floor execution systems may send data in one direction to [. . .] ERP, but rarely the other way around. While real-time enterprise concepts always will be far less stringent than the plant floor, the demand for real-time agility is driving a convergence with the world of corporate office solutions. What's driving [this] is the necessity to better service customers in a much tougher competitive environment.[170]

When a response is required from a human being, the reaction time is based on her ability to process the information, and its relevance and priority at the time. An effective data-based decision support process relies on the system to filter the information so that only critical exceptions are communicated according to their priority, presenting the information in a format that is most helpful at that moment.

Real-time event-driven management in a global economy requires a skillfully designed IT infrastructure. Adam Bartkowski, CEO of Apriso Corporation, a provider of supply chain integration solutions, recommends a "*bottom-out*" execution strategy:

The action is not on the 35th floor, where the business is ostensibly being run, but down on the production floor where things are actually being made. [The success of enterprise performance management] depends upon the event-driven, real-time dispatching of raw or processed information outwards. Any of these destinations could be nearby—or halfway around the globe. And at a destination node, a complementary means is required to analyze such information on the spot to make a quality or process oriented decision, archive it for regulatory or traceability reasons, or to aggregate or abstract it, perhaps instantly, for use by higher level management or control processes. In the bottom-out paradigm, any business or supply/manufacturing process—however granular—can be integrated with a relevant software application, and then— through the Internet—be made to influence or be influenced by any other process, anywhere in the world, for any reason. Until the Internet—one of the major drivers of the execution economy—such functionality would have been impossible.[171]

The boundaries of time and geography between the plant floor, the front office, the global supply chain, and the customer are becoming indistinct. With the rapidly growing sophistication of the Internet, relational database and software development tools, and Web Services, this creates a broad scope of opportunity for the application of information technology to manage the flow of materials and information. Dave Caruso, VP and Director of Research for AMR Research, suggests that a shift in perspective of data management is under way that will alter the practical boundaries and economics of event management:

The globalization of markets is accelerating the need for supply chain visibility and standardized processes. Likewise, new technologies like RFID could change our notion of transactions [. . .].[172]

The Impact of RFID

Radio Frequency Identification (RFID) technology may cause the most fundamental shift in supply chain data management since the widespread adoption of EDI in the 1980s. Although RFID technologies have been applied to internal automation applications for years, we are now witnessing the early stages of an explosion in the application of RFID technologies across entire supply chains. The initial drive behind RFID is Wal-Mart's North American Pallet, Carton, and Case Initiative (NAPCCI), a mandate for their top 100 suppliers to implement by January 2005.* As a testament to their leadership, the United States Department of Defense turned to Wal-Mart for assistance in the development of its own RFID supply chain initiative.

Although RFID got its first big push through major retail and Department of Defense mandates, the next phase is beginning in earnest—the staffing and organization of consulting groups targeting vertical markets. Now many other industries are receiving considerable focus, including automotive and transportation, retail, consumer packaged goods, point-of-sale, life sciences and pharmaceuticals, defense, and security.[173] Not surprisingly, the experimentation and adoption of RFID is gathering speed across entire supply chains, where the benefits are expected to be the most pronounced. The current emphasis is on pallet, carton, and case granularity, but many industries may realize considerable benefits once the focus turns to specific item identification and tracking. RFID is like bar coding, because you may capture information simply by pointing a device at the product, but the similarity ends there. Rather than a bar code label, an attached radio frequency tag communicates with a radio signal. The RFID tag may contain a wealth of information on the product, and in particular it may carry the Electronic Product Code (EPC)—a unique number that identifies a specific item anywhere in the global supply chain.

Unlike bar code labels, RF devices do not need to be in the direct line of sight to be read. In fact, multiple RF signals may be processed at once, so an RFID device may identify all packages in a single container, pallet, or shopping cart with a single interrogation. Because RFID must only be pointed in the general direction of the signal, this improves the feasibility of data capture in many challenging physical environments, or where products are always on the move. Furthermore, RFID applications generally require less human interaction than bar coding. Noninvasive RFID technologies and application software interfaces introduce the potential for capturing vast amounts of data for process control, feedback, decision support, and continuous improvement,

---

* Another 37 suppliers queued up to meet the challenge, although they were not required.

without introducing waste. This approach calls on a principle called "*0HIO*", which stands for *Zero Human Intervention Operations*. According to Sami Cassis of Factory Logic:

> RFID tags that are continuously tracked by readers allow operators to go about their usual work, un-encumbered by computers, while systems automatically track the movement, and hence status, of production. This technology allows computer systems to be "involved but not in the way". Such systems meet the requirements of Lean guidelines in making computers as unobtrusive as possible on the shop floor while allowing the Lean environment to track, assess and re-adjust factors such as kanban and buffer sizes as painlessly as possible.[174]

RFID will be an emerging technology for many years to come and will not replace bar coding in many prevalent applications. Bar coding is well entrenched in many business processes where no economic or practical benefit for its replacement would be realized. Beyond the cost factor, RFID faces many practical challenges for wide acceptance. An RFID tag contains a small semiconductor and antennae that must be attached to an item. Although this sounds simple, many peculiar challenges arise, and many environments are unfriendly to this technology. Increasing signal power often causes read-rates to decrease, because the signals drown each other out. Extreme heat, caustic chemicals, metals, liquids, strong radio frequencies, and cardboard containers are not the only enemies of RFID; even innocuous consumer products can be trouble. In addition to absorbing diaper messes, according to Mike O'Shea, Director of RFID Strategy for Kimberly-Clark, "the baby wipes absorb RF signals." Unilever is experiencing similar problems with other moisture-based materials.[175] Privacy is another critical concern, because a company or a consumer may object to the broadcast of sensitive product information from a passing truck, pallet, or shopping bag.

There is a Lean lesson to be learned from nascent RFID experiments. Overzealous attempts to automate apparently simple tasks often meet with unpredictable results. Do you remember the millions that were spent on the new Denver International Airport automated baggage handling system in 1995? Not only did the widely publicized failures regularly delay flights and mangle baggage, but they postponed the opening of the airport, causing international embarrassment to the City of Denver. After spending more than $230 million, in 2005 the system was shut down and baggage trucks were deployed. After a thorough postmortem, Cal Poly researchers identified *unnecessary complexity* as the primary culprit for the system failure.[176] This experience emphasizes that simplicity is paramount for the automation of any Lean operation.

Despite these initial challenges, which should be expected with any emerging technology, RFID offers many advantages over bar coding. Because data is stored electronically, RFID tags can store vast amounts of dynamic information. Unlike the information contained on a bar code label, some RFID tags also permit the interactive reading and writing of data, so they may be updated during each step of a process. The information contained on the tag may even store product specifications used as inputs for automated process control, carrying information from a prior operation that set processing parameters for the next. During each stage of assembly, the RFID tag may record the product characteristics and lot lineage required for quality certification and regulatory compliance. This suggests that an RFID tag can become an interactive component of the information and material flow. RFID tags may also be helpful in industries such as pharmaceuticals and electronics, for the prevention of counterfeiting and the identification of lost or stolen materials.

An endless variety of events may be triggered by an RFID signal interacting with a sensing device. Keep in mind that we're experiencing the first surge of practical interest in RFID; where might this technology possibly lead us in five or ten years? What changes will be required in the foundation logic of enterprise software and business practices to take advantage of its full capabilities? Will this lead to new sources of competitive advantage? At what cost?

Through their aggressive NAPCCI initiative, Wal-Mart is "basically pushing the burden of their logistics back on their suppliers," suggests Kara Romanow, consumer product goods analyst at AMR Research. "It's brilliant. Ultimately this is a revolutionary opportunity for the manufacturer, but based on what is available now, where the technology is now, and the cost that's going to be involved in becoming compliant with the Wal-Mart requirements—there is no ROI in thirteen months," she adds. "It's just cost."[177]

According to Wal-Mart spokesman Tom Williams, in 2003 the company moved 2.5 billion boxes through its distribution centers, and "RFID will dramatically improve the management of this inventory."[178] Dramatic improvements for Wal-Mart, but what of their suppliers? Can a value stream where one participant benefits at the expense of the others truly be called a "value stream"? It depends upon your long-term perspective. "These suppliers are being forced to implement the technology in a way that may not suit their business. The cost model just doesn't work right now," states Romanow. And the suppliers agree. "We don't have a business case for RFID," says one Wal-Mart supplier, speaking on assurance of anonymity. "I don't think RFID is a mature application at this point and time," adds another supplier, also asking to remain anonymous. Neither ventured to guess a dollar figure of the cost to meet this mandate.[179]

One month before the February 1, 2005 deadline, Wal-Mart CIO Linda Dillman predicted that all of the top 100 suppliers would meet the initial requirements. According to Romanow, however, "Most of the top companies will probably have less than 10 percent of their volume RFID-compliant by the deadline."[180] Now that the deadline is past, the fact that some or all did or did not meet the deadline is relatively unimportant. The competitive direction of industries is measured in years, not in months and weeks. In the long run, these pioneers hope to establish a powerful beachhead in the consumer products supply chain.

RFID Lessons Learned from EDI

To avoid giving RFID technology a bad rap, let's be sure to make an important distinction—RFID applied to our internal value streams is just another product identification and data capture technology like bar coding. Within an enterprise, the use of dynamic RFID tagging for process control and material handling applications can certainly produce immediate benefits, sometimes with flexibility and interactivity that bar coding cannot offer. But extending RFID across the supply chain, for real-time tracking of parcels and other product information, based on a global information infrastructure designed by a powerful few? Wisdom gained from the past suggests that we must be very cautious, and check our expectations for cost and near-term benefit.

EDI did not reach very far inside each trading partner's business processes, but merely provided external transaction touchpoints among them. Even so, EDI sent shockwaves through many companies that did not have the information systems and business processes to support the required information granularity and accuracy. And though EDI was standards based, standard EDI transactions were often tweaked according to the peculiarities of individual trading partners. Even today, hundreds of man-hours are usually required to implement a single "standardized" EDI transaction set between two trading partners. And many smaller companies still perform what is known as *rip and read EDI*—taking the order from the printer or fax machine, entering it by hand into their ERP system, and later manually generating Advanced Shipping Notifications and invoices—all to comply with their trading partners. These small companies are paying the cost of compliance, while missing the internal benefits of process improvement that should accompany such an initiative.

Will history repeat itself? We have seen many manufacturers that, to comply with the initial NAPCCI mandate, *slap and ship* outbound pallets with RFID tags simply to identify their contents. Although this satisfies Wal-Mart's requirements, it delivers no upstream benefits for internal production operations and materials handling. Although RFID allows hands-off operation, if even one manual touch point is introduced to bridge the new RFID with the old information system and material handling process, this will cause a high volume of narrowly concentrated workflow and data processing activities—a data management bottleneck.

"The Data Avalanche", an article in *Logistics Europe* summarizing the 2003 European Logistics and Supply Chain Forum meeting, suggests that technology, policy, and rights issues are of less immediate importance than basic logistics and business case questions:

> The immediate problem with RFID and the barrier to effective implementation is—what do you do with all this data? If you track a truck through in-cab telemetrics you have one lump of data. If you track each roll-cage, you have 40. If you track each tote, that's perhaps another 20-fold increase. If you track each item, you could have increased the volume of available information 80,000 times.

Most of this data is good but irrelevant. Management by exception is vital, but that is more than a technological change. All managers want to report the 99 per cent of activities that meet the plan. To focus purely on the one per cent that don't requires not a technological change, but a culture change. How are you going to identify and act on the error and event messages? That is the number one issue.[181]

RFID may create an event management windfall for many companies. On the other hand, as this article suggests, it may become a destructive avalanche. Regardless of these significant unknowns, the larger supply chain players are relentlessly pushing RFID tracking granularity down to the parcel, and soon to the item level in many industries. This may improve their supply chain visibility and cost, but how about *your* productivity? In a blunt article in June 2004 Gartner published *Prepare for Disillusionment with RFID*, proclaiming that:

The benefits of radio frequency identification have been oversold, and RFID cannot live up to the near-term promises that have been made for this technology. This means that RFID will soon be engendering a period of disillusionment, when at least 50 percent of RFID projects are likely to fail.[182]

However, Gartner remains positive about the technology for the long term, stating that RFID will be one of the most strategic technologies that enterprises will embrace through 2018. They suggest that companies should prepare for the coming "Trough of Disillusionment in the Hype Cycle" by:

- Distancing your RFID projects from others' projects to avoid getting caught in the downdraft.
- Getting a realistic message about RFID into your organization.
- Making sure that your vendors can survive a downturn in RFID spending.
- Ensuring that you can inexpensively support RFID-labeling projects.
- Having a written RFID assessment that demonstrates that your RFID strategy is based on sound, well-thought-out concepts.

Did Gartner say that 50% of RFID projects are likely to fail? And what's this about the year 2018? For many supplier organizations it appears that the lessons of the EDI era may be repeated. Compliance with RFID supply chain initiatives may not only fail to drive any internal process improvement in the near term, but could also increase operating costs and harm productivity. Unlike EDI, however, which only touches the surface of internal transactions and events, downstream RFID integration requirements may create disturbances that ripple throughout a manufacturing enterprise's internal value streams and upstream supply chains.

Is this how it has to be for many small and medium-sized companies? Are they going to recognize RFID's transformative power on the supply chain as a call to arms, to invest in making the leap to new efficiencies and through-put? Will they invest the time, money, and expertise to realize lasting bene-fits? Or are they going to simply paste RFID over the top of their existing systems, like slapping a new coat of paint on an old jalopy? My advice to the small or medium-sized manufacturer is to think strategically about RFID, focus on adding real value to the value streams, test each initiative thoroughly in collaboration with your trading partners, and make cautious initial invest-ments that you can afford to lose.

Event Management and Lean ROI

For an information system to be useful in a Lean organization, it must capture data in the most noninvasive manner possible, filtering massive quantities of data across the entire value stream, alerting human beings by exception when an event requires their attention. The information system should focus on the *controlling simplicities*[183] within a complex environment, the points of leverage where the most benefit is derived from the least effort and cost. From a Lean Enterprise perspective these are constraints: policy, market, material, and process; from a Lean Manufacturing perspective they are the demand signals that drive production through constraint and pacemaker operations. It is through the careful management of these critical leverage points that an entire value stream may be planned and controlled with minimum complexity and waste.

To prove its value within this context, an event management investment should be supported by an ROI justification. Three areas of cost and benefit should be considered for an event management ROI estimation: waste elimi-nation, throughput improvement, and demand management.

Waste Elimination. Waste elimination is the most obvious result of an event management solution and is typically measured as the reduction of costs including inventory, labor, and other operating expenses. In addition, signifi-cant (though less tangible) benefits may be realized, such as improved employee morale, improved accuracy and quality, fewer stock-outs, better schedules, lead time compression, and happier customers. When focusing on the reduction of waste with automated data capture and event management techniques, remember that it's important to ask first whether the task can be simplified or eliminated.

Throughput Improvement. An enterprise cannot save itself to prosperity. Although waste elimination and cost reduction are important, their benefits are amplified by an increase in throughput. When event management tech-niques are implemented skillfully they may increase throughput. By concen-trating real-time feedback where it counts, on the bottleneck and pacemaker operations, the throughput of the entire plant may be finely tuned.

For example, when a sudden constraint appears—a critical machine is down, or a quality problem arises on a bottleneck operation—an automated alert is immediately broadcast. Managers, specialists, and customer support representatives quickly converge upon the problem for a Kaizen Strike, drilling into the transactional, engineering, and process knowledge databases in real time with their wireless portable devices. They search the Web, log onto their supplier's self-service portal, searching the knowledge base, talking with and sending still and video images to the technician in Boston, Bangalore, or Beijing in real time. Problem identification, resolution, and prevention capabilities are enhanced, and throughput climbs another notch. Or at least throughput *capability* does.

Demand Generation. So how do we increase throughput? We improve throughput capability by making our processes more capable, able to produce more. Once we have the capability, we increase throughput by selling more. If we have a capable factory but no orders, or many production orders but for the wrong items, then throughput (measured by production that is sold, not accumulated in inventory) is not improved. This takes us back to the principle of demand pull: Only make what the customer wants, when he wants it. How can event management assist here?

Ask yourself: Where does your demand originate? Wherever that is, that is *the* Gemba where the chain reaction, the initial pull signal of demand, begins. Go to your *Customer Gemba* with friendly IT-enhanced tools, treating your customer quickly, fairly, and accurately.

In 1993 mobile device manufacturer Intermec (then Norand) published a report on the benefits of route automation.* During this early generation of handheld route automation devices, the obvious ROI emphasis was on cost reduction in the form of efficiency, accuracy, reduced errors, time savings, legibility, inventory control, reduced administration, and accounting costs. But the two greatest benefits, those that resulted in the highest dollar impact on the operations, were better marketing information (reported by 73% of the users) and improved customer service (70%).[184] These customer-focused benefits are the catalyst for increased sales and thus throughput.

This principle of automated and knowledge-enabled customer interaction applies regardless of how and where a manufacturer sells its products and services. When customer interaction is face to face, wireless handheld devices may be useful. When the relationship is long distance, a personalized Web portal stocked with helpful self-service features may do the trick.

Regardless of exactly how and where the interaction occurs, can you see the power of going to Customer Gemba, channeling highly personalized demand information into your CRM system and Lean planning process? Envision tapping into your customer's systems, analyzing sales transactions and

---

* Route automation is used when drivers deliver products and services such as parcels, food and beverage, retail grocery, laundry service, etc.

inventory status within minutes of your product being sold. Imagine value stream mapping your customer interactions, monitoring KPIs to spot important trends, with automated event alerts to notify you the instant something requires your attention. Customer Gemba is electronic poka-yoke for every customer interaction, enabling event-driven customer service.

Chapter **10**

Linking Strategy with Action: Performance Management

It was not enough to chase out the cost accountants from the plants.
The problem was to chase cost accounting from my people's minds.[185]

Taiichi Ohno

Imagine when Ohno questions a factory worker, "Why are you performing this task in this way?" the worker replies, "Because we have always done it this way, Ohno-san." Envision taking a Gemba walk through your executive offices and asking that same question. What answer will you hear?

Continuous improvement does not end at the shop floor. For an enterprise to truly be successful, continuous improvement must be applied to all tasks within the enterprise, from top to bottom. In this chapter we'll explore some of the challenges of performance measurement and explore ways to effectively lead and manage change. We begin with a look at the perplexing task of measuring the Return On Investment (ROI) of an IT project.

THE HUNT FOR ROI

The perceived value of IT has undergone several transformations during the past fifty years. At the dawn of the industry, computers were large and cumbersome machines used to manipulate massive amounts of raw data. Because these early automated processes were typically mechanical in nature, IT ROI was usually a straightforward computation of cost savings.

Lean Enterprise Systems: Using IT for Continuous Improvement, by Steve Bell
Copyright © 2006 by John Wiley & Sons, Inc.

During the 1970s, minicomputers brought mainframe-style computing power to midsize organizations. The development of business software advanced considerably, and more general business requirements—including many accounting, payroll, inventory, and production management functions—were automated at this time. Because the focus of these systems was usually limited in scope to a particular department or operation, the calculation of cost, benefit, and return were still relatively straightforward.

The 1980s brought personal computers, networking, and desktop productivity software. Isolated islands of data sprang up like mushrooms, providing users with local autonomy while often encumbering the overall business process. How to balance the benefits of local task improvement against the potential harm done to the value stream? Many shrugged their shoulders if anyone happened to question the financial justification as these new tools became increasingly popular.

The 1990s brought faster networks, powerful database engines, flexible programming languages, and open architectures. Personal computers were slowly brought back into the managed IT realm as nodes of larger client/server networks, and large-scale business software systems aspired to integrate multiple applications and databases to support entire value streams. The determination of cost, benefit, and ROI for IT investments naturally became more indirect and complex.

Then came the surge of information technology spending approaching the year 2000 phenomenon, accompanied by the rise of the commercial Internet. These events led to many popular but unrealistic expectations that the nature of commerce in general might be electronically transformed. Until this time, IT was generally considered a supporting function to business. Suddenly information technology was driving new products and services, creating new business models, opening new markets, driving outlandish market valuations, and promising to alter the business landscape forever. Or so it seemed to many at the time . . . but the boom became a colossal bust for most.

Nevertheless, something significant *had* changed. Comprehensive and relatively affordable enterprise information systems have evolved that not only automate the local tasks of the enterprise but orchestrate the flow of information across entire supply chains. The belief has evolved that IT can move beyond a traditional administrative support role, creating new business opportunities. The calculation of IT ROI is naturally far reaching, abstract, and difficult to measure.

We may forever look back at the period of 1995 to 2002 as a time of mass market information technology overindulgence, implausible expectations, and the failure of economic justification. On the bright side, many companies are now approaching IT investments with optimism, accompanied by renewed vigilance and healthy skepticism. An enterprise will not, and should not, consider an IT investment without a reasonable effort to determine ROI.

Beyond IT investing, an enterprise must develop an ROI model for measuring all investments, including Lean initiatives. An enterprise must therefore embrace a responsible approach to investing, with a balanced measure of value, while encouraging agility, team creativity, and educated risk taking.

The Components of ROI

The term ROI is used loosely and implies a valid financial approach to the justification of an investment. All ROI models share the basic factors of costs, benefits, and timing estimates. But how well can an enterprise predict? Are all the benefits clear? Are they direct? Are they tangible and measurable? If not, how can they be quantified?

Mohanbir Sawhney, professor at Northwest University's Kellogg School of Management, argues that:

> ROI measures only the returns that the company sees within its internal operations. ROI tends to favor projects that result in cost avoidance, at the expense of projects that promise revenue growth. However, the only way to grow the bottom line on a sustainable basis is to grow the top line, which is easy to ignore if every project is measured on tangible ROI. By ignoring the value created for partners and customers, ROI may be missing the real point. ROI requires that all benefits from a project be translated into financial terms. However, most e-business projects result in payoffs on multiple dimensions. For instance, a partner relationship management initiative may provide lower inventory costs (measured in dollars), faster order fulfillment (measured in time) and improved partner satisfaction (measured subjectively). Not all returns are financial returns in the short run, although they eventually may impact financial performance of the company.[186]

Continuous improvement efforts, and especially the IT systems that support them, often provide their greatest benefits in the form of improved quality, innovative products, and enhanced customer satisfaction. Collectively, these benefits may be the creative breakthrough initiatives that lead to competitive advantage. How do you predict and measure *that*?

The *CIO Magazine* article "Value Made Visible" contends that new approaches that attempt to account for the intangible benefits, although difficult, are quite necessary:

> Valuation's most crucial contribution to IT might very well be that it maps a clear cause-and-effect relationship between technology and the bottom line. "There isn't a first-order relationship between IT investment and financial outcome," points out David Norton, one of the two original developers of the balanced scorecard. "Investment in IT typically has a third-order financial effect," he explains, where, for example, technology improves some intermediate valuation, like customer service, which in turn boosts customer confidence, which finally results in increased sales for the company. What the balanced scorecard and other methods try to do is make visible those intermediate steps, in ways that can be quantified, measured, and tracked.[187]

Estimation of these future costs and benefits is not a simple exercise based on definitive financial variables found in any Finance 101 text. To estimate *all* the costs and benefits requires *intuition and judgment.* Let's look at the basic elements of the ROI model: costs, benefits, and timing, and explore how these elements can be adapted to reflect a more accurate picture of *value.*

Costs. Direct costs for IT investments are relatively simple to account for, to the extent that funds are paid for software and hardware acquisition, temporary personnel, outside consultants, training, implementation, ongoing maintenance and support expenditures, etc. Indirect costs are more difficult to measure, because they may be commingled with payroll and other internal operating costs such as the procurement process, system administration, facilities, analysis and design, training, education, quality assurance, software modification, and report development.

Even more difficult to measure are the less tangible costs (and risks) incurred during system implementation. How do you value the potential loss of a customer because of a misfire in the project? How do you measure the cost of lost transaction history during the data conversion process? What about key employees, along with the value of their knowledge and relationships, who may leave or choose early retirement as a result of the project? And what of the opportunity cost of other projects that are delayed or foregone? These and many other risks and indirect costs of an IT project should be identified in the early planning stage so they can be managed proactively. Various risks may also be used as weighting factors in the ROI model to anticipate the probability and severity of costly events.

Benefits. The traditional benefit driver for an IT project has usually been cost reduction, by reducing effort, time, and other forms of waste in a business process. However, cost reduction is only the beginning. When applied thoughtfully, IT can create new business opportunities, attract new customers, and increase throughput and profits. However, ROI models may have to be creative to quantify and measure these unpredictable, intangible, and indirect benefits. Although many cost reductions can be measured according to line items in the chart of accounts and financial statements, measures of increased market share, revenue, or competitive protection often appear as indirect factors in a sales and marketing forecast.[188]

Implicit in the justification process for an IT investment is the assumption that these costs and benefits will be measured on a periodic basis and used to assess the status and ultimately the success of the project. However, many enterprise software implementation projects fail to clearly identify measures of cost and benefit for the new system. Well-considered KPIs should correlate cause and effect relationships to the assumptions used to build the ROI model. However, project teams often neglect to take necessary baseline measurements before the project begins, so they are unable to perform a before-

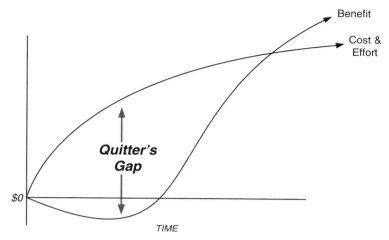

Figure 10-01. Quitter's gap—the lag between cost and benefit

and-after comparison. These omissions result in the inability to answer two apparently simple questions: *"Are we better off than before?"* and *"Was this project successful?"*

Timing. Timing must be considered carefully when calculating ROI for any project. Costs are usually frontloaded, paid out early in the project, whereas benefits, in addition to being difficult to measure, are usually realized later.

As the costs and difficulties mount, an organization may lose its resolve to follow through on a project. It's easy to stir up initial enthusiasm but far more difficult to sustain it. Observe in Figure 10-01 the period of negative benefit where people are doing two jobs: using the old dysfunctional system while developing a new one at the same time. This is a difficult time for a project team and the end users, with frequent meetings, planning and design sessions, documentation, education, training—all accompanied by plenty of stress. During this difficult time the project may be discontinued, after most of the costs have been incurred but before any benefits are realized.

Timing also complicates the ROI calculation because indirect benefits not only are difficult to measure, but are farther into the future and thus more unpredictable. Furthermore, although many cost reduction assumptions used to justify the investment are short-term tactical initiatives, the longer-range benefits are often related to higher-level strategic business goals. Many of these benefits are not only indirect but intangible, referred to as *soft* or *ncn-monetary benefits*. Although most people will argue that improved customer service, time to market, or product quality will add value to an enterprise, fewer will attempt to quantify them.

The Value of Intangibles

Although they are difficult to quantify, these intangible benefits *do* matter. We can agree that an ROI model is difficult to construct for a risky project with indirect and intangible benefits. We can also agree that some form of justification is needed for an organization to responsibly commit thousands or millions of dollars to such an initiative. Some form of ROI model is therefore necessary.

Recall the earlier assumption, that to estimate *all* the costs and benefits requires *intuition and judgment*. Who is best able to provide these insights? Definition of costs and benefits, along with the guidance of a successful initiative, must involve the individuals who are responsible for executing the business strategy, and who will realize the potential benefits: the managers and improvement teams. Who better to answer the questions: What will this system enable us to do better? How much better? When will these benefits be realized? Determination of IT ROI, project justification and measurement must be team activities, not solely the responsibility of the IT department.

Let us not forget the Y2K phenomenon followed by the eCommerce frenzy, when expectations were often irrational, and project justification was often no more than, "Because we have to". To avoid repeating this grave error, expectations, that intangible result of intuition and judgment, should be team-based for balance and validated against a thoughtful strategy.

This team-based participation in project planning and ROI justification may necessitate a cultural shift, accompanied by education and communication, to eliminate the aura of mystery that commonly surrounds IT initiatives. Effective company-wide IT project collaboration is one reason why many IT managers have been invited into the boardroom, and the executive role of Chief Information Officer was created, elevating the stature of IT from administrative support to a key role in value creation.

THE PAINFUL ANNUAL RITUAL

When performance is measured against expectations, they are often expressed as budgeted versus actual financial results. Near the end of each fiscal year, a painful annual ritual* begins across many companies. First, the finance department sends a directive accompanied by multilayered worksheets requesting input to construct the annual budget. Then each department manager is required to account for her current year performance, to project her goals and objectives into the future year, and then translate these predictions into numbers arranged in rows and columns, representing a vast array of bewildering revenue and expense accounts.

The instructions from the finance department often include executive-level mandates and targets based on mathematical formulas using the past as

* This name was given to the traditional budgeting process by Gartner in 1995.

reference ("decrease all telephone expenses by 5%" or "derive travel expenses as 10% of sales"). This type of historical formula budgeting is much like staring into the rearview mirror while driving a car forward.

Depending on how many sublevels of a particular business unit roll up to the departmental level, the traditional budgeting process of distributing, gathering, and recombining multiple layers of worksheets may take several weeks to complete and may be performed several times during a complete budgeting cycle. At each layer of combination, the differences between the mathematical calculations, the assumptions, and the actual logic used to manage the daily decisions become disconnected. So when the time comes to defend the assumptions behind any particular category of revenue or expense, the manager must search through layers of calculations and summarizations, looking for documented assumptions or patterns behind a decision made months before. The approval of these consolidated budgets may require several iterations, where the budget figures roll up to management, then down again for adjustments. This cycle creates yet more abstraction from the relevance of the actual decision-making at the departmental level. And it consumes more time.

Finally, these budgets are "approved," and for the remainder of the year they are used to measure performance. However, as the year continues and business conditions change, the budget becomes out of date and the detailed assumptions used to develop the budget become a vague memory. Periodic adjustment of the budget to adjust to these changing conditions may be impractical, because the iterative and clumsy process requires so much time and effort to complete. For this reason many enterprises are unable to repeat this process on more than an annual basis. And for that, most are grateful.

Fast forward through the year. The time to make a large budgeted expenditure arrives. Should you make the investment? Recall the budgeting process, completed late in the fiscal year, where requests were made based on data available at that time to calculate ROI. Are those assumptions still valid? What has changed? Should a new ROI calculation using current data be performed before making the previously approved expenditure? Many companies, and especially governmental organizations, perform a year-end rush to spend the budgeted money regardless of the benefit, or the money will be lost in the following year's budget. This behavior indicates a complete disconnect between budget and reality.

Let's also consider the implications of unforeseen opportunities and their effect on the budget. Often a manager will become aware of a new opportunity for which no funds have been budgeted. To capitalize on this opportunity, he may redirect funds allocated for another purpose. If the budget is not then updated based on these changing conditions, then late in the year the department may find its budget exhausted and funding unavailable for many of its originally planned expenditures.

Anyone who has been involved in such a process feels the frustration and understands the harm, not to mention the wasted effort, because little relevance exists between the budget and the actual decision-making processes. What is going on? Simply put, the top-down management view of the organization is driven by financial measures seen from a high level. This view contrasts with how each business unit is actually run, using intuition and judgment, basing plans on forecasts, life cycles, detailed operating plans, existing sales and purchase agreements, inventory policies, capacity plans, staffing decisions, etc. Decisions at the operating level are ideally made using relevant and predictive measures of those events that drive the business forward.

Planning is often perceived as good, implying forward thinking and proactive management. Budgeting, however, is often perceived as bad, because it carries connotations of bureaucracy, centralized control, and a recurring makework budget process that creates little value. In reality, the two should be inseparable—planning is done at the operational or line-of-business level, and the information directly impacts the budgeted financial decisions made throughout the organization. The harmful disconnect occurs when the financial measures are used to disregard or override the operating-level decision-making process.

This disconnect is similar to the difficulties experienced when financially focused ROI models are applied without flexibility or consideration of intangibles. This disconnect is a natural result of the financially centric way that many companies are managed because the stakeholders, whether they are capital markets, lenders, or private owners, understandably need a standardized set of financial measurements and controls to manage their investments.

This disconnect with traditional ROI models, illustrated by the painful annual ritual of the financial budgeting process, argues for a reasonable marriage between the business planning and the fiscal budgeting process. When business plans or conditions change, the budget should be adapted to reflect current assumptions and objectives. The budget must reflect the top-down strategic goals of the organization, as well as the bottom-up planning and control at the operational level. This marriage is the purpose of *activity-based budgeting** software and practices, which attempt to model the relationship between *business drivers*, the detailed operational planning decision factors, and financial results. For example, a driver of payroll expense could be the headcount plan at the functional level, which is ultimately driven by demand and capacity planning at the detailed operational level. Complex activity-based costing, planning, and budgeting models using specialized software have evolved to address this challenge.

Activity-based planning and budgeting software programs look and act much like spreadsheets. In fact, these tools are powerful programs with underlying relational databases and report writers, with the ability to query ERP and

---

* Activity-based budgeting (ABB) is related to activity-based costing (ABC) in that the costing assumptions within the budget are tied to the actual business drivers that cause these costs to occur. ABB extends this approach, connecting these low-level cost *and revenue* drivers to the high-level assumptions built into the fiscal budgets.

other enterprise applications to gather detailed operating data to incorporate within their calculations. When these models are standardized and automated, the budgeting process becomes faster and less burdensome, enabling rolling quarterly or even monthly budgets, which are prevalent in industries (such as high technology) that must be replanned frequently because of rapid change. A rapid planning and measurement cycle promotes agility, but a properly designed information system must be in place or frequent activity-based planning and replanning will create an enormous managerial burden.

Unfortunately, even when they are well implemented, enterprise-wide activity-based planning and budgeting tools are complex to build and maintain; they may be impractical for the smaller enterprise. Furthermore, activity-based planning and budgeting software is usually a separate application that must be attached to MRP II and other transactional systems, creating a dichotomy where work is performed in one system while planning is done in another. Wouldn't it make sense to apply the principles of activity-based planning and budgeting within the MRP II system where the activity takes place?

SALES AND OPERATIONS PLANNING

Which comes first, the chicken or the egg, planning or budgeting? An enterprise should focus on the drivers of the business, the delivery of products and services to meet customer demand. Although an enterprise may only produce one comprehensive fiscal budget each year, they are continuously planning at an operating level. Furthermore, the actual decisions that direct revenue and expenditures are executed at the operating level, so operational decisions drive financial results. This suggests that operational planning should drive the business, whereas fiscal budgeting should be used as a fiscal measurement and regulation mechanism. This effectively describes how the Sales and Operations Planning (S&OP) process works, as shown in Figure 10-02.

The S&OP process serves as the cornerstone of company-wide planning, a monthly exercise that rationalizes the plans for demand, supply, finance, and company strategy. S&OP provides the monthly reality check to executives, a review of how the business is running, and may also suggest replanning (or the questioning of fundamental assumptions) at the strategic business plan and fiscal budget level.

Any planning and budgeting process that does not provide useful input to the monthly S&OP process should be closely examined. Although executive management, shareholders, lending institutions, and regulators may always require top-down fiscal budgeting controls, these controls should not be allowed to take precedence over legitimate operational planning mechanisms. The fiscal budget should not override the operational planning process without a legitimate reason, such as limited working capital; this is the very reason that finance is the last stage in the S&OP process before executive approval of the production plan.

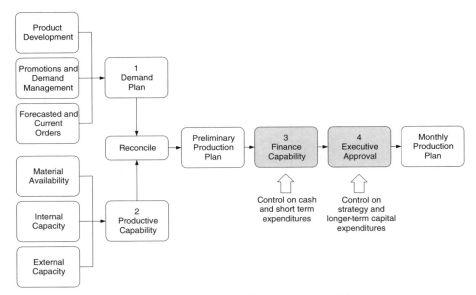

Figure 10-02. Financial Controls within the S&OP Process

Just as scheduling needs a focus on controlling simplicities and constraints, there should be a clear focus within strategic planning and budgeting as well. A company should reconsider its planning and control processes and develop new awareness of the *cause and effect linkages*, blending financial and operational measures, so that disconnected financial measures do not force inappropriate operational decisions. This points to the discipline of Lean accounting.

LEAN ACCOUNTING

Lean Accounting is comprised of two elements:

Administrative Process Improvement—The improvement of administrative processes with the same principles of flow and waste reduction that are applied on the shop floor. Improved administrative process design eliminates waste in many forms including process time, unnecessary transactions, postprocess controls, and audits.

Performance Measurement—The accounting of an operation to appropriately measure and improve Lean performance, while providing financial stakeholders assurance that their assets are being managed properly. Lean accounting acts as a counterbalance to the traditional focus on standard financial measures and cost accounting.

Our focus within this chapter is on performance measurement. As Taiichi Ohno suggested in the quotaion at the beginning of this chapter, no greater or more persistent impediment may exist for an enterprise embarking on the Lean journey than entrenched cost accounting assumptions.

Traditional cost accounting produces results that aren't just misleading; they can be entirely contradictory to Lean principles. Consider that many managers are measured or compensated by traditional cost accounting notions such as efficiency, utilization, and cost per unit. Failure to modify these practices means, at best, that the Lean initiative must constantly struggle upstream against a strong current, never achieving satisfactory results because they are attempting to satisfy mutually exclusive objectives. At worst, these attitudes and policies will not only cause a sincere Lean effort to fail but may also create a strong disincentive for further attempts. For example, I have heard of several instances where the Lean champion or project manager lost their job after a successful Lean initiative where inventory levels and production costs were reduced. In at least one of those cases, Lean improvements quickly backslid to the former practices. Why? Because a large quantity of inventory carried at a higher absorbed cost was suddenly purged from the balance sheet and replaced by a smaller quantity at a lower cost. This caused a sudden and unanticipated financial loss (when the excess and overvalued inventory was expensed), which created heartache for executives (especially the CFO, who must answer for unanticipated losses) and a potential loss of stock price for the shareholders. This illustrates how short-term market valuation considerations can inhibit healthy long-term decisions.

Traditional financial accounting is so deeply entrenched that you can't afford to mince words. Take, for example, what Brian Maskell and Bruce Baggaley have to say in *Practical Lean Accounting*:

> Traditional accounting, control and measurement systems [. . .] motivate people to use non-lean procedures. Traditional systems are wasteful. Standard costs can harm Lean companies because they are based on premises grounded in mass production methods. The methods are complex and confusing to generate, they provide a misleading understanding of cost, and they lead to wrong management decisions on important issues, such as make/buy, profitability of sales orders, rationalization of products or customers, and so forth.[189]

So what is the monster lurking in the bushes, the source of this erroneous mass-production thinking? It's the widely held belief that the more of something you make, the less each unit costs. We have been taught the principle of *economy of scale* since we were children, that it was the basis for the Industrial Revolution, and it has become a core assumption of most economic theory. Economy of scale is so intuitively obvious that to think otherwise is, well . . . counterintuitive.

Let's start by suggesting that this principle is correct, but only within a narrow definition of cost. In traditional thinking, cost is comprised of fixed and variable elements. Variable costs, such as direct materials, increase as volume increases. Fixed costs, such as certain costs of plant, equipment, labor, utilities, and so on, do not directly increase with volume*. We attempt to allocate a portion of these fixed costs to each product, treating them as if they were direct costs. For example, if fixed costs are $100 and we produce 100 units, then we allocate $1 per unit. If we produce 50 units, then we allocate $2 per unit. This makes intuitive sense: If you're planning to make a lot of something, you might buy a big machine, spreading the costs over a large volume. You certainly wouldn't buy a big machine if you planned to use it just once a month, would you? This logic sounds simple enough.

As we have learned, though, these traditional cost accounting assumptions can be contrary to Lean principles. Why? Because they encourage *bigness*; traditional cost accounting favors large batch sizes, long production runs, long lead times, and large inventories. Furthermore, many manifestations of the seven forms of waste are difficult to measure from a traditional cost accounting standpoint. And Lean practitioners insist that rather than measuring them, which compounds the waste, we must simply eliminate them.

Although Lean practitioners argue that they are counterproductive, traditional cost allocations have become deeply entrenched in our minds and our laws. The fundamentals of cost accounting are a pillar of every accounting class taught in our schools, and of Generally Accepted Accounting Principles (GAAP) that guide financial accounting practices in the United States. The Security and Exchange Commission (SEC), the Financial Accounting Standards Board (FASB), the American Institute for Certified Public Accountants (AICPA), banks, lenders, and the public financial markets must all play by these rules. The valuation of companies, and the stability of entire economies, depend on the consistency of these regulatory rules. Conservative standards for compliance are necessary, so we should expect that these measurements will remain at the financial reporting level, but these "financial views of the data" should not rigidly guide operational decisions. According to Dr. Richard Schonberger in "Kanban at the Nexus":

> It is common sense that in any process sequence the non-bottleneck processes product at a rate no faster than the bottleneck process. To run any faster will just produce idle inventories that cannot get through the choke point. This is the main idea of [. . .] theory of constraints. Reasonably managed companies would surely have practiced this brand of good sense from the beginning.

> Or would they? Our microlevel management accounting systems muddy the water. Typically, they drive managers to strive for maximum outputs at every process, and hang the common sense.[190]

This highlights a fundamental conflict between Lean and traditional accounting. In traditional accounting, inventory is an asset, placed right next to

---

* This is not quite accurate, because at some point a fixed asset such as a building or a machine must be expanded to provide additional capacity; some fixed costs increase in relation to volume, but as a step-function rather than smooth growth.

cash and investments on the balance sheet. Inventory contributes to the overall value of the enterprise. But to Lean accountants, excess inventory is considered a *liability*, consuming cash, materials, and productive resources that may be better utilized elsewhere. The contrast could not be more striking, that of an asset versus a liability. However, mass-production assumptions can be so deeply entrenched that they can quickly eradicate a Lean initiative, so reeducation across all levels of the organization is essential for sustained Lean success.

On the shop floor, Lean Accounting provides feedback to Kaizen teams, while in the boardroom it neutralizes the negative inertia caused by traditional cost accounting and compliance measures. With a properly designed information system, all information consumers may be served by a single set of fact-data. However problems arise when using the right views of the data for the wrong purpose, for example when making operational decisions based upon financial or compliance views.

Maskell and Baggaley explain that during Lean transformation an organization may evolve through several stages of Lean Accounting. At the outset, traditional financial controls are left in place since it would be irresponsible to remove them too quickly. At the same time, focused Lean measures such as inventory valuation and cell performance guide emerging improvement initiatives. As Lean transformation spreads throughout the enterprise, Lean Accounting measures should focus on overall value stream cost and effectiveness, with decreasing emphasis on traditional department performance and cost accounting. As processes are simplified and become self-regulating, many traditional measurements, transactions, controls, reports, and meetings (and the information systems required to support them) may be reduced or eliminated entirely.

THE BALANCED SCORECARD

> Balance suggests a steadiness that results when all parts are properly adjusted to each other, when no one part or constituting force outweighs or is out of proportion to another.
>
> *Webster's Third New International Dictionary*[191]

Lean Accounting balances operational and financial measures by eliminating the distortions caused by measuring productivity and throughput with inappropriate methods. Even so, Lean Accounting still favors a financial perspective, and this can present an unbalanced view of the overall value of the organization, an incomplete picture of the organization's health and future potential. According to authors Michael Cowley and Ellen Domb in *Beyond Strategic Vision*:

> Financial measures tend to be lagging indicators, that is, they really measure the result of actions taken by the company in the past. The financial indicators are necessary for any business, but they are not very good indicators of things to come, which will be the result of how good a job is being done now on devel-

oping competitive products, attracting and retaining customers, entering new markets, and so on. [Performance measures including] Innovation, Customer Satisfaction, and Customer Loyalty are better predictors of the future.[192]

Recall the central position occupied by the financial core of ERP illustrated in the Copernican view of the enterprise software universe. This unbalanced emphasis on financial measurement has pervaded every discussion in this chapter—from ROI, to planning and budgeting, to traditional cost accounting, and finally—to some extent—even with Lean Accounting. This singular financial bias must be exorcised from management and shareholder thinking for Lean initiatives to thrive. For lasting competitive advantage an enterprise must focus its resources on creating real value, not accounting "book" value.

Economies of scale and the power derived from concentration of assets may actually create a competitive *disadvantage* as markets and supply chains favor time-based competition, innovation, quality, customer satisfaction, and agility over lowest cost. This strategic shift requires mastery of the flow of information across the Lean Network, where the focus is on both intangible and measurable financial value. As companies around the world transform for competition that is based on information, their ability to exploit intangible assets has become far more decisive than their ability to invest in and manage physical assets.[193]

In recognition of the value of nonfinancial measures, in 1992 Robert Kaplan and David Norton published the ground-breaking *Harvard Business Review* article, "The Balanced Scorecard—Measures That Drive Performance". Kaplan and Norton suggested that financial measures are not flawed, just incomplete. They proposed a set of four measures that, when kept in balance, can assess the overall health of an organization:

Financial—To succeed financially, how should we govern and protect our financial interests? How must we present information to our stakeholders and regulatory reports?

Operational Effectiveness—How do we measure and improve our business processes to deliver the best value to our customers?

Value to the Customer—How do our customers perceive us? How do we add value in their eyes? How can we enhance satisfaction and loyalty? Recall that the Value Stream, as defined by Womack and Jones in *Lean Thinking*, begins with value *from the customers' perspective.*

Innovation—How do we sustain our ability to continuously improve? How do we discover and implement the right new things? Are we adding value to our customers through new products, services, and relationships, in a manner that builds competitive advantage?

It is important to emphasize that if any one element is out of balance then the health of the enterprise is compromised. For example, an enterprise may have strong finances, good customer relationships, and a handle on operational

Balanced Scorecard Norton and Kaplan	Finance	Value Creation		
		Operations	Innovation	Customer
Lean Thinking Value Stream Womack and Jones		Physical Transformation	Problem Solving	Customer Value
Discipline of Market Leaders Treacy and Wiersma		Operational Excellence	Product Leadership	Customer Intimacy
Enterprise Software Component	ERP	MRP II	PLM	CRM

Figure 10-03. The elements of value creation

efficiencies. But if the enterprise is not innovative, then competitors may perceive a ripe opportunity and aggressively pursue their customer base with innovative products and services for which the incumbent has no response. Likewise, the enterprise may have excellent customer relationships, strong research and development in collaboration with their customers, and operational efficiencies, but if they aren't profitable then the underlying assumptions and strategy must be reconsidered.

We appear to be getting close to a *comprehensive* set of Lean performance measures that balance the focus of the enterprise, integrating tangible and intangible measures of value through the eyes of the customer. Just how comprehensive, how all encompassing, how relevant are these balanced measures? One way to suggest the validity of any theory is to provide evidence that the theory is supported by other accepted theories and practices.

In chapter 6 we explored the Copernican view of the enterprise software universe and demonstrated the correlation with Womack and Jones' *Lean Thinking* and Treacy and Wiersma's *The Discipline of Market Leaders*. We also learned that the primary enterprise software components of ERP (finance), MRP II, CRM, and PLM were consistent across these categories.

You can see in Figure 10-03 that when we add the Balanced Scorecard, this framework aligns perfectly. These complementary approaches point to the same conclusion regarding the fundamental elements of value creation, and it is not by coincidence that these elements align with the core capabilities offered by ERP, MRP II, CRM, and PLM enterprise systems.

The lesson here is that Lean operational excellence, innovation, and customer focus drive value, while financial measures are an important control mechanism. With such a perfect correlation, perhaps we have arrived at a general theory of enterprise performance management?*

---

* With all respect to Professor Stephen Hawking, we are only exploring a general theory of the *software* universe, not the real one.

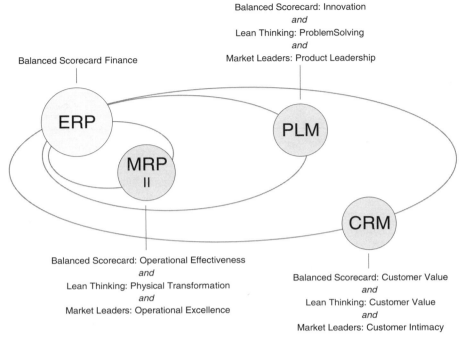

Figure 10-04. Balanced scorecard and the Copernican view of enterprise software

HOW PERFORMANCE MEASUREMENT LEADS TO PERFORMANCE MANAGEMENT

By exploring the shortcomings of traditional ROI, fiscal budgeting, and cost accounting models, we have learned the importance of balanced performance measures in a Lean Enterprise. Now we turn our attention to the mechanics of *leading* and *managing* an organization by those measures, so that the actions within each value stream, department, team, cell, and individual are consistent with the organization's strategic goals and objectives. We must also develop effective feedback mechanisms to alert executives when front-line reality diverges from the assumptions behind top-line strategy.

There is a traditional distinction between leading and managing. Although both are necessary, managing sometimes carries a negative implication, as something that does not add value and should be eliminated. Reduced and simplified perhaps, but managing cannot be eliminated entirely, because it is the mechanism that guides and controls the organization. According to Warren Bennis in *On Becoming a Leader*, the traditional manager/leader dichotomy is characterized by the following statements:

- The manager administers; the leader innovates.
- The manager maintains; the leader develops.
- The manager asks how and when; the leader asks what and why.
- The manager has a short-term view; the leader has a long-range perspective.
- The manager does things right; the leader does the right thing.
- The manager focuses on systems and structure; the leader focuses on people.

Contrary to these traditional distinctions, in a Lean Enterprise *all* managers must strive to lead, focusing on people doing the right things, innovating and continuously improving with a long-range perspective. In *Thriving On Chaos*, Tom Peters describes a flexible environment in which people are not only valued, but encouraged to develop to their full potential, and treated as equals rather than subordinates, making their own suggestions to initiate change. To accomplish this, executives must transform the organization from a rigid pyramid to a *fluid circle*, guiding an ever-evolving network of autonomous units. This requires reshaping the corporate culture so that creativity, autonomy, and continuous learning replace conformity, obedience, and rote; and long-term growth, not short-term profit, is the goal. This organization must be self-correcting, identifying weak links in the chain and repairing them. And this organization must encourage innovation, experimentation, and risk taking. In sum, Peters describes a world of people who are leading—not merely managing.[194]

From a mechanical point of view, management is the administration of policies and procedures, moving in a direction that is guided by leadership. To the extent that IT can simplify, structure, and automate the communication and control mechanisms, then NVA management activities may be reduced and managers will have more time to lead. When decision-making flows in a fluid top-down/bottom-up circle enabled by information systems (shown in Fig. 10-05), then managers spend less time managing and more time guiding and empowering initiatives.

This leads to an important precaution on executive leadership. A Lean Enterprise guides by strategy from the top down, whereas ideas and initiatives for improvement should flow from the bottom up, through managers, teams, and individual employees. The top-down strategic view, empowered by powerful reporting systems and "drill-down" software tools, may tempt executives to fiddle with the fine controls, overriding the actions of their managers, teams, and employees—but this can be dangerous, causing unintended consequences while harming the culture of empowerment. According to William Christopher, co-author of the *Handbook for Productivity Measurement and Improvement*:

> Drilling down may give the senior level information it shouldn't have, and, worse still, shouldn't use to decide on an intervention. A "cause", an executive-level intervention, doesn't have an "effect"; it has consequences that may rumble over a large territory.[195]

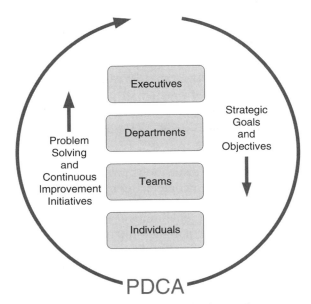

Figure 10-05. The fluid circle of leadership and management

In accordance with Lean, as well as other contemporary management philosophies, many enterprises have flattened their organization structure. By emphasizing fewer non-value-added layers of command and control, managers give more responsibility and authority (leadership) to those performing the work because they best understand how the process may be improved.

At the executive level, the time horizon is very distant—typically a strategic plan looks ahead three to five years, while many executives have ten- or twenty-year visions in mind. At the managerial level the time horizon is usually bounded by the annual plan, although contracts and performance agreements with customers and suppliers may span multiple years. At this level, most performance measurements are made in annual and monthly increments. At the team level, the time horizon is quite short, measured in months and weeks, and on the production floor in days and hours.

As you would expect, the scope of responsibility also narrows as you travel down the hierarchy. According to Rother and Shook in *Learning to See*, improvement of the overall value stream (flow kaizen) is management's responsibility, whereas process improvement and waste reduction (process kaizen) are the responsibility of those at the front lines.[196] Executives should be concerned about front-line process improvement and waste reduction, because the success of their business strategy results from these actions. However, they should limit detailed top-down interventions as much as possible, encouraging the advancement of improvement initiatives by those doing the work. In the *Harvard Business Review* article "*How the Right Measures Help Teams Excel*", Christopher Meyer suggests four guiding principles for the design of a performance measurement system to maximize the effectiveness of empowered teams:

1. The overarching purpose of a measurement system should be to help a team, rather than top managers, gauge its progress. A team's measurement system should primarily be a tool for telling the team when it must take corrective action. The measurement system must also provide top managers with a means to intervene if the team runs into problems it cannot solve by itself. But even if a team has good measures, they will be of little use if senior managers use them to control the team. A measurement system is not only the measures but also the way they are used.

2. A truly empowered team must play the lead role in designing its own measurement system. A team will know best what sort of measurement system it needs, but the team should not design this system in isolation. Senior managers must ensure that the resulting measurement system is consistent with the company's strategy.

3. Because a team is responsible for a value-delivery process that cuts across several functions (like product development, order fulfillment, or customer service), it must create measures to track that process. While such measures are extremely important, teams still need to use some traditional measures, like one that tracks accounts receivable to ensure that functional and team results are achieved.

4. A team should only adopt a handful of measures. The long-held view that "what gets measured gets done" has spurred managers to react to intensifying competition by piling more and more measures on their operations in a bid to encourage employees to work harder. As a result, team members end up spending too much time collecting data and monitoring their activities and not enough time managing the project [or process].[197]

The Importance of Alignment

> **Align-ment**: An arrangement of groups or forces in relation to one another
> Merriam-Webster Online Dictionary[198]

Let's carefully consider point #2 from Meyer's list. For each team to make *appropriate* process improvement and waste reduction decisions in support of strategic goals and objectives, there must be a clear alignment and communication of these measures from the top down and from the bottom back up again. According to Lean management consultant Bob Kerr:

> Winning today demands the achievement of results through people which can only occur when there is alignment of action. Such alignment is only possible when a clear direction exists. It begins with an appreciation that only vision provides direction. The secret of successful alignment lies in the ability to select those measures that will align all behaviors with the vision. Simply stated, the right measures are those that align all activity with a company's corporate vision, or future desired state.

It is tragic to see good people honestly working to achieve higher output levels of a given measure that is wrong or misguided. Such measures tend to be those imposed upon people without their input or involvement. Such measures they truly do not understand since they have had no involvement or hand in its definition, and therefore can hold no feeling of ownership. Hence, the challenge for management is to ensure that everyone understands the company's vision, and the short term goals to be attained, and what their personal roles must be to achieve the corporate business plan. For there to be alignment there must be clarity, understanding and involvement.[199]

This alignment must begin at the top with articulate vision and strategy, because executive leadership charts the course and steers the ship. Unfortunately, strategic vision is sometimes articulated as a lofty, vacuous, feel-good mission statement about being "world-class" and "customer-centric." As one senior executive described his company's strategic plan, "It's where the rubber meets the sky."[200] Similarly, the challenges of the traditional budgeting process demonstrate the disconnect that may occur when strategic and operational directives are communicated and/or controlled as visionless and irrelevant financial measures.

Because each successive layer of an organization focuses on progressively more detailed actions, there should be a step-down process where each goal and objective is translated to the next level as a more specific measurement. When these layers of objectives and measurements are aligned, and integrated from a database perspective, this permits a manager to drill down, across, and back up at will, examining the appropriate detail to address whatever question is being asked at the moment. To facilitate such fluid analysis there must be clear cause-and-effect relationships among the measurements and the underlying data.

Recall the distinction between a result measure and a process measure: A result measure quantifies the outcome of a process, whereas a process measure assesses the inner workings of the process to identify the root cause of the result. Linking multiple layers of result and process measures creates a clear chain of cause and effect.

Figure 10-06 illustrates the effectiveness of the *Five Whys* in problem solving. We ask *Why?* because something is not working properly; the result measure of our action does not agree with our desired objective. The answer to the first *Why?* leads to one or more process measures that suggest the probable causes of the problem—recall the Ishikawa Fishbone cause and effect method from the previous chapter. We choose a particular process measure that seems to be a likely cause of the unsatisfactory result, then ask *Why?* again. In such a progression the process measure from the last *Why?* becomes a result measure for the current *Why?,* which in turn leads to one or more underlying process measures. We ask *Why?* as many times as necessary, until we finally end up with the root cause(s) of the problem.

The causal chain of process and result measures may circle around a particular spot in the organization, or causalities can travel up and down the

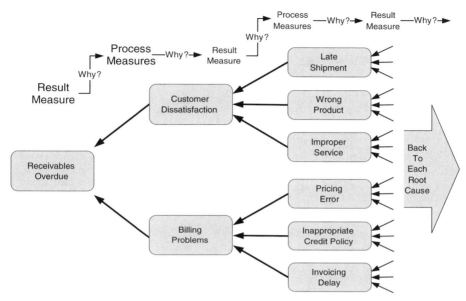

Figure 10-06. The Five Whys in action

hierarchy and across departmental and functional boundaries. Although it's a natural assumption that most cause and effect relationships move down the hierarchy, from less to more detail, from plans to specific actions, that is not always the case. Figure 10-07 illustrates the causal chain of a disconnected financial budgeting process. Note that the root cause of the problem stems from the top-level budget restricting a necessary capacity expansion.

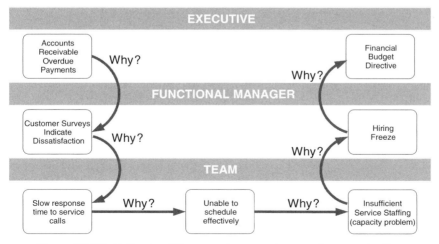

Figure 10-07. Following Whys up, down, and across the organization

HOSHIN PLANNING

We have learned that an effective performance management system should link strategy with specific action across the organization hierarchy. This suggests that an information system should integrate these hierarchal layers into a coherent set of fact-based views and exception-based reports. Does such a system exist, one that can translate strategic goals into consensual team actions, with the flexibility to learn and adapt as conditions change, without creating a substantial administrative burden? The enterprise communication, collaboration, business intelligence, and knowledge management software tools exist to create such a system. More importantly, there is a proven management framework to guide a Lean Enterprise, which may be implemented with the appropriate IT tools. *Hoshin Planning* (also known as Hoshin Kanri or Policy Deployment) has been employed extensively by Toyota, as well as many other organizations known for their management prowess, including Hewlett-Packard, Intel, Milliken, Zytec, and Proctor and Gamble.[201]

Hoshin Planning techniques evolved from Management By Objectives (MBO), a popular approach introduced in the 1950s. Using MBO, management established objectives that were communicated throughout the organization and translated into lower-level departmental and individual targets. MBO evolved into Hoshin Planning when it found its way to Japan. Hoshin Planning built upon the hierarchal foundation of MBO, while inviting the fluid interaction across the entire organization to develop, test, and implement linked plans. In Figure 10-08[202] you will notice a distinct similarity to Deming's PDCA cycle—a sure indication that a continuous improvement process is at work here.

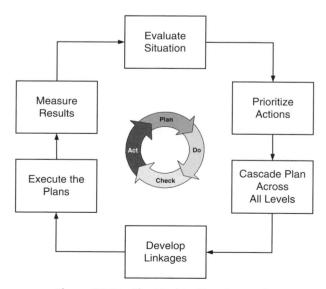

Figure 10-08. The Hoshin Planning cycle

Hoshin Planning begins with the strategic planning process, resulting in a limited number of strategic (breakthrough) goals and objectives. These are communicated downward through all hierarchal levels of the organization, spanning all functional areas. At each level collaboration occurs to determine the appropriate actions to achieve the desired objectives, before the plan is passed downward to the next level. Focus is critical to Hoshin Planning: Too many priorities dilute the energy of an organization. Determination of the appropriate KPIs to support a few critical objectives at each level requires *all* levels of the organization to clearly distinguish symptoms from true root causes.

Hoshin Planning creates agreement down and across the entire organization on key issues to be addressed, with initiatives for improvement *communicating upward* from the lower levels of the organization. While the strategic goals are owned by executive management, the means for their achievement are the responsibility of the teams and individuals. This iterative and cascading consensus-building process is accomplished through a technique called *Catchball*, described by Pascal Dennis in *Lean Production Simplified* and illustrated in Figure 10-09:

1. Company officers develop a vision of what the organization needs to do, and capabilities that need to be developed. They "toss" the vision to senior managers.

2. Senior managers "catch" the officers' vision and translate it into Hoshins [individual plans]. Then they toss them back to the officers, and ask, in effect, "Is this what you mean? Will these activities achieve our vision?"

3. Officers provide feedback and guidance to senior managers. The Hoshins may be passed back and forth several times.

4. Eventually a consensus is reached. Officers and senior managers agree that, "These are the Hoshins that our company will use to achieve our vision."

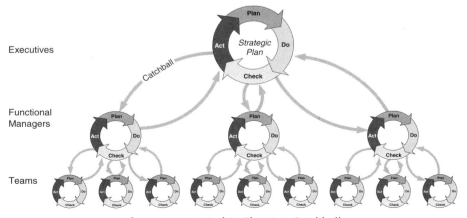

Figure 10-09. Hoshin Planning Catchball

5. Senior managers now toss *their* Hoshins to middle managers, who catch them and translate them into activities. These in turn are tossed back to senior managers who provide feedback and guidance. Eventually, a consensus is reached. Senior and middle managers agree that, "These are the activities (Hoshins) we will use to achieve the senior managers' Hoshins, which in turn will achieve our company vision."

6. Middle managers will in turn toss *their* Hoshins to their subordinates. The process culminates with the performance objectives of individual team members.[203]

Each Hoshin Plan must be communicated simply and associated with a limited number of well-defined KPIs, avoiding the proliferation of complex and nonstandard reporting formats. *A3 Reports*, originally used at Toyota in the 1960s, were given this name since the entire report would have to fit on an A3 size engineering paper measuring roughly $11'' \times 13''$. Although a consistent decision-making process is more important than the particular report format, a consistent format encourages simplicity and economy. For this reason there should be a standard Hoshin format for each purpose, such as for strategic planning, problem solving, and project proposals, illustrated in Figure 10-10 by an example for a Setup Time Reduction initiative.[204]

An enterprise may compile all of the completed Hoshin reports into a binder, creating a concise and chronological history of strategic initiatives and accomplishments across the entire organization. This comprehensive Hoshin record may be used to review the effectiveness of the overall strategic planning and performance management process and is a valuable tool to orient a new employee to the strategic initiatives and organization of the enterprise.

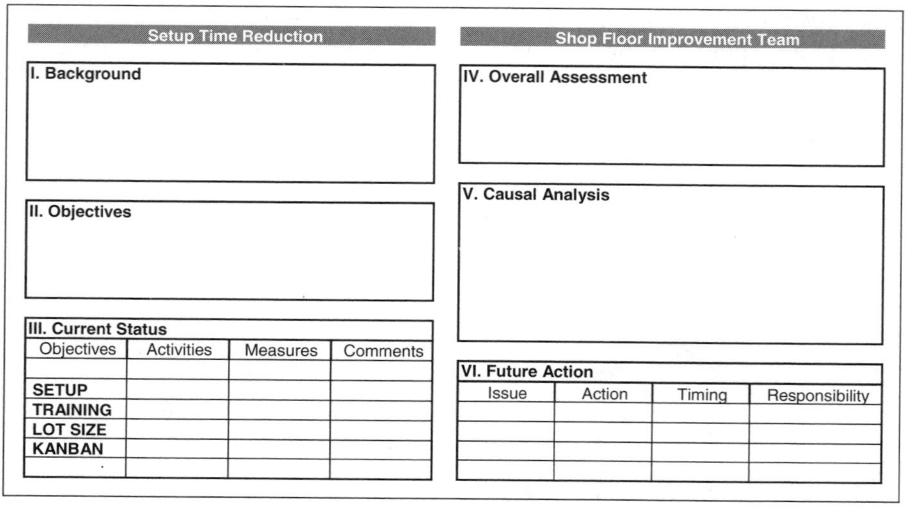

Figure 10-10. Hoshin A3 Report for a Setup Time Reduction Initiative

The development of each Hoshin Plan can be a time-consuming process as it cascades through the levels of an organization, and too many Hoshin Plans can dilute focus and dissipate energy. Hoshin Planning is therefore recommended as a tool to implement a short list of strategic *breakthrough* strategic initiatives, not as a mechanism for managing the many tactical continuous improvement activities that naturally result from kaizen team efforts.

In *Lean Thinking*, Womack and Jones describe the initial breakthrough transformation (Kaikaku) an enterprise may experience on the journey to Lean:

> Converting a classic batch-and-queue production system to continuous flow with effective pull by the customer will double labor productivity all the way through the system while cutting production throughput times by 90 percent and reducing inventories in the system by 90 percent as well. Errors reaching the customer and scrap within the production process are typically cut in half, as are job-related injuries. Time-to-market for new products will be halved and a wider variety of products, within product families, can be offered at very modest additional cost. And this is just to get started. This is the *kaikaku* bonus released by the initial, radical realignment of the value stream. What follows [are] continuous improvements by means of *kaizen* en route to perfection. Firms having completed the radical realignment can typically double productivity again through incremental improvements within two to three years and halve again inventories, errors, and lead times during this period. And then the combination of *kaikaku* and *kaizen* can produce endless improvements.[205]

This distinction between Hoshin Planning breakthrough kaikaku and Continuous Improvement kaizen is described in the table in Figure 10-11, from Cowley and Domb in *Beyond Strategic Vision: Effective Corporate Action with Hoshin Planning.*[206]

Despite the emphasis of Hoshin Planning on a select few strategic breakthrough initiatives, the Hoshin communication framework (supported by IT tools such as Business Intelligence and EIS systems, dashboards, scorecards, and portals) can be useful to focus enterprise-wide continuous improvement. At the higher levels of the organization, the goals and objectives are abstracted from specific actions, requiring the aggregation of information upward through the organization to assess progress—these may be presented as objective-oriented dashboards. At the team level, although numerous continuous improvement initiatives may be underway, individual KPIs may roll up to a single result measure that is called out on a particular Hoshin A3: this is shown in Figure 10-12.

By linking detailed team initiatives to high-level strategic goals and objectives, the Hoshin Plans focus each continuous improvement team. This linkage between Hoshin and continuous improvement is important: Although all team-based improvements are beneficial, some have a greater impact on competitive advantage and strategic success than others. For example, if the enterprise strategy is focused on improving lead time, which has been identified as

Continuous Improvement - Kaizen
• Many small incremental improvements
• Tactical
• Teams and individuals
• Systematic improvement methodology, mostly analysis
• Deals with existing systems and methods
• Part of daily process management

Hoshin Planning - Kaikaku
• Breakthrough, not incremental change
• Focus on a few things
• Strategic
• Must usually be driven from the top
• Frequently involves invention of new systems and methods
• Planning process may only involve certain employees (if not taken to the individual level) but entire organization is aligned to Hoshin objectives

Figure 10-11. Kaikaku and kaizen working together

a current weakness, initiatives that affect lead time reduction should be emphasized over those having less influence on this objective.

This strategic linkage to continuous improvement helps create what Richard J. Schonberger calls an *economy of control*, since an overabundance of controls and measures is wasteful. In *Let's Fix It! Overcoming the Crisis in Manufacturing*, Schonberger describes the result of "data-based operator-centered improvement":

> Various kinds of transactions and reports—the trappings of conventional heavy-handed control—may fall by the wayside. Finally, middle-managerial operational controls and executive-level financial controls prove to be redundant, except for scorekeeping purposes and long-term indicators of business health and success.[207]

This illustrates how Hoshin Planning can guide the continuous improvement efforts of teams and individuals; front-line continuous improvement should ultimately lead to fewer transactions and greater focus using event-driven, exceptions-based, simple, and often visual feedback. Hoshin Planning should not interfere with general continuous improvement, but focus its energy. This requires a clear commitment to design a comprehensive performance management and decision-making process, supported by appropriate information systems, with a determination to keep it simple and to use it regularly.

Hoshin A3 – Lead Time Reduction Objective

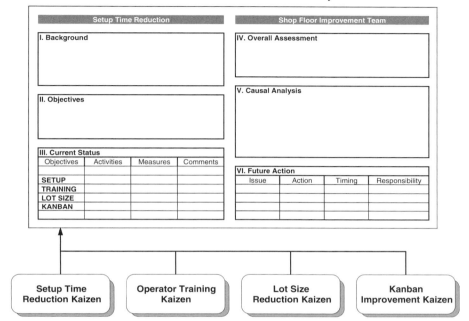

Figure 10-12. Hoshin linkage to kaizen activities

THE MATRIX ORGANIZATION

Hoshin Planning emphasizes the vertical linkages of the hierarchal organization, but we must also coordinate the *horizontal* linkages that follow the flow of value streams across functional and departmental elements within the organization. To create an effective Lean Enterprise we must link strategy with action from the top down, from the bottom up, and *across*. Three forms of measurement must be coordinated throughout the enterprise: department, process and result, and project.

1. **Departmental Measures**—align with the traditional organization structure and may include performance measures for the finance, sales and marketing, product development, operations, human resources, customer service, and other departments. Departmental measures are vertically oriented, inward-looking measures of cost and operational effectiveness that are required to manage the department in a responsible fashion. However, an exclusive focus on departmental measures leads to suboptimization of the overall value streams.

2. **Process and Result Measures**—follow value streams as they cross departmental boundaries. It is important that these measures be associated with overall value to the customer and as suggested by the balanced scorecard, that they evenhandedly measure financial, operational, innovation, and customer perspectives. An organization that has a cross-functional continuous improvement initiative under way, mapping their current- and future-state value streams, has the team framework already in place to develop and manage these holistic measurements. In fact, these process and result measurements should be a natural outcome of the future-state mapping process. If teams do not have relevant *process* measurements in place, then they must rely only on value stream *result* measures, which limit their ability to identify root causes.

3. **Project Measures**—cross departmental boundaries and may impact several value streams. By definition a project is temporary, with a distinct beginning and end, whereas value streams are ongoing. Examples include the construction of a plant or cell, the implementation of a software system, or the development of a training program. Each project should have measurable goals and objectives, a work breakdown structure, and a project plan that identifies organization, responsibilities, tasks, timelines, phasing, milestones, resource requirements, risks, and costs.

The interrelationships among departmental, process, and project activities often lead to a matrix style of organization, illustrated by Figure 10-13.

The mantra "Think global, act local" challenges an enterprise to be more centralized, while simultaneously becoming more decentralized and thus able to react to local demands and markets. This dichotomy led consulting firm A T Kearney to survey more than 200 executives and managers from seven major U.S.-based corporations in six industries. Companies selected for the study had operated within a matrix structure for anywhere from three years to more than twenty years. Although the matrix organization allows a company to address multiple business dimensions with various command structures, the survey results indicate many natural pitfalls as well:

- A company that adopts a matrix structure gains agility and is able to react more quickly to market and customer demands.
- Successful matrix organizations are grown over time, not abruptly installed. Successful organizations must tailor the matrix to meet their own unique needs, and a copycat matrix almost always guarantees failure.
- A truly balanced matrix boasts the following three attributes: information flows freely, power and authority are equivalent in all dimensions, and multiple business objectives are pursued with equal importance.
- A primary challenge of operating in a matrix organization is aligning goals among many different dimensions. Confusion over responsibilities is a

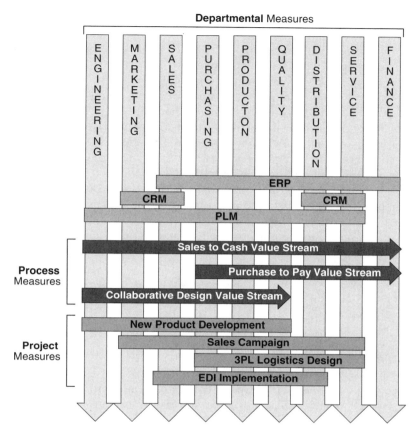

Figure 10-13. The matrix organization

problem in almost all matrix organizations. Leaders can have responsibility without authority; this can give rise to ambiguity and conflict.

- Poor planning aggravates the situation. When organizations make the transition to a matrix structure, they generally do a good job of establishing roles and responsibilities at the top levels but fail to address the roles and responsibilities at the middle and lower levels of the organization. Executives expect employees at these levels to adapt to change as necessary and often expect employees to simply take the initiative when a new situation calls for a reaction. Employees, on the other hand, expect clarity from senior managers during times of change. This disconnect creates confusion and ambiguity, which is exacerbated if organizational goals are unclear, constantly changing, or misaligned.

The study concludes that a matrix organization requires clear guidelines, accountability, training, cascading goals and measures, *and a climate for information sharing.* Senior leaders say the secret to success is to constantly communicate their objectives to employees.[208]

Effective matrix-style management may hold the key to either managing complexity or becoming mired in it. Here is the interesting part—this study (and prevailing attitudes) suggests that an enterprise may *choose* to manage with a matrix approach, or it may not. In fact, any enterprise pursuing continuous improvement naturally evolves into a matrix-style organization whether or not they choose to formally acknowledge the structure. The moment continuous improvement initiatives assemble cross-functional teams and focus on improving value streams, there arises a cross-functional process orientation accompanied by cross-functional performance goals and measurements. The enterprise is also likely to have a variety of projects: education, team building, kaizen initiatives, software implementation, and so on—each of these requiring a cross-functional perspective. A Lean Enterprise naturally evolves into a matrix organization, but if they do not recognize this evolution and reorganize their information flows and decision-making processes accordingly, then conflict and confusion ripple throughout the organization, hindering sustained change efforts.

THE IDEAL PERFORMANCE MANAGEMENT SYSTEM

We have painted a broad landscape in this chapter, integrating several related subjects—ROI, Activity-Based Budgeting, Lean Accounting, the Balanced Scorecard, Hoshin Planning, and the Matrix organization, placing them all within a coherent framework to support continuous improvement and the real-time, event-driven organization. Broad brushstrokes have been applied to illustrate process and reason. More detailed discussions of these topics can be found in a number of sources listed in the bibliography. The purpose here is to help you understand the integrative framework of continuous improvement throughout a Lean Enterprise and how IT can support its realization.

Most importantly, if by adopting continuous improvement an enterprise naturally develops the inherent challenges of a matrix organization but does not directly confront them with the appropriate leadership, management, and information systems, the outcome should be obvious—confusion, conflict, and waste. So the operative question is this: If some form of hierarchal, cross-functional, event-driven, exceptions-based, and balanced performance management system is vital to the advancement of a Lean Enterprise, how could you *possibly* accomplish this without a well-designed information system?

In Chapter 8 we examined knowledge management and its vital components: fact-based decision-making, automated exception notification, drill-down reporting, business intelligence, EIS, content management, dashboards, scorecards, and portals. Then in Chapter 9 we explored the notion of an event-driven organization, where teams and individuals manage the complexities through a focus on critical exceptions within each value stream. The virtuoso orchestration of these elements are encompassed within Schonberger's definition of the ideal performance management system:

> The ideal system of performance management, which perhaps does not exist, even among the world's best-managed firms, goes something like this: All employees are dedicated to intensive, data-based management of processes. Direct results of those efforts show up as weekly, daily, hourly, or, in some cases, real time. Those metrics, therefore, are tracked that often, displayed on visual signboards in all the work centers, and summarized in main trafficways. They constitute the workforce's and management's time-relevant scorecard.[209]

William Christopher points out that we must look beyond Lean Manufacturing shop floor performance measures to achieve the goals of the Lean Enterprise and the Lean Network:

> While Schonberger's definition is good for shop floor related measures such as quality and productivity, it does not very well cover organization capability, customer creation and satisfaction, innovation, profit improvement, government and community relationships, environmental relationships, or outcomes management. Few companies presently have data-based management of processes in these key performance areas.[210]

To be holistic, a performance management system must be multidimensional, looking beyond the traditional financial and operational measures and mechanisms, looking outward to trading partners and the global economy, to manage the key causal relationships that drive the business. In *Software Systems that Support Performance Management*, Brian Maskell and Gay Gooderham summarize their requirements for such an ideal system, representing a blend of management and IT acumen:

- Communicate strategy clearly and consistently throughout the organization.
- Link strategy to action for process managers, departments, and teams through the entire organization, capturing detail at the level appropriate to each user. The cause and effect linkage allows the users to see how their actions support the critical success factors to achieve the strategic goals of the organization.
- Create multidimensional views including process views, modeling the company and its related performance measurements in more than one way. This includes traditional department, region, division, process, project, team, and individual perspectives.
- Link actions to people accountable.

- Trace balanced sets of measures, including finance, operations, innovation, and customer service.
- Present scorecards at various levels of the organization.
- Promote focus on key drivers and critical results that link actions to strategic objectives; a key contribution of an information system is to sort the wheat from the chaff, presenting exceptional and actionable information to each individual.
- Make results readily accessible with appropriate security, the system must become an everyday part of people's work—easy to use, intuitive, accurate, and visual. The system must be widely available, and since it's key to corporate strategy, it must also be highly secure.
- Provide tools for analysis of results, scenarios, and measure relationships. Analysis creates insight into the company's operation and the changing marketplace.
- Support team collaboration and rewards, closing the loop on a performance management system is the link to a rewards or gain-sharing program. It is important that the concepts and calibration of the performance management approach are sound before moving into the delicate areas of compensation, otherwise unintended and harmful consequences may result.
- Integrate with corporate information systems so there is no time wasted in gathering, entering or reconciling data from multiple sources.[211]

To their list I add the following recommendations:

- Reduce transactions and focus a limited number of measurements on key causes and constraints.
- Focus on root causes rather than symptoms, establishing clear cause and effect relationships with result and process measures.
- Continuously evaluate measures for relevance, simplicity, and economy.
- Direct activities by event-driven measurements whenever possible, focusing near-real-time attention on key exceptions and business drivers.
- Encourage teams to develop their own measurements, according to clearly communicated strategic goals and objectives.
- Extend the boundaries of value stream measurements across the Lean Network to eliminate all waste and improve customer value.

The Need for a Performance Management Champion

Do information technologies now exist to create such a system? As we explored in Chapter 8 on Knowledge Management, the answer is clearly yes. However, IT is necessary but not sufficient to enable strategic performance measures to drive effective continuous improvement throughout the organization. Continuous improvement emphasizes that change must be owned at the individual and team levels. This is supported by Hoshin Planning methods, where goals are communicated downward and solutions are developed by the teams. Yet with so much decentralization, with so many moving parts, horizontal and vertical linkages, potential conflicts of matrix-style responsibility and authority, cultural, language, geographic, and time boundaries, parochialism, self interest, and corporate inertia—we cannot expect this all to come together without leadership. An integrated performance management system needs an architect, a champion.

In their study on matrix organizations, A T Kearney discovered that several organizations use either a process guardian or a committee to monitor performance of their matrix. If the monitor of choice is a process guardian, the person appointed should be in a position of influence and well respected within the organization. At one Japanese manufacturer, process guardians are well-respected executives nearing retirement. Another top matrix organization designates the process guardian position as a direct report to the CEO. Says one manager, "The process guardian needs to be fireproof."[212]

For example, our firm assisted a respected one hundred-year-old manufacturing enterprise to develop a program of continuous improvement and an IT strategy to support their strategic plan for the generations to come. We began by forming and educating teams and process mapping the current state value streams. This led to the development of future-state objectives, the definition of system requirements, and the selection and implementation of new ERP and related systems. To accomplish change of this magnitude, this enterprise asked their Vice President of Operations, a well-liked and respected 35-year veteran approaching retirement, to accept a newly created role as *Vice President of Business Effectiveness*. This individual reports directly to the CEO with the following objectives:

- Creation of a company-wide education program in continuous improvement
- Formation and nurturing of cross-functional teams
- Design of a comprehensive performance management system
- Creation of the project management office to oversee major initiatives
- Collaboration with IT to ensure appropriate functionality and usability
- Development of a team-based continuous improvement framework that will sustain itself long after he retires and passes the baton to the next generation

Call this leadership role what you will: architect, orchestrator, process guardian, champion, coach—for lasting transformation, a respected individual or team should rise above the traditional organization structure, ensuring that people, processes, and technology act in unison. To encourage breakthrough and continuous improvement, executives must clearly articulate the vision and strategy, and the champion may then take the lead role in spreading this message throughout the organization. Finally, the IT architecture supporting the performance management system must be simple, agile, and intuitive, so that it becomes an integral and value-adding part of the culture. With that challenge we turn to our final chapter.

Chapter **11**

Lean IT: Applying Continuous Improvement to Information Systems

We seem to spend more time fighting with our software than working with it.
Can I really trust this data?
Why doesn't our IT staff seem to understand how our business works?
Why do we invest in systems that don't solve our problems?
We seem to waste half of every meeting arguing over whose data is correct!
But we just replaced that software five years ago. . . .
After all that money and effort, and everyone still relies on spreadsheets?

Do these lamentations sound familiar? The unfortunate fact is that many companies feel they are held hostage by their information systems: They can be unreliable and overly complex, not suited to the business needs, while consuming vast resources to purchase, implement, and maintain.

Furthermore, business managers are often expected to accept responsibility for systems they don't understand, can't manage, don't trust, and possibly even fear. It would be unacceptable for the production, marketing, sales, or finance departments to operate in this fashion; why should IT be any different?

Fortunately, IT has reached a significant evolutionary milestone, and we now have the opportunity to change this untenable condition. Since the birth of the computer industry change has been revolutionary, with each generation effectively replacing the last, despite expensive attempts to integrate old and new. With an established foundation of standards in hardware, communications, software, and database technologies, change can now be evolutionary. And with the maturity and consolidation of the ERP industry, an enterprise

Lean Enterprise Systems: Using IT for Continuous Improvement, by Steve Bell
Copyright © 2006 by John Wiley & Sons, Inc.

application foundation may be established upon which we can build for the future. We can now strive for the continuous improvement, rather than the continuous *replacement*, of our enterprise information systems.

In the first ten chapters of this book we have explored how IT may aid in the continuous improvement of a Lean Enterprise. In this final chapter we turn the tables, exploring how continuous improvement, and the lessons learned from decades of Lean Manufacturing evolution, may enhance the performance and longevity of IT. To understand this new approach to managing change we begin with a brief examination of the past; then we'll explore the future of Lean IT. This chapter is thus organized into four sections:

The Challenges of Traditional IT—explores some of the pitfalls of traditional IT change management practices.

What is Lean IT?—examines the tools of Lean IT, the significance of an enterprise software ecosystem, and lessons learned from Lean Manufacturing.

Guiding Change with Lean IT—explains why information technology is *not* the solution, merely a tool in the hands of people to improve processes.

Applying Lean IT to the Lean Enterprise—illustrates how Lean IT and the Lean Enterprise may work together for sustained continuous improvement.

THE CHALLENGES OF TRADITIONAL IT

Pain, Chaos, and Project Failure

Although they won't readily admit this, and you won't see this word appearing in any glossy brochures or fancy websites, IT sales and marketing professionals spend considerable time talking about their customer's *pain*. What is it? Why does it happen? What does it cost? Has the pain reached a critical threshold to stimulate a buying decision? If not, what can they do to elevate this perception? Who is the decision-maker that is motivated to relieve this pain? What is it worth to him or her? And most importantly, how can they position their "solution" to eliminate this pain? Astute sales and marketing professionals know that we all live with pain—it's simply a fact of existence. They understand that people are able to ignore most pain for long periods of time with a variety of clever avoidance techniques. They also know that when pain reaches a critical point, people react quickly and often through emotion. Being at the right place and time with a quick remedy for the pain is often more important than having a legitimate and long-term solution to the problem. In fact, most business problems are solved by changes in policy and process enacted by people—the information technology is simply a tool to facilitate the change.

But because it is human nature to wish for the easy way out, the unfortunate fact is that many reactive information technology "solutions" create new problems as they solve the old ones. In fact, a new solution may introduce more pain than it relieves, or it may shift the pain from one part of the organization to another. In any case, the introduction of a new information technology often creates a new cycle of pain that leads to yet more information technology acquisition.

IT is all about managing change, and change can be unpredictable. The *APICS magazine* article "Give Change a Chance" observes:

> IT projects create value by creating organizational change. The greater the scope and impact of the technology, the more change it creates. ERP touches every area of the company, and many people don't necessarily perceive a problem in their own area. Why is change—and workers' natural inclination to resist it—so prevalent in most ERP implementations? It's the nature of the beast, both the ERP beast and the human one. Bring those two beasts together and you've got what could be a volatile, expensive, and time-consuming situation.[213]

IT projects are indeed volatile, expensive, time-consuming . . . and risky. The Standish Group has been conducting surveys on all types of IT projects since 1994. Their research, published in their annual CHAOS report, reveals that in 2001 a staggering 31.1% of projects were cancelled before completion. Further results indicate 52.7% of projects cost 189% or more of their original estimates. The proportion of successful projects completed on time and on budget is only 16.2%. And, even when these projects are completed, many are a mere shadow of their originally specified requirements. Projects completed by the largest American companies reported achieving only 42% of the originally proposed features and functions; and while smaller companies [and smaller projects] do better, there is still a pretty large gap.[214]

The pain of failed IT projects is felt by large and small enterprises alike, but it is the failures of the largest that have become legends. Many well-publicized ERP and SCM failures have cost large enterprises hundreds of millions of dollars. The widely reported losses include only immediate revenues, project costs, and market valuation, but the damages extend beyond dollar figures to the immeasurable loss of customers, employees, and reputation. Perhaps large enterprises can recover from hits like these, but small and medium-sized companies cannot.

In addition to the losses suffered by individual companies, there is a massive economic impact from poor enterprise software quality, according to a *The Economist* article titled "Managing Complexity":

> A 2002 study by America's National Institute of Standards (NIST), a government research body, found that software errors cost the American economy $59.5 billion annually. Worldwide, it would be safe to multiply this figure by a factor of two. So who is to blame for such systematic incompetence?

Cost overruns and delays are common in numerous industries—few large infrastructure projects, for instance, are completed either on time or on budget. But it is peculiar to software that billions of dollars can be spent only for nothing useful to result.[215]

Large IT projects are dangerous territory, introducing significant business risks and seducing us with high expectations for new technology, while confounding us with assertions of intangible ROI. The practice of IT is often permeated with a sense of magic and mystery, where businesspeople suddenly find themselves at a loss of confidence. And when large and complex information technology projects meet with the culture and unpredictability of organizations and people, they often fail.

The Traditional Approach to Enterprise Software Management

Let's examine the traditional methods of how a company selects, implements, and maintains enterprise software, and later we'll build on these concepts to show how a company can achieve Lean IT.

An enterprise must first match their needs against the capabilities of enterprise software, and it is commonly accepted wisdom that most horizontal* ERP applications should satisfy 80% to 90% of the requirements of any enterprise. Depending on how sophisticated and unusual the operations, they may call for some customization, or the integration of specialized vertical software, to exceed 95% of these requirements. Beyond 95% most find there is a decreasing marginal cost/benefit for a software application investment, and often the *gap* (the remainder of unfulfilled requirements) is satisfied by a mixture of process redesign, manual work, offline spreadsheets, and disconnected desktop databases.

Herein lies a vital question that many fail to ask: Are the unique requirements in the '95% and above' zone (that are not satisfied by an off-the-shelf software application) simply remnants from obsolete legacy systems and outmoded practices, or are they what create distinction and real competitive advantage for the enterprise? The answer to this question may determine whether these practices deserve investment or should be thrown out with the legacy system bathwater.

How do we arrive at this measure of 85%, 90%, or 95% fit to our requirements? First, we must identify those requirements; approaches can include extensive investigation by consulting firms using elaborate requirements management software tools and checklists, or internally led initiatives with varying degrees of formality. In any case, let's assume that we have identified a list of 1000 specific software capabilities needed to run the business, weighted by importance. We'll discuss the practical implications of managing such a large

---

* A 'horizontal' application is generalized software designed for most types of businesses, whereas a 'vertical' application is designed for a specific industry or purpose.

list in a moment, but for now we must develop a plan for the implementation of these capabilities when we install the new system.

In our practice we rarely see a company that can implement all of its requirements in a single "big-bang" implementation. Even those that claim to implement in this way choose to leave a few low-priority capabilities on the table to address later—thus even so-called big-bang implementations are phased to some degree. Let's assume for this example that a company selects an application that satisfies 90% of its requirements (900 out of 1000 from their list)—are they going to realize the full 90% potential of the application after the first phase of implementation? Probably not. Can they get by with fewer than 50% of these capabilities? Again, probably not. So what results is a set of necessary and target zones somewhere between 50% and 90% of their requirements, with the 90% line representing the *system potential*.[216] The necessary zone defines those capabilities the enterprise should have to run the business effectively, and the target zone indicates the nice-to-have capabilities that will boost performance. There is also a danger zone below which the enterprise cannot function; these zones are illustrated in Figure 11-01.

Should an enterprise lack the skills or resources to rise above the necessary zone and into the target zone, it may be perpetually trapped by manual workarounds and dis-integrated systems that compromise business effectiveness. And if the meager capabilities fall into the danger zone, the application is probably doing more harm than good.

Traditionally a project team should select (or develop) a new application that provides a path to the target zone, implementing the application in phases. In the example illustrated in Figure 11-02, this particular enterprise intends to implement phase 1 beginning with 65% of the functional requirements, proceeding to phase 2 implementing another 20%, which results in an 85% fit. Finally they plan to creep over 90% with a combination of incremental refinements to the software and processes, supplemented by spreadsheet and manual workarounds.

Following this rigorous methodology, companies may succeed with the implementation, delivering a functional application. According to the Standish

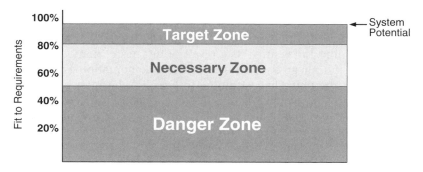

Figure 11-01. Software Fit Zones

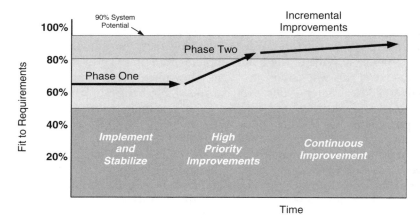

Figure 11-02. Phased Implementation Plan

Group CHAOS report, however, quite often this story does not have a happy ending. There are many ways a company can be trapped by the complexity of an IT project and become a statistic. For example, a company may struggle with a poor application, or perhaps an adequate application implemented poorly, for years. One day the pain becomes too great and they react, selecting another application that they believe will suit their needs better—hoping for a target of 80–90% fit. But they don't do a good job identifying their requirements, because they lack the time, the skills, or a vision for the future; they simply define their requirements as an extension of the capabilities of their current application.

This company chooses an application that can provide 80% of the requirements they can articulate. Because the company is in crisis, struggling with their current application while trying to keep the business running, they decide to implement 70% of the system potential as quickly as possible. Unfortunately, the project soon begins taking longer and costing more than expected. As they learn of new capabilities they did not consider when defining their initial requirements, they may increase the scope during the project. The timeline and budget slip further.

As they continue falling behind they begin to hurry, devoting insufficient attention to training and testing. As a result of this haste, many unanticipated problems arise during go-live. By now the implementation has become a chaotic dash for the finish line, straining to keep the business running while fighting brush fires in every direction. Finally someone throws "the switch" and the new system is live. However, because not all the desired functions work as planned, they end up achieving only 60% of their new system potential. Everyone in the company is exhausted, the project team is disenchanted, and management is frustrated. The project did not go well, and no one is looking forward to phase 2.

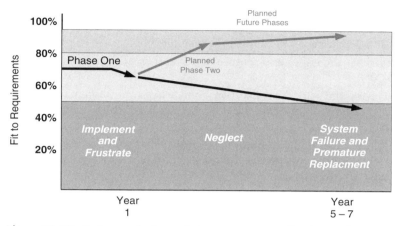

Figure 11-03. Train wreck: the path to premature application replacement

No surprise—phase 2 never happens. In addition, the business requirements continue to change, and while the vendor issues updates that improve the application potential, the use of the application is not improved. The effort to upgrade the software is a substantial project in itself, and is thus avoided, so the software upgrades stay on the shelf. Because of the lack of training and proper user documentation, combined with employee turnover, system performance continues to degrade, until one day someone hollers, "This software doesn't work, we should replace it!" And the cycle begins again. This unfortunate drama is illustrated in Figure 11-03.

Whatever the specific causes, ultimately this is a failure of change management. The business processes and the underlying systems are not continuously improved, and the situation follows the natural path of entropy. To manage change in the traditional way, the team must manage the multitude of issues and requirements that arise during an IT project. During the early discovery phases of a project, while the team is conducting interviews, process analyses, mapping sessions, etc., hundreds or even thousands of distinct issues will arise—problems, questions, variables—that lead to specific requirements. Some newly discovered issues will be critical, whereas others will not be. It is important, however, to record all issues and requirements as soon as they are discovered, to prevent them from becoming lost.

At some point, this list develops into a prioritized set of issues and requirements that may be used to select software, to direct the phases of its implementation, and to measure the results. Earlier we suggested a hypothetical list of 1000 requirements. In fact, there may be many more, but it is not necessary to manage all of them with equal diligence. Here is a traditional approach that focuses effort on the most important issues:

- Select a tool to manage the issues and requirements. Although sophisticated software tools are available for this job, a simple spreadsheet or desktop database may suffice for a smaller project.
- Prioritize the issues and requirements, and carefully manage the top 10% to 20%:
  - During the selection of software these key requirements are called *critical differentiators*, because they distinguish one application's suitability from another.
  - During the implementation, as well as ongoing maintenance of the system, these key issues should be prioritized by the impact they will have on the business.

Using this approach, the project team may whittle a list of 1000 down to one or two hundred critical issues and requirements; these may be assigned to various subproject teams, where the volume becomes manageable. Because these issues and requirements are managed in a database, they can roll up into a consolidated list that is used to measure overall system performance by *percentage of fit* to total requirements.

The Seven-Year Itch

Despite the fact that the traditional change management tools and methods just described have existed for many years, enterprise software applications are replaced with surprising frequency. In particular, it is well known that ERP systems historically have a lifetime of no more than five to seven years. And this frequent replacement cycle of individual applications is only the tip of the iceberg; the fact is that most enterprises maintain *several* critical enterprise software applications, each with their own challenging implementation and life cycle management issues. Each of these applications must also be integrated and managed collectively to support the smooth flow of processes and value streams. As the statistics suggest, the chances for success in the face of such instability and complexity aren't encouraging.

The enterprise software industry clearly knows about this five to seven year replacement cycle; in fact, it has come to depend on it. Doug Burgum, formerly CEO of Great Plains Software, is now Senior Vice President of Microsoft Business Solutions. In 2002, *Red Herring*, a Silicon Valley technology magazine, asked Burgum when he expected the ERP market to rebound:

> He not only skips the 'poor visibility' jargon, he names specific years. "This market is going to get a lot stronger in '04, '05 and '06," says Burgum. Call it the Y2K echo or the seven-year itch. Whatever the name, it's the cornerstone of Microsoft's stealthy plan to storm the business applications market. In the late 90's, companies loaded up on business apps to upgrade systems that couldn't withstand the year 2000 date change. Since then, sales have nosedived. But companies typically refresh their application software about every seven years; hence Mr. Burgum's prediction.[217]

Following the Y2K boom and bust, the enterprise software market in general, and ERP suppliers in particular, have seen some hard times. In late 1999 ERP sales suddenly plummeted, like someone turning off a light switch. In the post-Y2K aftermath, the number of viable ERP vendors has dwindled from hundreds to a few dozen. Some of the ERP products simply disappeared, leaving their customers racing to find a replacement. Many more were acquired by ERP software companies bent on acquisition* to capture the ongoing maintenance revenue stream and customer base.

These parent companies are consolidating multiple ERP products into a portfolio, hoping to leverage their large customer base and service revenue stream to fund the development of the "next generation" of ERP system. With that said, the likelihood of the emergence of many *completely new* ERP systems is very unlikely for two reasons: 1) ERP software has become extremely broad and complex, creating a significant barrier to entry, and 2) the ERP growth boom is over, and most publishers are now incrementally improving their systems primarily funded by ongoing maintenance and professional service revenue streams. The top tier ERP vendors are striving to move their complex systems downmarket because most of their largest customers have long since purchased ERP systems. After garnering 50% or more of the global ERP market for large enterprises, SAP's revenue growth has shifted from product sales to professional services.

But what about the next seven-year replacement cycle? If anyone has the muscle to introduce a revolutionary new ERP system, wouldn't it be Microsoft, with their +$3 billion annual R&D budget? Even Microsoft's *Project Green*, their initiative to develop a comprehensive new ERP system to replace their assortment of acquired systems to meet this next replacement cycle, was placed on the back burner in 2004. An *InfoWorld* article, "Microsoft Puts Brakes on Next Business Apps", states that:

> Microsoft plans to build completely new business applications on a single code base that will eventually replace its existing offerings. Microsoft originally had planned to ship the first results of Project Green as early as late 2004. Because the first products now won't be out until 2008 at the earliest, the number of developers assigned to Project Green is being reduced from 200 to 70. "We have made a decision to move resources off Green and back on the core product lines to strengthen those product lines because we realize now that it is going to take much longer," Burgum told a federal court in testimony in the U.S. Department of Justice's case to block Oracle's takeover of PeopleSoft.[218]

Could it be that even the powerful Microsoft is experiencing difficulty managing the overwhelming complexity of several ERP systems? Then in

---

* These hungry ERP acquisitors include SSA Global, Epicor, Infor Global Solutions, Sage Group PLC, Exact Software, Intuit, Microsoft, and Oracle.

March 2005 Microsoft restated their strategy, suggesting that they would maintain the separate product lines for much longer, while investing in their enhancement with the latest developing technologies.[219] We may expect Oracle to experience the same challenges as they assimilate PeopleSoft; shortly after the acquisition in early 2005 Oracle announced Project Fusion, the intent within three to four years to combine three massive ERP products (Oracle, PeopleSoft, and JD Edwards) into a unified code base[220]—a seemingly impossible feat.

Have we arrived at a critical milestone in the evolution of the enterprise software industry, where managing complexity, and not the mixture of promise and chaos of emerging technology, is our greatest opportunity and nemesis? What does this mean to the company that is planning to purchase a new ERP system? What does it mean to a company striving to improve the one it already has? Is an enterprise destined to continue replacing its core enterprise system every seven years?

This crisis of complexity does not just affect ERP systems; it's a challenge that the entire IT industry must confront. The imperative question that results from all of this pain, chaos, cost, risk, complexity, and failure is this: *If a manufacturing enterprise cannot compete in the global market without IT, then how can they make IT manageable?*

WHAT IS LEAN IT?

Lean IT is practical, manageable, agile, and team-based, and it must add value to the enterprise. That sounds easy enough, but don't expect that achieving it will be. Jim Womack speaks of the natural inertia of human nature working against the idea of Lean IT:

> I'm not naive about getting the world to embrace Lean information management. We're not quite yet at the end of thinking that more information is always better and that if we just had all possible information, perfect algorithms, and lightening fast central processors, life would be easy. For example, despite 50 years of evidence that this isn't true, we are now embarking on a new experiment with RFID in which every item in every process can be tracked individually.[221]

So how do we compensate for the natural tendencies toward overcomplication, overautomation, and rigidity, to realize the benefits of Lean IT? Recall that in Chapter 1 we contrasted the rigidity and long-range planning of traditional IT with the agility of Lean thinking; this is shown again in Figure 11-04.

For IT to become Leaner it must:

- Manage change incrementally and continuously
- Organize and execute with cross-functional teams

Attribute	Lean	Traditional IT
Change Management	Organic, incremental and continuous	Engineered and planned large events
Organization	Cross-functional teams	Central command and control
Measures	Top-down and bottom-up performance measures linking improvement initiatives to strategic goals	Cost containment and uptime
Knowledge Management	Generalization	Specialization
Education	Process focus	Task focus
Definition of Success	Speed and Flexibility	Stability

Figure 11-04. The attributes of Lean and *traditional* IT

- Measure performance in a holistic and relevant manner
- Encourage the general development and sharing of knowledge
- Focus education and improvement initiatives on processes and value streams
- Measure success by speed and flexibility, without causing chaos
- Be accepted and used properly by the user community
- Enable users to be more effective at their value-added activities

Lean IT is attainable, but it requires similar effort and resourcefulness as an enterprise striving to transform its traditional manufacturing operations to Lean. The fact that we're dealing with computers rather than drill presses and assembly lines makes little difference.

Michael Hugos, author of *Building the Real-Time Enterprise: An Executive Briefing*, stresses that:

> IT can be a big part of what makes a company agile, or it can be a big part of what makes it a clumsy, slow-moving bureaucracy. One of the major determinants of this is the way your company answers the question, "Should we build our systems fast, or should we build them good?" The agile answer is to build them fast and good enough for now.
>
> What does "good enough for now" mean? In a fast-paced, competitive world, opportunities arise quickly and then either fade away or evolve into something else. The advantage goes to companies that can develop systems that are ready when the business needs them and don't cost more than the opportunity is worth.

The best way to do this is to create systems out of combinations of simple build-ing blocks and repeatable processes.[222]

On our path to Lean Manufacturing we must transform the factory with flexible and standardized processes; similarly, to craft Lean IT we must develop agile and standardized software, development, integration, training, and support tools and methods. More importantly, however, to achieve Lean Manufacturing we must change the way we think—new attitudes accompa-nied by effective change management and continuous improvement practices are equally vital to the development of Lean IT. First we'll explore the tools of Lean IT, then in the next section we'll turn our attention to the essential issues of change management and continuous improvement.

The Future of Enterprise Software

IT professionals now have the opportunity to do more than just keep their heads above water. This change from revolutionary to evolutionary, from continuous replacement to continuous improvement, offers a real opportunity not just to achieve a system's potential but, more importantly, to continue improving upon it indefinitely.

Gartner, who is known for spotting and naming emerging trends, has coined the term *ecosystem* for the new industry model of enterprise software, and *ecosystem vendor* for those large entities around which the supporting players cluster. *ComputerWorld* notes in the article "Gartner Sees Shift to Bite Size Business Software":

> Gartner believes that makers of software [. . .] such as SAP, IBM, Oracle, and Microsoft must carve smaller pieces out of their large packages to make it possible to adapt them more quickly. And they must also make sure that those pieces can also plug into competing products, as companies cherry-pick more specialized programs from different vendors but want to stitch them together seamlessly. This increases the agility of the software because it's now easier to arrange the process or determine who will actually perform each step in the process.
>
> In this way, business process is shifting IT projects from large multi-year marathons to rapid deployment gap applications [incrementally improving capa-bilities up to and beyond the 90% fit zone]. Instead of continuing to sell com-prehensive products, software makers are trying to create what Gartner calls "ecosystems"—realms where they shape the environment and create frame-works and standards within which others operate. Increasingly a [software] company is destined to become part of an ecosystem in order to survive. On the other hand, software buyers need to switch from one large purchasing decision to picking the right product for individual tasks.[223]

This ecosystem model suggests that enterprise software life cycles will lengthen and there will be fewer new entries into the marketplace for the *core* enterprise systems: particularly ERP, and to a lesser degree CRM, PLM, APS, MES, WMS, and others. Why the distinction between ERP and all the others? As I described in Chapter 6, ERP is the backbone of enterprise software, the core around which all others integrate. The broad scope and complexity of an ERP system, although on one hand desirable, on the other hand is costly to implement. Once an ERP system is well-established, it is dreadful to replace because a mosaic of supporting applications have united around its framework; the entire ecosystem now relies on the survival of its host.

With the continuing advance of integration technologies, these host software vendors strive for evolutionary (not revolutionary) advances, so they do not risk losing their existing customer base. In this new model the ERP system now sets the pace of evolution through incremental functional and technical changes—the upgrade paths and future development cycles of the entire ecosystem depend on their plans. Ideally the entire ecosystem will evolve in a smooth fashion, ensuring that the core ERP system will persist, gathering new developments and partners that emerge as markets shift and requirements change.

According to Ray Lane, former President of Oracle, speaking before 1100 software industry executives at the illustrious Software 2004 conference in Silicon Valley, "Software innovation on a grand scale is dead." The chances of revolutionizing the software market today, on the scale of what SAP or Siebel have done, are slim. Consolidation rules. The real change taking place is in software as a service: delivering applications—faster, cheaper, and more nimble—to an enterprise's Web Services-based architecture, rather than offering packaged software.[224]

This new ecosystem model means that companies can no longer hope to solve their problems with the replacement of their ERP software every few years. Of course they can try, but with the remaining ERP suppliers improving their products to the point where they are functionally similar, what's the point? Because the enterprise software market has changed, so must the approach to selection of a partner. Once there were hundreds of ERP vendors competing for business; now there are only a handful of viable candidates. This suggests a thorough consideration of the partner as well as the product. An enterprise should ask: Do we trust these people? Do we buy in to their vision for business and technology advancement? Forget the seven-year replacement cycle, this is a long-term marriage. The software function, look, and feel that evolves ten years from now may be quite different, but we'll still have the same partner.

The bottom line is that an enterprise should expect to use its ERP system for many years to come, so the continuous improvement of the entire enterprise software ecosystem surrounding the ERP core is essential. Recall Michael Hugos' earlier suggestion in "*Agility Is a Frame of Mind*", "The advantage goes to companies that can develop systems that are ready when the business needs them and don't cost more than the opportunity is worth. The best way to do this is to create systems out of combinations of simple building blocks and repeatable processes."[225] This is the very essence of what Web Services and Service-Oriented Architectures offer, and how the enterprise software ecosystem may be created.

The Role of Web Services and Service-Oriented Architectures

In the Lean IT model, although the core components of the ERP system provide the stable structure and transactional framework for the enterprise, changes in requirements may be managed by implementing relatively small and standardized pieces—sometimes called *components, objects, building blocks*, or *granules*. This Lean change management approach is similar to an Assemble or Configure to Order production environment. Although there are strong competitive pressures to move quickly in this direction, these are counterbalanced by great industry inertia. Most ERP suppliers have been investing in components and Web Services for some time now, granularizing their systems to some degree. However, many of their organizational structures, consulting methodologies, licensing structures, maintenance policies, pricing, compensation, and revenue recognition models are still managed as a monolithic framework.

Then there is the Open Source software movement, where the underlying source code and intellectual property are available to a community of developers and users. Open Source involves a rapid, communal, and democratic development approach with frequent interaction between developers and users. The popular Open Source principle of Free (capital F) software shows the characteristics of a social movement. Free software is accessible via a license that grants users permission, in perpetuity, to copy, modify, study, and distribute the software's source code. It is a philosophy about the development, distribution, and accessibility of software, namely, the freedom involved to that end—Free does not refer to price.[226] Whatever the underlying motivation, economics or social movement, Open Source is a rapidly growing phenomenon that commercial software publishers are being forced to confront and embrace.

The article "Demand at the Fount for Open Source" argues that we are seeing early signs of a significant shift in how companies think about software development:

> [Within the Open Source community] the features of the software may literally be developed by a party other than the one that originally provided the software, and that development may actually be incorporated into the original source of the software itself. This means that if, for example, a company needs something from its Open Source software which is not supported, it can then develop or sponsor development for that functionality in the product. The functionality can further the growth of the product as a whole. Thus the software's entire user base can benefit and the primary development team of the software may not have to devote as much in the way of resources to creating new functionality on its own.

Companies sponsor such development because they have the opportunity to get what they need at a lower cost, and via an efficient process. This development process signals a shift in how the software industry does business.[227]

The Open Source community proudly demonstrates their large-scale commercial viability, pointing to Amazon and Google, whose infrastructures are deeply rooted in Open Source. Erik Keller originally coined the term ERP while at Gartner, and is the author of *Technology Paradise Lost*. Keller predicts that the Open Source movement will cause a significant and irreversible disruption to the enterprise software industry:

> It used to be assumed that either you outsource a business process/application, build it yourself, or buy software for it. Open Source changes these assumptions dramatically, as it brings back the potential to build and own an application; or to blend methods by means of a consortium, in-house development, or contracting with offshore providers. Thus the economics of how IT gets deployed is turned on its head and is ready for reevaluation.

> Buyers of technology now have choice, and with that choice a large amount of bargaining leverage. This is one of the reasons why the software market has yet to recover to historical growth patterns. To cope with Open Source, many sellers of technology will need to switch their investment and revenue strategies from the front end of selling a piece of software to the back end of supporting a process. What a seller may lose in hardware or software license fees, it will need to pick up in long-term support contracts and consulting. These factors and the maturing landscape of Open Source products and initiatives will permit buyers to play software license vendors against Open Source service vendors, forcing margins and long-term pricing downward.[228]

The Open Source movement is gaining strength, with the major *infrastructure* players (including IBM, Hewlett Packard, Sun, Oracle, and perhaps even Microsoft) enthusiastically signing up. However, it is unlikely that the mainstream enterprise systems (ERP, CRM, PLM, and others) will be quickly replaced, because these massive applications represent many man-centuries of development. In fact, as ecosystem hosts, many enterprise software publishers are now embracing (or are being forced to embrace) the potential for Open Source component integration. As Gartner pointed out, the customer must select an ecosystem, then form a fluid relationship with its constituencies, adapting to changing requirements. In the future, many of the vital components of an enterprise architecture may be Free.

Agile Software Development

The use of these new tools, techniques, and relationship models lead to an approach called *Agile Software Development*. This approach reduces development lead time while improving flexibility, helping to avoid the massive software debacles experienced by large and small enterprises alike. Until now, complexity has been our biggest constraint to producing timely, flexible software tools. According to *The Economist*:

> There are five steps involved in creating a piece of software: enumerating the requirements; designing the program; actually writing the code; testing it; and then deploying it. Traditionally and naturally enough, this was seen as a sequential process. However, John Swainson [formerly in charge of software development for IBM Corporation, and now CEO of Computer Associates, one of the world's largest software companies] points out that by the time an organization gets around to deploying a piece of software, its requirements have often already changed. This, he says, means that an "iterative" model, in which an organization continually cycles through the five phases, makes more sense than the traditional "waterfall" which puts them in sequence.
>
> The main principle of agile programming is that developers must talk to each other often, and that they must talk to the business people setting requirements equally often. Combine this with a short time-scale—ideally agile proponents seek to deliver a working bit of software every few weeks—and you have an accelerated, informal version of the iterative model. This means that no project can go on for years and produce nothing—a fatally flawed project will be caught sooner.[229]

Figure 11-05 contrasts the traditional waterfall approach to the iterative (spiral) model, where phases are much smaller in scope (measured in days rather than months), using smaller development teams and frequent interaction with the users, to deliver workable solutions faster. Requirements that are

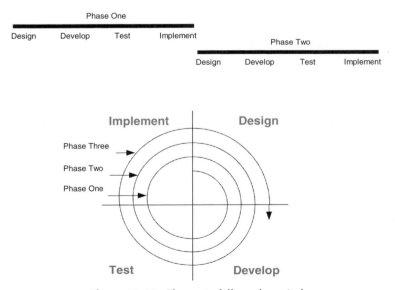

Figure 11-05. The waterfall vs. the spiral

not prioritized for the current phase are set aside for a short while, until the next cycle comes along.

The iterative development approach favors smaller projects, which means that development teams are not required to forecast requirements far into the future, so the system is able to quickly adapt to change. Furthermore, the teams themselves may be kept smaller and tightly focused. A fundamental challenge for a traditional software-development organization is Brook's law*: Adding more programmers to a late project makes it later. More generally, Brook's Law predicts that the complexity and communication costs of a project rise with the square of the number of developers, while work done only rises linearly.[230] In other words, several small teams work faster and better than one large team. The corollary to Brook's Law happens to be one of my favorite resource management principles: Nine women cannot make a baby in a month.

Agile software development replaces long, arduous phases with rapid and continuous improvement cycles. Requirements forecasting is reduced, lead times for delivery are shortened, waste is eliminated, and quality is improved. By the way, have you noticed that this is the same iterative diagram used to illustrate Dr. Deming's PDCA cycle? And did I just say "continuous improvement cycles"? It's beginning to sound like we're a software factory, producing on an Assemble to Order basis. Work is pulled by real-time customer demand, utilizing concurrent product development methods, small teams, standardizing work, eliminating waste, resulting in reduced lead time, improved quality, and agility. This is the essence of Lean IT.

GUIDING CHANGE WITH LEAN IT

Technology is NOT the Solution

The tools to support IT change management are important; but as we emphasized with Lean Manufacturing in Chapter 3, the tools are *necessary but not sufficient*. Just as Lean Manufacturing requires a fundamental change in thinking, Lean IT requires effective change management attitudes. Agility begins as a frame of mind.

Despite what much sales and marketing literature gushes forth, information technology is just a tool, it is not "the solution." To deliver value to the customer, Lean IT must aid in the solution of a business problem by developing effective and standardized procedures, thereby enabling continuous improvement.

To get to the heart of the business problem, to find the controlling simplicity, the point of greatest leverage, we must identify and eliminate the *constraint*. According to the Theory of Constraints, we should begin with policy constraints, because they embody the habitual attitudes and behavior of the organization. In the latest installment of TOC novels, *Necessary But Not*

* Frederick Brooks is author of *The Mythical Man-Month.*

Sufficient co-authors Goldratt, Ptak, and Schragenheim emphasize that in order for IT-induced change to be effective, the policies (rules) of the organization must come first:

> ". . . now we install some new technology. Let's assume successful installation occurs; the limitation has been diminished. But what happens if as part of the implementation of this new technology, we neglected to address the rules? What happens if we still operate with the old rules, the rules that assume the existence of the limitation?"
>
> "In that case, the rules themselves will impose a limitation," Lenny says.
>
> "Exactly. And then what benefits will we gain from the new technology?"
>
> "I don't know," Lenny answers. "It depends on the technology and what it does. But I see your point. If we don't also change the rules, we can be assured that we will not realize the full benefits."
>
> Scott looks at the sky, still pretending to smoke his imaginary pipe. "You see, Watson, technology is a necessary condition, but it's not sufficient. To get the benefits at the time that we install the new technology, we must also change the rules that recognize the existence of the limitation. Common sense."[231]

This may be common sense; most people understand the old phrase *garbage in/garbage out*. If we invest in automating a broken process, we only speed up the creation of waste, while at the same time cementing the problems into place. But if this is common sense, then why do software implementation teams commit this blunder so often?

Even when information technology appears to be focused upon the policies and processes of the organization, this may not lead to an effective or enduring solution. *People* must drive change. During a presentation titled *"Run the Business, Grow the Business, and Improve the Capabilities,"* Roger Brooks, President of Oliver Wight North America, illustrated the challenge of sustained change management with the diagram shown in Figure 11-06.[232]

Brooks' message is poignant:

- A technology and process improvement solution that does not involve the hearts and minds of people leads to alienation, depersonalization, and turnover; all significant threats to sustained continuous improvement.
- A technology and people solution that does not include process improvement simply automates chaos and inefficiency; this is the GIGO (garbage in/garbage out) principle.
- A people and process improvement solution that does not include technology may indeed work just fine—information technology is not inevitable. A process should *first* be simplified before it is automated. But although a Lean Enterprise should initially focus on process improvement and simplification, continuous improvement efforts may eventually

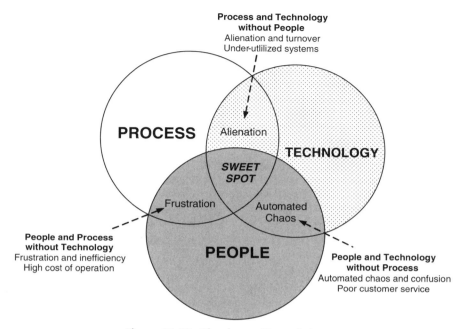

Figure 11-06. The three pillars of change

lead to complexity through increased volume, velocity, and variety. The situation may then call for the skillful application of information technology.

This three-way interdependence of people, process, and technology requires coordination, or as I prefer, "orchestration." Coordination implies centrally planned and controlled behavior, whereas orchestration suggests a leader providing guidance, setting the pace, while encouraging individual creativity and inspiration.

Orchestrating Change Through Project, Program, and Portfolio Management

How does an enterprise manage change involving people, process, and technology?

Through the Project Management Office, which orchestrates the disciplines of strategic planning, program and project management, project portfolio management, and system lifecycle management.

The Project Management Office. Many organizations, even large ones, have a localized view of project management. Not only do large projects require dedicated, cross-functional teams, but many medium and large organizations may have dozens of projects running simultaneously (IT and other types) competing for scarce capital and human resources. Without central planning and management, these resource battles starve some projects while feeding others, the inevitable result is a number of failed (or at best underserved) projects. This suggests the need for not only central allocation and management of resources but a decision-making process to establish priorities. Any projects that do not justify sufficient resource commitments should be cancelled or delayed. How do we determine these priorities? They must be aligned with strategic goals and objectives.

In their study on enterprise software project failures, the Boston Consulting Group found that initiatives based on a clear strategic vision had positive outcomes 53% of the time vs. only 22% for projects lacking such vision. The study concludes that smaller, more focused projects have better chances of success than broader ones, and companies should focus on "smaller, high-value chunks" of their business, where big returns can be gained from modest IT investments.[233]

The instrument required for such clear prioritization is called the Project Management Office (PMO), which performs both Program and Project Management activities, defined by the Project Management Institute:

Project Management—is the application of knowledge, skills, tools, and techniques to project activities to meet project requirements. The work typically involves competing demands for scope, time, cost, risk, and quality; stakeholders with differing needs and expectations; and identified requirements. Operations and projects differ primarily in that operations are ongoing and repetitive whereas projects are temporary and unique. The project life cycle serves to define the beginning and the end of a project.

Program Management—a program is a group of projects managed in a coordinated way to obtain benefits not available from managing them individually.[234]

Program management involves a collection of projects that are *interrelated*; they compete for shared resources while working toward shared goals. Programs and projects may involve a variety of participants and stakeholders from within and outside the boundaries of the enterprise, including providers of products and services, customers, suppliers, funding sources, regulatory agencies, and so on. Programs include multiple projects, and each project may include multiple phases and subprojects. The competition for resources can be intense, and the management of all of these interrelated parts requires great skill and proper tools.

Just like managing the allocation of resources on a manufacturing shop floor, periodically the need will arise to make trade-off decisions among projects due to time or resource constraints. How are these prioritization and trade-off decisions made? Within a single project this may be within the scope of the project manager. Within a program consisting of multiple related projects, this may be within the scope of the program manager. But what happens when an enterprise has many unrelated programs across various departments and locations, all competing for the shared pool of capital and human resources?

Project Portfolio Management. The Project Management Institute defines Project Portfolio Management (PPM) as "the selection and support of projects or program investments. These investments in projects and programs are guided by the organization's strategic plan and available resources."[235]

The Chief Financial Officer of an enterprise is often ultimately responsible for the oversight of *all* projects, and the portfolio of IT projects is accountable to this authority. From a corporate governance perspective the CFO is particularly attentive to controls upon cost and risk. The *Business Finance* magazine article "Project Portfolio Management Goes Mainstream" describes the value of PPM from this point of view:

> Few of the challenges CFOs face are as complex and demanding as managing a portfolio of enterprise projects. Dropping dollars into one initiative affects all of the others. Shifting time and resources among ongoing projects directly impacts risks and returns. A typical medium to large organization has dozens, even hundreds, of projects under way at any given time, and finance executives are increasingly hard-pressed to establish priorities and ensure that initiatives stay aligned with corporate goals.
>
> Just because a project is important doesn't mean we have resources to do it. In this era of tight budgets and limited resources, a haphazard approach to project management is no longer workable. Companies are looking for a better way to analyze the risks, costs, and returns that their enterprise initiatives generate. Many organizations are turning to software that enables them to manage a broad range of efforts holistically: Project Portfolio Management tools.[236]

As you might expect, PPM has much in common with investment or product portfolio management. Although PPM software applications are diverse, most provide a combination of project management, content management, collaboration, portals, reporting, and analysis tools that are fine-tuned to PPM tasks. Many offer sophisticated and expensive tools that may be appropriate only for larger enterprises with dedicated PMO resources. However, many small and medium-sized companies also have the need for such capabilities, because they have similar functional requirements and project management complexities as their larger counterparts, but with fewer resources to manage them. The *Industry Week* magazine article titled "Made for Midsize" stresses this important point:

> "It's a major misconception that smaller companies have less complex technology needs," observes Rod Johnson, a vice president at AMR Research. "While larger companies can afford to pursue innovation around technology, smaller firms typically have to pick their spots and make sure that they're investing in solutions that will provide a significant payback. They face some tough decisions."[237]

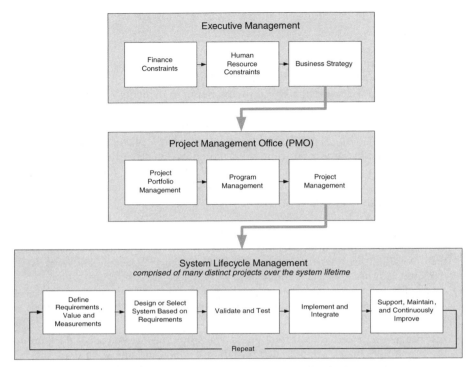

Figure 11-07. PMO linkages from overall strategy to individual project management

If they are resourceful, a small or medium-sized company may practice the disciplines of Project Portfolio and Program Management without sophisticated software tools. In fact, earlier in this book we explored an array of knowledge management tools that may be useful. Furthermore, Project Portfolio and Program Management principles share the same organizational framework as enterprise performance management described in the previous chapter, emphasizing the hierarchical linkages between business strategy and ground-level initiatives illustrated in Figure 11-07.

More important than the software tools is the necessity for every enterprise, large or small, to develop *competency* in portfolio, program, and project management. The modern enterprise exists in a dynamic environment; operational excellence (the performance of ongoing and repetitive activities) has become a basic requirement for viability. However the ability to plan, execute, and control a variety of interrelated projects with a wide variety of stakeholders, and according to a clear strategic direction, may distinguish top competitors from the also-rans. From shop floor kaizens, to internal IT projects, vital constraint-breaking kaikaku initiatives, and elaborate global supply chain programs, the enterprise must juggle numerous projects, utilizing scarce resources to best advantage.

But there's more to the PMO than just balancing the project portfolio, programs, and projects of the enterprise. The *EAI Journal* article "A Guide to ERP Success" suggests that there are five vital roles that the PMO and its staff may contribute to institutionalize effective change management practices:

- **Project Management Solution Architect**—The PMO assumes a leadership function in defining the combination of processes, technologies, and standards required to meet strategic and tactical project management needs.
- **Process Champion**—The PMO develops, implements, and continuously improves project management processes based on organizational feedback, management requirements, and industry best practices. Implicit in this role is the need to provide value to project and senior management stakeholders.
- **Mentor and Coach**—The PMO assumes an active role in promoting knowledge, understanding processes, and achieving buy-in from stakeholders. The focus is on promoting an understanding of relevant PMO processes but may extend to an understanding of general project management knowledge that's relevant to the stakeholder. This role also includes developing and implementing project management training.
- **Facilitator**—This role includes working directly with project teams and conducting project workshops designed to gain consensus on key parameters such as scope, resource requirements, plans, and schedule dependencies.
- **Knowledge Broker**—In this role, the PMO ensures that all project-critical management data and information necessary for process implementation and decision-making are available to all stakeholders. This includes the analysis and reporting of project metrics, including performance and risk metrics and quantitative and qualitative analyses, including variance analysis, critical path analysis, and trend analysis.[238]

System Life Cycle Management

Finally we arrive at ground level, where systems are selected, developed, implemented, and continuously improved in the five basic stages described in Figure 11-07:

1. **Define Requirements, Priorities, and Measurements**
   During this stage, a company should form cross-functional teams, map current-state processes, and identify desired future states and the gaps between current and future states. The team should identify linkages to the strategic plan and then establish a value for each gap, to prioritize their closure. Finally, the team should define particular software requirements

to address each gap, remembering that a balance of people, process, and technology is required at this point.

Now here's the kicker—if continuous improvement efforts are already underway, shouldn't this current/future state gap analysis, strategic linkage, and prioritization process already exist? An enterprise shouldn't perform these analyses only when they're shopping for software; this should be an ongoing process.

2. **Design or Select a System Based on Prioritized Requirements**
   Although this stage may include the development of software, our focus in this narrative is on the selection of a commercially available software product. The team should be armed with a prioritized set of requirements based on their desired future state. A team selecting software without a clear definition of its needs, and without predefined criteria and methods for evaluating, weighting, and selecting the right software based on those requirements, is just conducting a fashion show. Either the cheapest or the best-looking software, presented by the most persuasive sales team, that happens to say the right things at the right times, is likely to win. Or, if the selection team is imbalanced, with too much emphasis on a particular function or department, then an imbalanced selection may result. A *cross-functional* team ensures the development of a balanced set of criteria through the future state definition process, linking priorities to strategic goals and objectives, ensuring that the selected application provides the best overall fit. As Rother and Shook point out in *Learning to See*, without a guiding future state, improvement efforts are just wasteful.

3. **Validate and Test**
   Software has now been selected (or designed) based on a preliminary definition of key future state requirements. But before the final system design is determined, the team should undertake several rounds of prototyping to validate key assumptions and design decisions. This testing process should be iterative and repeated in multiple rapid cycles and include tight interaction between the design team and the user community. In addition to system design, testing also validates training effectiveness, user proficiency, data accuracy, documentation, and system performance, before a go-live decision is made.

4. **Implement and Integrate**
   Skillful planning, project management, and a committed team are essential to a successful implementation; otherwise delivery of a successful project is pure luck.

5. **Support, Maintain, and Continuously Improve**
   As we have already established, it is critical that selection, implementation, and maintenance decisions be made with the entire ecosystem in mind. Once a system is in place, changes made within the entire collective of systems and processes (the ecosystem), including rapid and fine-grained adaptation to changing business practices, software upgrades,

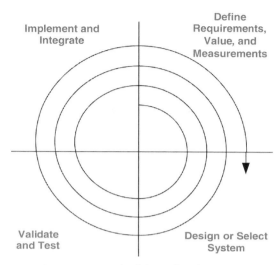

Figure 11-08. The life cycle of a system

customizations, and integrations, must be carefully orchestrated in conjunction with cross-functional training and continuous improvement initiatives. Period.

This process describes the initial selection and implementation of a new system; however, the very same approach applies to the ongoing improvement of any system in a rapid cycle as shown in Figure 11-08.

APPLYING LEAN IT TO THE LEAN ENTERPRISE

Focusing Change

To realize their full potential, Lean Enterprise and Lean IT initiatives *together* must encompass the entire organization, guided by effective strategy, focusing on specific initiatives. According to Goldratt, Ptak, and Schragenheim, "Software adds value only to the extent that it overcomes limitations." So what are your limitations? What are your Strengths, Weaknesses, Opportunities, and Threats? Where are your constraints? How do you improve throughput? How do you satisfy your current customers and grow your market? How do you nurture and leverage the collective knowledge of the enterprise? The answers to these questions must focus the initiatives of the Lean Enterprise, supported by Lean IT.

Leaders must develop a strategy, identify the constraints that limit the achievement of that strategy, and focus the energy of the entire enterprise on breaking those constraints to achieve breakthrough improvement. Does this focus on constraints mean that an enterprise should disregard the abundance of incremental improvements that naturally arise from team-based continuous improvement activities? Of course not. If something is broken, fix it. If something has fallen on the floor, pick it up. Individuals and cross-functional teams should be empowered to make incremental improvements whenever they find them. But should these random incremental kaizen improvements be the focus? No. The often-cited Lean ideal of the "pursuit of perfection in everything" may sound virtuous, and is a worthy aspiration for tactical improvement teams, but it's not practical from a strategic perspective. To be effective we must focus on priorities.

We began this chapter by exploring *traditional* tools and techniques for managing IT change. This included guiding the selection and maintenance of enterprise software by identifying gaps between the current and future states, managing hundreds (or even thousands) of distinct requirements, and measuring overall system performance by percentage of fit to those requirements. This conservative methodology of rigorously prioritizing and managing a portfolio of countless programs, projects, issues, requirements, and gaps may lead to reasonably effective information systems, but many aspects of this process are wasteful. Furthermore, this approach is *not* assured of creating a system that will propel a Lean Enterprise to the next level of performance. This is because the burden of managing the many small details may bog the change process down, restricting the team's focus, leading to incremental but not breakthrough change.

In this chapter we also explored the benefits of modern enterprise software tools, which may deliver increased agility through granular component architectures and a rapid spiral PDCA process. Although this approach is likely to succeed in creating more flexible information systems, it may not deliver breakthrough business performance; good tools do not guarantee good results.

Forget guiding the business forward by managing thousands of detailed issues and requirements—which are the few critical issues that drive Lean Enterprise success? Where is the business going, and what are the system capabilities required to get there? Whichever enterprise software application best satisfies these *critical requirements* will probably perform satisfactorily on the hundreds or thousands of others.

That is not to say the selection/development team shouldn't keep this list of the top 20% requirements in mind, because failure to satisfy any one of these may create a new constraint. And an internal application development and support organization should track *all* of the issues and requirements in a database, because that is how they manage their activities and support the end users. But from a system life cycle management perspective, even if there *are* noncritical shortcomings with the chosen system, with the flexibility of Web Services and rapid deployment methodologies, creating a strong ecosystem foundation upon which to grow and adapt is more important than focusing on the many insignificant details.

So how do you determine the critical issues that prevent you from achieving breakthrough performance? Suppose that an enterprise forms a *functional team* comprised of nine leaders representing engineering, marketing, sales, purchasing, production, quality, distribution, service, and finance; these are the nine functional areas illustrated in the matrix organization (Fig. 10-13).

Each functional team leader then forms cross-functional *process teams* to focus on the elements of the value streams for which they are responsible. For example, the production functional team leader may form process teams for scheduling, capacity management, setup time reduction, and other vital production processes. Remember that this is a cross-functional (matrix) approach, so engineering and quality members must be involved on the setup reduction process team, because their actions impact setup time reduction effectiveness.

In this example, suppose that the nine functional teams are each responsible for ten process teams; these process teams are responsible for the results of their processes within the overall value stream. Functional team leaders communicate strategic goals and objectives downwards to the process teams, and ideas and initiatives percolate upwards. Through catchball, each process team develops a list of (for example) five improvement initiatives that support enterprise objectives; 450 separate kaizen initiatives are now under way (9 functional teams × 10 process teams × 5 initiatives), each with its own targets, activities, and measurements. The results of these initiatives roll up to the functional team leaders who direct and encourage the teams. The aggregate results of these initiatives at the functional team level then roll up to the executive level as a handful of KPIs.

Although 450 separate process team improvement initiatives are guided by the top-down communication of strategic goals and objectives, how many of these initiatives are truly strategic? How many directly address a critical constraint? Most likely, just a few. And who decides which initiatives are strategic, directing the focus of enterprise-wide resources to break critical constraints? Are the nine functional teams or the ninety process teams responsible? Can they determine whether any of their improvement initiatives address a critical constraint? Possibly not, because they do not have a perspective on the overall *portfolio of* projects. Only executive management has the necessary top down perspective to determine where the critical constraints exist; thus strategic constraint elimination efforts should be coordinated through the Project Management Office.

Does this mean that the process team improvement initiatives are unimportant? Absolutely not. From the bottom up, continuous improvement should pursue perfection in every process. Each employee should feel a sense of ownership, responsibility, and empowerment for making incremental improvements each and every day. But from a top-down perspective most relatively minor initiatives do not merit specific visibility.

How is the fabric of top-down and bottom-up objectives and initiatives to be woven? The answer lies in a blending of Hoshin Planning mechanics and the Project Management Office leadership, enabled by a fabric of IT tools such as alerts, portals, and scorecards, orchestrating enterprise-wide kaikaku and kaizen initiatives.

Hoshin Planning concentrates enterprise resources on breakthrough change, by focusing a few strategic kaikaku initiatives that break constraints. The Hoshin Catchball process reaches down and across the organization to the individual kaizen teams, focusing resources on the critical enterprise constraints. And when a strategic improvement initiative requires software support, this process also guides the management of strategic requirements for software selection, implementation, and life cycle management.

And finally, the PMO may also be responsible for orchestrating the KPIs that direct the numerous team-based, continuous improvement initiatives within the enterprise. With responsibilities for guiding both breakthrough and incremental change of processes, and supporting enterprise software requirements, perhaps the Project Management Office deserves to be renamed the Office of Breakthrough Change and Continuous Improvement.

Sustaining Continuous Improvement with Lean IT

The underlying message of this book is this: Continuous improvement and IT are complementary disciplines. Leveraging IT tools and methods to enhance Lean Enterprise performance, and using continuous improvement techniques to enhance Lean IT performance, are two sides of the same coin. Both aspects must focus on constraints to limit complexity and optimize results.

As the enterprise change management process charts a direction for focused process improvement, to the extent that IT can enable these improvements, systems should be planned, tested, and implemented quickly and decisively. Once the systems are in place, they should be measured and continuously improved during their entire life cycle. As processes are improved and the need for transactions and controls diminishes, the systems may be simplified and perhaps eliminated.

This symbiotic relationship among people, processes, and information technology requires a new way of thinking about enterprise information systems. In the traditional IT paradigm, a system became a monument that embedded itself deeply within the minds and processes of the organization, often requiring a meltdown to bring about change. By contrast, Lean IT is proactive and agile. This contrast between the old and the new is shown in Figure 11-09.

For breakthrough results and lasting change, the Lean Enterprise and Lean IT must work hand in hand. Throughout this book we have explored a variety of PDCA cycles for both Lean improvement and IT implementation and life-cycle management. *Are these really separate cycles?*

Consider these two statements: "Continuous improvement is a cyclical process" and, "The flow of materials and the flow of information are two sides of the same coin." The continuous improvement of the Lean Enterprise and the continuous improvement of Lean IT are two aspects of the same cycle. Not two, but one integrated cycle is required. Figure 11-10 illustrates such an integrated cycle, where each step contains an aspect of process *and* information.

Attribute	Traditional IT	Lean IT
Organization	Centralized	Centrally Managed Team-Based Functionally Driven Initiatives
Focus	Tactical Techology-Driven Solutions Prioritize Local Optima	Strategic Business-Driven Solutions Value Stream Constraints
Change Management	Rigidly Planned and Long Term	Dynamic
Primary Success Drivers	Stability and Cost Containment	Value Creation
Knowledge Management	Functional Silos	Fluid
Education	Specialized	Cross-Functional
Speed	Long Term	Rapid Cycles
Agility	Discourage Change	Encourage Rapid Adaptation
User Involvement	Periodic or Episodic	Continuous
Phasing	Waterfall	Spiral

Figure 11-09. Attributes of traditional and Lean IT

People, Process, and Technology merge in this one great cycle:

1. **Develop Strategy and Identify Constraints**—Executive management should evaluate value streams at a strategic level. Measure value creation through Lean operational improvement, innovation, and customer service. Identify the sources of competitive advantage. Develop a clear marketing and production strategy, positioning the overall mix along the product/process diagonal. Identify constraints that limit the achievement of the strategy, pursue constraint elimination initiatives aggressively, and monitor them carefully.

2. **Manage Goals, Objectives, Measures, and Project Portfolio**—Articulate the strategy, then develop measurable goals and objectives to support it. The PMO should manage the portfolio of projects and improvement initiatives carefully and continuously, focusing scarce resources on the elimination of constraints. Targets and measures should be communicated from the top down, encouraging cross-functional teams to generate ideas and initiatives for their achievement from the bottom up. Teams should own their KPIs.

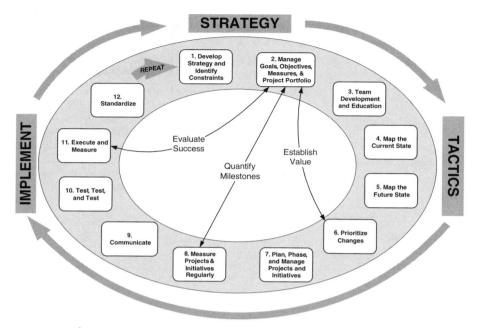

Figure 11-10. Sustaining continuous improvement with Lean IT

3. **Team Development and Education**—Invest in the development of cross-functional teams throughout the enterprise; include an IT representative on each team. Provide the teams with ongoing education and coaching in continuous improvement techniques. Guide the teams with vision and strategy, and encourage them to develop their own initiatives and measures. Focus teams on creating value and eliminating constraints; remove obstacles from their path, and discourage inappropriate management interventions.

4. **Map the Current State**—Map the existing value streams, processes, and sub-processes so that cross-functional teams have a clear and holistic understanding of all activities and interrelationships within the enterprise. This mapping should include the complementary flow of materials and information. Manage these maps and supporting documents as vital enterprise knowledge.

5. **Map the Future State**—Develop a vision for the future state that is consistent with strategic goals and objectives. Do not bog down in the details; focus on simplicity and economy. Carefully map and analyze the constraints, and verify that they are legitimate and not merely symptoms of underlying problems. Ask "why" often.

6. **Prioritize Changes**—Perform a simultaneous gap analysis of business processes and their supporting information systems, identifying the incremental changes required to achieve the future state. The PMO

should focus the majority of effort on eliminating strategic constraints, while the improvement teams discover and execute tactical improvements along the way. Simplify processes first, then determine where IT may be applied to enhance performance. Identify how data should be captured, according to how it adds value. Remember that although spreadsheets and other quick technology fixes may be used sparingly, information should flow as smoothly as production; information disintegration causes waste.

7. **Plan, Phase, and Manage Projects and Initiatives**—Using a rigorous project management methodology, develop a plan for each project and initiative. Think PDCA. Strive to reduce project cycle times and improve agility through tightly controlled scope, regular interaction between users and designers, the use of small and standardized components, and a rapid deployment spiral. Deliver quick wins to build confidence.

8. **Measure Projects and Initiatives Regularly**—Establish measures that link strategy to action. Develop clear relationships between cause and effect, establishing appropriate result and process measures across the entire enterprise. Make sure these measures are balanced, reflecting finance, operational effectiveness, customer satisfaction, and innovation.

9. **Communicate**—Communicate the project plans and progress reports widely, ensuring that everyone in the company understands the purpose of each initiative, how it may impact them currently and in the future, and how the project supports strategic goals and objectives.

10. **Test, Test, and Test**—Design and test new processes and systems rigorously. Verify design and structure, user proficiency, data accuracy, user documentation, and system performance. The team should make a clear go/no-go decision to proceed beyond testing to implementation.

11. **Execute and Measure**—Execute the new processes and supporting systems, and measure the results. Use the monthly Sales and Operations Planning process to regulate all aspects of planning, execution, and control, and to perform a reality check against executive expectations and strategy.

12. **Standardize**—Now it's time to act, the "A" in PDCA. As the changes to processes and systems prove effective, institutionalize them through standardization, best practice documentation, and frequent education, training, and cross-training. Standardization does not mean rigidity; it is a pillar of continuous improvement—processes must be reliable so they can be consistently measured and quickly improved.

After each improvement cycle, celebrate your accomplishments, and praise judicious experimentation and risk-taking. Determine whether a constraint has been broken. If it has not, then what must be done next? If it has, then identify the next constraint, and start again.

There's *always* a next constraint. Repeat the cycle and keep the momentum going! Although it takes great effort to get this cycle rolling, it takes much less effort to keep it moving.

Keep improvement cycles small in scope and short in cycle time—the protracted waterfall change management approach must be replaced by a swift spiral. Just like Lean Manufacturing, Lean IT replaces long lead times and workorder-based production with a level schedule and standardized components. This approach provides the flexibility to shift resources and alter the project scope in near-real-time according to changes in demand. Changing system design quickly is not scope creep, nor should it create instability—short project cycle times allow for rapid changes, just like a short takt time allows for a flexible product mix. Using this approach, Lean IT delivers agility and stability at the same time.

This holistic approach continuously improves teams, value streams, and information systems, aligning company strategy through the design of future-state processes, with an unwavering focus on constraint elimination and value creation. Lean IT creates competitive advantage by accelerating and amplifying the continuous improvement of people and processes. Lean IT, as well as IT in support of Lean initiatives, requires the effective change management of people, processes, and technology, *in that order*.

Postscript

Zen and the Art of Lean

This divorce of art from technology is completely unnatural.

Technology presumes there's just one right way to do things and there never is. But if you have to choose among an infinite number of ways to put it together [...] the art of the work is just as dependent on your own mind and spirit as it is upon the material of the machine.

It is this identity that is the basis of craftsmanship in all the technical arts. And it is this identity that modern, dualistically conceived technology lacks. The creator of it feels no particular sense of identity with it. The user of it feels no particular sense of identity with it. Hence, it has no Quality.

The craftsman isn't ever following a single line of instruction. For that reason he'll be absorbed and attentive to what he's doing even though he doesn't deliberately contrive this. He's making decisions as he goes along. His motions and the machine are in a kind of harmony.

Robert Pirsig
Zen and the Art of Motorcycle Maintenance[239]

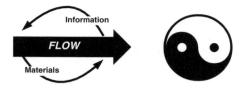

I first read Robert Pirsig's classic *Zen and the Art of Motorcycle Maintenance* shortly after it was first published in 1974. It is the story of a man and his young son riding across the Western US, retracing the events of a personal tragedy that concluded with an interlude in a psychiatric hospital.

During this journey the author is moved by the natural beauty of his surroundings and absorbed by his intimate experience with the motorcycle—the act of riding it, maintaining it, and keeping it in balance. His experience contrasts with his riding companions, who constantly struggle with their own machines. The author realizes that their frequent mechanical mishaps are preventable, and that most of their problems and frustrations, although they attribute them to the equipment, are the natural result of their lack of awareness. This becomes a metaphor for the author's own struggle with the complexities and challenges of modern life that led to his crisis.

While researching this book I again read Pirsig's story. I was inspired by how he reconciled the forces of science and humanity, and by the relevance of this story to the essence of Lean.

For many years, authors and researchers have attempted to show a connection between the culture, sociology, and education of Japanese society and their collective talent for quality and adaptability. During a session our firm recently facilitated on employee-led continuous improvement, one participant insisted that it was "easier for the Japanese to think and act this way, since it was part of their culture." I find this to be a common belief held by many Westerners; however, I believe these qualities can be developed by anyone, with practice.

Zen has been a powerful influence in East Asia for over 1400 years, and in the nineteenth century there were over 470,000 temples in Japan alone. Although certainly not all contemporary Japanese practice the religious or spiritual aspects of Zen, its influence on individual and collective behavior can be found, if you know what to look for. Perhaps there is something to be learned about Lean through a better understanding of Zen. Many of the principles of Zen—respect for the individual, embracing change as an ally, and acting with patience, awareness, simplicity, and in harmony with the surroundings—are consistent with Lean and continuous improvement. These are not religious practices, but merely behaviors that help us to become more centered human beings.

A common theme in Zen meditation is the flow of water, which is never in conflict, effortlessly seeking balance and moving around every obstruction.

This is the very same metaphor used by Taiichi Ohno to describe the flow of materials in a Lean operation, and the same idea may be applied to the flow of information.

Zen is semantically inseparable from the practice of Lean: *Kaizen*, the Japanese word for continuous improvement, is derived from the Japanese roots *kai* meaning "to take apart", and *zen* meaning "to make good."[240] This suggests the art of reducing a system to its components, understanding the inner causal relationships so that its performance may be improved. *Beginners Mind* is an important idea in Zen, suggesting a similar clarity of thought: letting go of old habits and assumptions, looking at the familiar with a fresh set of eyes as if for the first time, thereby discovering new solutions.

Most people naturally make situations too complicated, interpreting and judging them based on past experiences, attitudes, and personal bias. According to Shunryu Suzuki, the master who brought Zen to the United States in 1958, *"In the beginner's mind there are many possibilities, in the expert's mind there are few."*[241] Although the development of experience and judgment is the purpose of education and socialization, it can also hinder creative problem solving. The Zen ideal of beginners mind releases the uninhibited and inquisitive mind of the child while harnessing an adult perspective.

THE SEARCH FOR QUALITY

In *Zen and the Art of Motorcycle Maintenance*, Pirsig muses on a principle he finally chooses to call *quality*. Pirsig's message of quality is the blending of art and science, the combination of analytical left brain and creative right brain into a meaningful and holistic experience. This is not the empirical quality so often associated with Total Quality Management, Statistical Quality Control, or Six Sigma. Perhaps we have overanalyzed the idea of quality, losing sight of what it truly means. When asked to define quality, Deming stated simply that *"quality is pride of workmanship."*[242]

Pirsig suggests that art and science are two aspects of the same reality, and either taken alone is not whole. There must be a balance, a harmony of these apparent opposites. Zen found its way to Japan from China around 600 A.D. Buddhism originally traveled from India to China, where it merged with Taoism to become Cha'an, or Zen. Taoism contributed to Zen the principle of complementary opposites, the balance of forces known as the Yin and Yang; their visualization is shown below. According to Taoists, everything is a balance of opposites. One cannot know light without dark, heat without cold, or hard without soft. This leads to a comprehension of each situation in terms of the balance between opposites, striving for a natural harmony in every situation. According to Taoism and Zen, extremes are unhealthy, and we should always strive for a comfortable middle path.

Yin/Yang

As a balance between extremes, every situation can thus be expressed as a continuum, and paradoxically also as a natural cycle with no beginning or end. It should be no surprise that you have seen many thematic continuums and cycles illustrated throughout this book. On a continuum the emphasis is not on an absolute right or wrong, but on the appropriate and balanced response to a particular situation, while the cycle represents the dynamic energy flow of process.

This sense of balance moves effortlessly with circumstance, so a situation is rarely at rest for long. Rigid thinking and rules-based policies can hinder the natural balance that lies within every situation. But when the individual worker is offered goals and guidelines (not inflexible rules), along with an intuitive comprehension of the complete process, he or she is ideally suited to make appropriate and timely decisions. This way of thinking is consistent with the empowering principles of continuous improvement.

So the Zen ideal is to naturally and effortlessly find the right balance and harmony within each moment, letting go of inappropriate habits and thought patterns. How difficult can this be? How often do we lose our balance when a situation becomes difficult? How many of us act and react habitually and compulsively? Much of the chaos within a manufacturing plant is caused by predictable reactivity: The same problems arise again and again, and we respond as we always do—thus the situation never really changes. We hastily respond to symptoms while leaving the root causes unchanged. How often, after having asked *Why?* five or more times, do we discover that the real source of a problem is our own rigid thinking and policies—that *we've always done it this way?*

In some cases this rigidity may be caused by an IT failure, because we don't have the right information to make proper decisions, or because poorly designed, obsolete, or inflexible systems impose inappropriate behavior. But more often this rigidity is caused by a failure of human and organizational nature, repeating the same old behavior patterns through habit and inertia.

Imagine how our sense of quality and balance would improve if we could simply drop the old habits and predispositions, looking freshly at every moment with beginners mind? Why is it so difficult for us to do this? From the perspective of a Zen practitioner, an ordinary human being *thinks too much.* Each one of us lives with the constant background noise of our minds, repeating the past and dwelling on the future, chewing on habitual thought patterns like a dog gnawing on a bone. The sum of these familiar thought patterns comprises our personality. For example, do you know someone who always seems a little bit angry, perhaps with a quick temper? Likewise, do you know someone that is cool and calm, responding sensitively to each situation? What makes these people so different? A Zen master would suggest that these tendencies naturally flow from their habitual states of mind, which are constantly reinforced by the persistent dialogue in their head that triggers predictable patterns of behavior.

If allowed to run out of control, this mental clutter can makes us poor listeners and observers. How often do you find yourself tuned out of a conversation, thinking your own thoughts while the other person chats away? The fact is that your mind—and its accumulated thoughts, experiences, likes, and dislikes—filters your every experience. The Zen master would say that you don't really live reality, you live *your* reality. To prove this point, ask several people to recall the same event and they will describe it differently; this is because each has his or her own unique perceptual filter. It's important to understand that this perceptual filter is critical to our survival in a complex and potentially dangerous world, yet it clearly inhibits creative thinking and problem-solving. Is it possible to control this compulsive thinking and reactivity, to discipline this filter rather than be controlled by it? Yes, but it's not easy. One technique that works for many is meditation, and this happens to be the foundation of Zen practice.

It is no surprise in this overstimulated world that various forms of meditation have become popular in the West. Through meditation a Zen practitioner endeavors to still the mental background dialogue, quieting the mind's nagging perceptual filter so the practitioner may pay close attention to what is *really* going on, thus achieving the state of *no mind*. It is easy to misunderstand this mental state—the individual does not go to sleep or become unconscious, nor does he try to negate the experience and fall into a numb state. The goal is to eliminate the compulsive background dialogue so his real experience becomes more vivid.

Anyone with an experience of the meditative state will attest to a briefly heightened sense of awareness. Most of us have experienced this heightened awareness at some time, perhaps while enjoying a quiet moment in nature, listening to music, practicing an art, holding a sleeping child in our arms in the dark of night, or perhaps through sustained physical exercise. When we focus deeply on an experience without conceptualization, we become momentarily absorbed. The colors are brighter, the sounds are sharper—we're no longer an observer, we're actually an integral part of what is around us. And then it's gone, leaving only a memory impression that the mind tries to analyze and verbalize. This momentary perfect awareness the Zen master calls *Satori*.

Is this perfect awareness so different from the ideal problem-solving process of a Kaizen event? Taiichi Ohno once said, "*Observe the production floor without preconceptions and with a blank mind.*"[243] This is an instruction you would hear from a Zen master. Many stories are told that Ohno would take a new engineer out into the shop, draw a circle on the floor, and instruct the new employee to stand within it. Ohno would then leave this person standing there for the entire day. Why? Simply to watch and observe carefully. Why didn't Ohno offer any guidance other than "stand inside the circle"? It would only create a preconception and thus bias the outcome.

Similarly, Jeffrey Liker shares an interesting story in *The Toyota Way*,

> Reflecting back on the early days when Fujio Cho was the first president of the Georgetown (Toyota) plant, the stories begin with the managers' visits to the factory floor in the morning. On the way in, they notice Cho standing and watching an operation. They pass nearby him, expecting Cho to notice and greet them, but he doesn't respond. He just stands and stares, as if off into space. They walk even closer. He continues to stare.

> They go about their business, then happen by 15 minutes later. Cho is standing and staring. They wonder if he is ill or frozen to the ground at that point. Finally, Cho relaxes, as if coming out of a trance, notices he is not alone, and says, "Good morning" with a smile. Later there are some orders from the president's office to tighten up some part of the Toyota Production System in the plant.[244]

This spectacle must have been startling and perhaps amusing when it was witnessed on the floor of a manufacturing facility in Kentucky. But Liker is describing behavior identical to a Zen master deep in concentration. A Zen master may sit motionless for hours, looking directly at the petals of a flower, or simply at a blank wall, evenly taking in all the sensory input that surrounds him. He does not mentally label or judge; he just sits. As he falls deeper into meditation, he becomes an integral part of the situation and the sense of separateness is lost, thus the inner workings of the situation become intuitively apparent.

Meditation is difficult; meditation is easy. In truth, meditation is more *not* doing than doing—ceasing the busy mind requires effort of a different sort, the effort to relax and stop thinking. The archetypal signs hanging in the workplace that exhort employees to "THINK" or to "Work Smarter, Not Harder!" seem comical, misguided, and somehow sad. Thinking isn't an act of will or obedience. Thinking, or more accurately, creative thinking, is a spontaneous act which happens when we're relaxed and in harmony with our surroundings.

According to Zen masters and scientists alike, the feelings of love, beauty, and creativity, to name three extremely powerful and ineffable inner experiences, don't arise from thought. This is the essence of the left- and right-brain dichotomy clearly established by brain science decades ago. The essence of direct experience of beauty cannot be put into words; the more you try, the more the experience eludes you. Ask any artist where her creative inspiration comes from, and she will tell you that it just happens, that she somehow become absorbed in activity. Likewise, sports psychologists and coaches have learned to help athletes attain peak performance by shifting from the analytical left to the creative right brain through practices that involve relaxation, meditation, and visualization.

Creativity and inspiration come from beyond rational thought, or perhaps rather from the silent gaps in the midst of rational thoughts. This is true for scientists as well as artists. Albert Einstein reported leaps of intuition after awakening from his dreams. Of course, he didn't just dream up his profound theories, he worked very hard on them—knowledge and experience *are* necessary. Likewise, we must study, learn, think, and test our ideas. There must be effort on a rational level to set the stage for a cognitive-intuitive leap to occur. But at some point, when we get stuck, when we've thoroughly analyzed a situation and there is no sense of forward progress, then it's time to *let go*. Have you ever chewed on a difficult problem and then become distracted, and suddenly the solution appears to you as a flash of inspiration?

Inspiration arises naturally under the right conditions. When you run a machine at 100% capacity with no rest, it will overheat, tire, and wear out. The same happens with people and organizations. Frantic behavior leads to poor quality and burnout—not creative thinking. People need time to stop, look, listen, and contemplate what is going on. With a quiet mind, with an open mind, inspiration is invited.

A PRESCRIPTION FOR LASTING CHANGE

All manufactured items, violins, muskets, and wagons, were once made one piece at a time by craftsmen. The skills, techniques, and handmade tools were passed down through generations, nurturing a mix of science and art that required careful attention to every step. The transmission of knowledge and skill had to be delivered personally from one individual to the next through experience. Long after the Western industrial revolution, post-war Japan faced the challenge of rebuilding with limited resources. They developed an approach to compete with the industrialized West on their own terms, adapting mass production techniques, focusing on craftsmanship, teamwork, simplicity, quality, and flow. The Sensei was not just a teacher or boss, but a master with insight.

Many individuals in Western society have begun to react with simplicity against the frantic pace of their lives, having learned that "the harder they work, the behinder they get." Lean has also developed a strong foothold in the industrial West, emphasizing principles of simplicity, yet paradoxically improving quality, throughput, and profitability where traditional mass production methods could not.

As we evolve from mass production to mass customization, the focus shifts to the adaptability of the individual worker, who must have the ability and awareness to flex as circumstances change. A shared belief of Zen and Lean is that change must be an ally, that we must balance with change rather than fight against it. This is the essential difference between push and pull—both in our personal lives and on the shop floor.

So what is the point to all this? Should each individual engage in meditation, Yoga, Tai Chi, or other Eastern practices? Not necessarily, just do whatever works for you: golf, gardening, playing music, creative play with your children, or a slow walk in the woods. But do something. It is important to find time in a busy life for daily personal restoration; your doctor would agree with this prescription.

To sustain Lean performance, an enterprise should behave the same way. It is often said that a Lean shop should never be pushed beyond 80–85% capacity on a regular basis. There needs to be slack time to accommodate sudden changes: unexpected machine downtime, rescheduling of an order, or a quality problem. When this extra 15–20% capacity isn't used for production, then the machines sit idle. The workers perform preventative maintenance, participate in a Kaizen event or education workshop, brush up on a skill that's become a little rusty, or walk the plant to lend a hand wherever it's needed. These are the fertile gaps between thought and activity where creativity seeps through.

Companies that provide each worker with a little downtime, encouragement (not fear or guilt for being underutilized), and a framework for the direction of the resulting creativity (Kaizen teams and initiatives guided by vision and strategy) often achieve profound results. Can these results be measured in strictly financial terms? Perhaps, though a more balanced measurement system is appropriate. It is no surprise that the balanced scorecard has become increasingly popular, indeed the concept of balance in our lives, our workplace, our schools, governments, and environment, is long overdue if we are to realize enduring performance improvement. Sustainability is a long distance run, not a sprint.

Managers cannot directly cause inspiration to happen, but they *can* create the conditions for it to arise on its own. Don't misunderstand; this should not be an individual free-for-all. Managers must set clear and attainable goals that are consistent with company strategy, while teams and individuals must be accountable to them. But individuals should have the flexibility to suggest the methods by which these goals are attained. To nurture this sense of creative problem solving, the environment must be conducive to growth and experimentation. If we spend all day working under stress, then is it any wonder at the end of the day we are left tired and uninspired? Each of us needs a little time during the workday where we let go, opening the door for creativity and inspiration. Inspiration can be elusive at times, yet inspiration is surprisingly effortless, it all depends: are you going with the flow or struggling against it?

As a result of his many visits to Japan, along with his workshops, his many publications through Productivity Press, and the creation of the Shingo Prize, Norman Bodek was instrumental in introducing Lean thinking to the West. He brought not only the techniques of Lean, but the stories of individual persistence and inspiration behind their conception:

> I look at one magical day that kick started The Toyota Production System. It was when Taiichi Ohno came over to Shigeo Shingo and said, "We have to reduce the setup time on this press from four to two hours." And Shingo, brilliantly said, "OK." Who else would have looked at a process that had been taking four hours for many years and simply said, "OK"?

A little time later Ohno came by again and said, "Two hours is not good enough, it has to be done in less then 10 minutes." And Shingo again said, "OK." Then Shingo just sat and watched the changeover process for days until the light bulbs went off in his head.

Shingo recognized the difference between inside and outside setups—that which could be done while the machine was running, and that which could be done only when the machine stopped—thinking always what could be moved from inside to outside. One day he thought about how quickly a tape could go in and out of a music player and asked himself if these huge dies could also be changed over as quickly.

I learned primarily from both Dr. Shingo and Mr. Ohno that the key to the success of Lean Manufacturing is simplicity. Maintain focus on continuous improvement, getting all workers consciously and continuously coming up with improvement ideas to eliminate waste. Instead of always telling workers what to do, you stop, make a shift, and empower your employees to implement ideas on how to shorten the lead time, reduce defects, improve safety, and reduce costs, thus making their lives easier and more interesting.[245]

Be mindful of Deming's definition of quality: *pride of workmanship*. We are most effective when we truly care about what we do. As an employee, a manager or an executive—what can *you* do to create a sense of ownership, individual involvement, and personal satisfaction in your workplace? Do it now. When everyone cares, powerful things happen.

To Build a Home

After a long flight and a six hour mountainous bus ride, I met eleven new friends, all Habitat for Humanity volunteers, in the small village of Totonicapan in the Western Highlands of Guatemala. The jobsite, dusty and rocky, lay perched on the side of a steep hill, with occasional running water and no electricity. All work was to be done with hand tools, muscle, and local methods, encouraging community knowledge preservation and sustainability. We were a mixed crowd, male and female, young and old, some with construction skills and others clumsy but willing to learn, and most of us speaking only broken Spanish.

On the first day, the Jefe (our nineteen-year-old boss who had been building homes since he was eleven) introduced his crew (in Spanish, of course) and the lessons began. The first day was awkward, while everyone developed the basic building and language skills: mixing mortar with shovel and hoe in the mud pit, digging trenches, shaping concrete blocks with chisel and machete, cutting and bending steel rods with hacksaws and pliers to form rebar, and fashioning straight walls with a plum-bob on a string.

By the second day people gravitated to their preferred tasks, cross-training each other, and helping out wherever needed. Every time the heavy wheelbarrow needed to be eased down the slippery slope, hands magically appeared, and there was never a fall. Water pressure was unpredictable, yet without water, production would instantly stop. Fortunately the Jefe had foreseen this problem and had hauled in a large metal bathtub, which sat nearby, constantly filling with water dribbling from a pipe: a constraint buffer.

With the foundation poured, next we began to form rebar from 1/3"-diameter steel rods. Within moments we realized that cutting all the needed rebar, with only two hacksaws, would consume at least one if not two days, leaving the majority of our team idle. One of our team members, recalling his days working in a sawmill more than 50 years before, jumped into action, fashioning clever devices out of materials laying around the jobsite and breaking the process into steps. Soon we had a high-volume rebar cutting and assembly line set back in the shade. What would have taken days, and delayed the pouring of concrete required only a few hours and much less effort, after an investment of a little ingenuity on physical movement and setup time reduction.

When the time came to begin laying the rows of concrete blocks, another constraint appeared. The heavy blocks lay in stacks, far up the steep, slippery hill. To move each block, or even several at a time in the wheelbarrow, would slow us down and make for dangerous work. I turned to the cluster of children who each day gathered at the edge of the jobsite to giggle at the dusty foreigners. Through pantomime, we invented a game: Who could carry the most blocks down the hill? We made it fun! Soon our blocks lay in tumbled stacks at the edge of the job site, and for the rest of the week, the entire team engaged in play with the local children.

By the third day, and for the remainder of the project, everything went smoothly, without mishap or injury, and though we exerted considerable physical labor, at over 9000 feet of elevation in one hundred degree afternoon heat, this labor was not *work*. Eyes and hearts were open, learning, helping, caring, with a purpose.

How can we make ordinary work a more valued part of each person's life? Lean, continuous improvement, quality, leadership . . . these are just concepts, empty words. A Zen master would suggest that humans focus on words and concepts like a dog staring at the finger rather than the object to which it points. So many companies we encounter read the books and repeat the phrases, but they can't seem to change their attitudes and behavior.

Do not overintellectualize. If you cannot truly feel what it means to transform the culture of your organization, to empower your staff to make continuous improvements, then you may be making it too complicated. Walk around, ask questions, listen, and learn. As Stephen Covey says, "seek first to understand, then to be understood."[246] Put into practice the celebrated

Hewlett-Packard MBWA: *Management By Walking Around*, another way of saying *Gemba*.

Learning from the experience and mistakes of others is important. But even more important is to take your own calculated risks and learn from them. Treat every employee as if they know more about the situation than you do—in many cases they will. Let go of the ego that insists you should be smarter since you're the manager. And forget the old lessons you learned about how to manage people.

Be curious, open, patient, and playful.
Constantly invest in developing people's skills.
Gather, nurture, and share knowledge generously.
Keep an open mind.
Reward experimentation.
Learn through direct experience.
Lead by example and inspiration.

What are you waiting for? As an old Zen master relentlessly questioned, *"If not now . . . when?"*

Acknowledgments

Life is precious and short; it challenges us to take what knowledge our predecessors have nurtured, add some measure of value, and pass it on to others. I am grateful to the pioneers of Lean: Ohno, Shingo, Womack, Jones, Rother, Shook, Schonberger, and many others listed in the bibliography for what they have begun. I am also grateful to many people for their time and effort in the development of this book, which I hope will add value in some way:

To my wife of twenty years, Karen, whose love, support, writing skill, and most of all patience, have sustained me through this effort. To my little muse, Biscuit, who stayed close during those early morning hours of research and writing.

To my friend and business partner Mike Orzen, your tireless effort added so much value. Though your name is not on the cover of this book, what is within represents our deeply shared vision and principles.

To Carol Ptak, your 'goal' is to improve the world in your own special way. You've been warm and generous with your criticism and encouragement.

To Norman Bodek, through Productivity Press and the creation of the Shingo Prize you helped introduce Lean to the West. Your recent books, and your generous assistance with mine, helped me to appreciate the human side of Lean.

To Jeffrey Liker, though I must have tried your patience at times, you always stuck with me and challenged my ideas. In the end you helped me to re-write the introductory chapter, lending a fine sense of balance between the disciplines of Lean and IT.

Lean Enterprise Systems: Using IT for Continuous Improvement, by Steve Bell
Copyright © 2006 by John Wiley & Sons, Inc.

To Bill Christopher, my literary agent and coach, thanks for taking me under your wing.

To my supporters at John Wiley and Sons publishing: George Telecki, Rachel Witmer, Dean Gonzalez, Danielle Lacourciere, and Melissa Yanuzzi; I couldn't have asked for better.

And finally my gratitude goes to many others who have helped along the way: Tony Baer, onStrategies; Adam Bartkowski, Apriso; Sami Cassis, Factory Logic; Tim Costello, BuilderHomesite.com; Diana Crossley, Tauber Manufacturing Institute, University of Michigan; Mark Douglas; Kevin Duggan, Duggan and Associates; John Gillam, Boyd Coffee Company; Tony Gorski, JCIT International; Sarah Harrington; Laila Hirr; Bob Kerr, High Performance Solutions; Terri King; Tom Knight, Invistics; Marsha Kremen; Elizabeth Lyon, Editing International; Brian Maskell, BMA Inc.; Leon McGinnis, Georgia Institute of Technology; Trevor Moore, Shop9000; Brad Paris, Harry's Fresh Foods; Ian Percy; John Pierce; Travis Pierce, WARN Industries; Roger Resley, eNSYNC Solutions; Kevin Rosell, Epicor Software Corporation; Andrew Sage, George Mason University; Dennis Severance, Accenture Professor of Computer and Information Systems, University of Michigan; Scott Simpson; Arleigh Taylor, Microsoft Corporation; Dave Turbide; and many others thank you!

Acronyms

0HIO: Zero Human Intervention Operations
3PL: Third Party Logistics
ABB: Activity Based Budgeting
ABC: Activity Based Costing, or ABC Inventory Classification
AICPA: American Institute of Certified Public Accountants
API: Application Programming Interface
APICS: The Association for Operations Management (formerly the American
　Production and Inventory Control Society)
APS: Advanced Planning & Scheduling
ASCII: American Standard Code for Information Interchange
ATO: Assemble to Order
ATP: Available to Promise
B2B: Business to Business
B2C: Business to Customer
BI: Business Intelligence
BOM: Bill of Materials
BOR: Bill of Resources
BPEL: Business Process Execution Language
BPR: Business Process Reengineering
C/O: Changeover Time
C/T: Cycle Time
CAD: Computer Aided Design
CAE: Computer Aided Engineering
CAGR: Compound Annual Growth Rate

Lean Enterprise Systems: Using IT for Continuous Improvement, by Steve Bell
Copyright © 2006 by John Wiley & Sons, Inc.

CAM: Computer Aided Manufacturing
CCR: Capacity Constrained Resource
CFR: Code of Federal Regulations
CIM: Computer Integrated Manufacture
CIRM: Certified in Integrated Resource Management
CNC: Computer Numerical Control
CORBA: Common Object Request Broker Architecture
CONWIP: Constant Work In Process
COM: Component Object Model
COM+: Component Object Model +
COOL: Country Of Origin Labeling
CPA: Critical Path Analysis
CPFR: Collaborative Planning Forecasting and Replenishment
CPG: Consumer Packaged Goods
CPIM: Certified in Production and Inventory Management
CFPIM: Certified Fellow in Production and Inventory Management
CPM: Certified Purchasing Manager
CRM: Customer Relationship Management
CRP: Capacity Requirements Planning
CTO: Configure to Order
CTP: Capable to Promise
DBR: Drum Buffer Rope
DCE: Distributed Computing Environment
DCOM: Distributed Component Object Model
DMAIC: Define Measure Analyze Improve Control
DRP: Distribution Requirements Planning
DSD: Direct Store Delivery
DSS: Decision Support System
DTD: Document Type Definition
DTS: Data Transformation Services
EAI: Enterprise Application Integration
EAM: Enterprise Asset Management
ECC: Engineering Change Control
ECM: Engineering Change Management
ECO: Engineering Change Order
EDD: Earliest Due Date
EDI: Electronic Data Interchange
EIS: Executive Information System
EODD: Earliest Operation Due Date
EPE*x*: Every Part Every *interval*
EPM: Enterprise Performance Management
ERD: Entity Relationship Diagram
ERP: Enterprise Resource Planning
ETL: Extract Transform and Load
ETO: Engineer to Order

FAS: Final Assembly Schedule
FASB: Financial Accounting Standards Board
FCFS: First Come First Served
FCS: Finite Capacity Scheduling
FDA: Food and Drug Administration
FIFO: First-In First-Out
GAAP: Generally Accepted Accounting Principles
GIGO: Garbage-In Garbage-Out
GMP: Good Manufacturing Practices
GPS: Global Positioning System
GUI: Graphical User Interface
HACCP: Hazard Analysis and Critical Control Point
HTML: Hyper Text Markup Language
ICS: Infinite Capacity Scheduling
ISO: International Organization for Standardization
IT: Information Technology
J2EE: Java 2 Enterprise Edition
JIT: Just-In-Time
KPI: Key Performance Indicator
LAN: Local Area Network
LLSF: Largest Lot Size First
MES: Manufacturing Execution System
MIS: Management Information System
MOM: Message Oriented Middleware
MPS: Master Production Schedule
MRO: Maintenance Repair and Operations
MRP: Material Requirements Planning
MTO: Make to Order
MTS: Make to Stock
.NET: Microsoft Web Services
NIST: National Institute of Standards and Technology
NNVA: Necessary Non Value Added
NVA: Non Value Added
ODBC: Open Database Connectivity
OEM: Original Equipment Manufacturer
OLAP: Online Analytical Processing
OLE: Object Linking and Embedding
PC: Personal Computer
PDA: Personal Digital Assistant
PDCA: Plan Do Check Act
PDM: Product Data Management
PLM: Product Lifecycle Management
PMO: Project Management Office
POLCA: Paired Overlapping Loops of Cards with Authorization
PPM: Project Portfolio Management

QFD: Quality Function Deployment
QRM: Quick Response Manufacturing
RCCP: Rough Cut Capacity Planning
RF: Radio Frequency
RFID: Radio Frequency Identification
RMA: Return Merchandise Authorization
ROCE: Return On Capital Employed
ROI: Return on Investment
RPC: Remote Procedure Call
RRP: Resource Requirements Planning
S&OP: Sales & Operations Planning
SAA: Systems Application Architecture
SCADA: Supervisory Control and Data Access
SCM: Supply Chain Management
SCOR: Supply Chain Operations Reference
SEC: Securities and Exchange Commission
SFA: Sales Force Automation
SLSF: Smallest Lot Size First
SMART: Synchronized Material Availability Request Tickets
SME: Society of Manufacturing Engineers
SMP: Simplified Market Pull
SOA: Services-Oriented Architecture
SOAP: Simple Object Access Protocol
SPC: Statistical Process Control
SPT: Shortest Processing Time
SQC: Statistical Quality Control
SQL: Structured Query Language
TOC: Theory of Constraints
TPM: Total Productive Maintenance
TPS: Toyota Production System
TQM: Total Quality Management
UDDI: Universal Description Discovery and Integration
VA: Value Added
VAN: Value Added Network
VMI: Vendor Managed Inventory
VSM: Value Stream Mapping
WIP: Work In Process
WMS: Warehouse Management System
WSDL: Web Services Description Language
XML: eXtensible Markup Language

Endnotes

Chapter 1

1 Reprinted with permission from Apriso Software, Inc.
2 Jeffrey K. Liker, The Toyota Way, McGraw-Hill, New York, NY, 2004, p. 226.
3 George Plossl, Orlicky's Material Requirements Planning, Second Edition, McGraw Hill, New York, NY, 1995, p 7.
4 Doug Bartholomew, Lean vs. ERP, Industry Week, July 19, 1999.
5 Geoffrey Moore, Inside the Tornado, HarperCollins Publishers Inc., New York, NY, 1995, p 65.

Chapter 2

6 Burger King Corporation advertising jingle.
7 Peter Hines, Richard Lamming, Dan Jones, Paul Cousins and Nick Rich, *Value Stream Management, Strategy and Excellence in the Supply Chain*, Financial Times-Prentice Hall, London, UK, 2000, p. 60.
8 Michael Hammer and James Champy, *Reengineering the Corporation*, Revised Edition, HarperBusiness, New York, NY, 2001, p. 42–47.
9 Norman Bodek, *Kaikaku, The Power and Magic of Lean*, PCS Press, Vancouver, WA, 2004, p. 32.
10 Terence T. Burton and Steven M. Boeder, *The Lean Extended Enterprise*, J. Ross Publishing, Boca Raton, FL, 2003, p. 100.
11 Pascal Dennis, *Lean Production Simplified*, Productivity Press, New York, NY, 2002, p. 102.
12 HH Dalai Lama and Howard C. Cutler MD, *The Art of Happiness*, Riverhead Books, New York, NY, 1998, p. 199.

Lean Enterprise Systems: Using IT for Continuous Improvement, by Steve Bell
Copyright © 2006 by John Wiley & Sons, Inc.

13 Rafael Aguayo, *Dr. Deming, The American Who Taught the Japanese About Quality*, Fireside, New York, NY, 1991, p. 6.

14 Ibid., p. 17.

15 Ibid.

16 Norman Bodek, *Kaikaku, The Power and Magic of Lean*, PCS Press, Vancouver, WA, 2004, p. 59.

17 James P. Womack and Daniel T. Jones, *Lean Thinking, Banish Waste and Create Wealth in Your Corporation*, Second Edition, Free Press, New York, NY, 2003, p. 179.

18 Dr. Richard J. Schonberger, Lean and Fat Factories, *The Manufacturer*, November 2002.

19 Edward De Bono, *De Bono's Thinking Course*, Revised Edition, Checkmark Books, 1994.

20 Steven W. Thompson, Lean, TOC or Six Sigma, Which tune should a company dance to?, *SME Online* electronic article.

21 Mike George, Dave Rowlands and Bill Kastle, *What is Lean Six Sigma?*, McGraw-Hill, New York, NY, 2003, p. 34.

22 Pete Pande, The Six Sigma–Lean Blend, *APICS—The Performance Advantage*, September 2003.

23 Rafael Aguayo, *Dr. Deming, The American Who Taught the Japanese About Quality*, Fireside, New York, NY, 1991.

24 Norman Bodek, *Kaikaku, The Power and Magic of Lean*, PCS Press, Vancouver, WA, 2004, p. 28.

25 Robert T. Parry, President, Federal Reserve Bank of San Francisco, Economic Outlook presentation to the Rotary Club of Portland, November 25, 2003.

26 David Drickhammer, Lean Manufacturing: The Third Generation, *Industry Week*, March 1, 2004.

27 James P. Womack and Daniel T. Jones, *Lean Thinking, Banish Waste and Create Wealth in Your Corporation*, Second Edition, Free Press, New York, NY, 2003, p. 289.

28 Patricia Panchak, Lean Health Care? It Works!, *Industry Week*, November 1, 2003.

29 Linda Riebel and Ken Jacobsen, *Eating to Save the Earth, Food Choices for a Healthy Planet*, Celestial Arts, Berkeley, CA, 2002, p. 20.

30 James Womack, Is Lean Green?, Lean Enterprise Institute Electronic Article, November 4, 2003.

Chapter 3

31 Jeffrey K. Liker, *The Toyota Way*, McGraw-Hill, New York, NY, 2004, p. 12.

32 Ibid., p. 7, 10, 12.

33 Tim Costello, Lean Software Does Not Mean No Software, Presentation, Shingo Prize Conference and Awards Ceremony, May 12–16, 2003.

34 Professor Leon McGinnis, Georgia Institute of Technology, interview, August 10, 2004.

35 Tim Costello, interview, May 14, 2004.

36 James P. Womack and Daniel T. Jones, *Lean Thinking, Banish Waste and Create Wealth in Your Corporation*, Second Edition, Free Press, New York, NY, 2003, p. 19.

37 Michael Hammer and James Champy, *Reengineering the Corporation*, Revised Edition, HarperBusiness, New York, NY, 2001, p. 53–86.

38 Mike Rother and John Shook, *Learning To See*, Lean Enterprise Institute, Brookline, MA, 2003, p. 5.

39 Ibid., p. 9.

40 Ibid., p. 42.

41 Mark Francis, Lean Information and Supply Chain Effectiveness, *International Journal of Logistics: Research and Applications*, Volume 1, Number 1, 1998.

42 Peter Hines, Richard Lamming, Dan Jones, Paul Cousins and Nick Rich, *Value Stream Management, Strategy and Excellence in the Supply Chain*, Financial Times-Prentice Hall, London, UK, 2000, p. 67.

43 Dan Jones and James Womack, *Seeing the Whole—Mapping the Extended Value Stream*, Lean Enterprise Institute, Brookline, MA, 2003, p. 3.

44 Terence T. Burton and Steven M. Boeder, *The Lean Extended Enterprise*, J Ross Publishing, Boca Raton, FL, 2003, p. 11.

45 Norman Bodek, *Kaikaku, The Power and Magic of Lean*, PCS Press, Vancouver, WA, 2004, p. 35.

46 Beth Bacheldor, Never Too Lean, *Information Week*, April 10, 2004, p. 40.

47 Erik Keller, ERP is the only system of record, *Manufacturing Business Technology*, April 2004.

48 Dave Caruso, Throw ill-advised caution to the wind, *Manufacturing Business Technology*, May 2004.

49 Robert Kennedy, Globalization—Transforming the Way America Works, *Dividend*, University of Michigan Business School, Spring 2004, Vol. 35, No. 1, p. 20.

50 John M. Hill, Toward Perfect Postponement, *Manufacturing Business Technology*, April 2004, Vol. 22, No. 4, p. 16.

51 Edward Teach, Working on the Chain, *CFO*, September 2002.

52 Michael Tanner, Inter-enterprise applications growth: where IT meets marketing, *Manufacturing Business Technology*, May 2004, Vol 22, No. 5.

53 David Drickhamer, EDI is Dead! Long Live EDI!, *Industry Week*, April 1, 2003.

54 A Perfect Market, *The Economist*, May 15–24, 2004, Volume 371, Number 8375, p. 14.

55 *Logistics Today*, Ten Best Supply Chains of 2004, December 1, 2004.

56 *Logistics Today*, Ten Best Supply Chains, December 1, 2003.

57 Merriam-Webster Online Dictionary, www.m-w.com.

58 Clayton M. Christensen, *The Innovator's Dilemma: When New Technologies Cause Great Firms to Fail*, Harvard Business School Press, Boston, MA, 1997, p. xv.

59 Nicholas G Carr, *Does IT Matter?*, Harvard Business School Press, Boston, MA, 2004, p. 7–11.

60 Ibid., p. 83–86.

61 Michael Schrage, Wal-Mart Trumps Moore's Law, *Technology Review*, February 28, 2002.

Chapter 4

62 APICS Online Dictionary, http://www.apics.org/resources/dictionary/.

63 Robert H. Hayes and Steven C. Wheelwright, Link Manufacturing Process and Product Life Cycles, *Harvard Business Review*, January–February 1979.

64 Robert H. Hayes and Steven C. Wheelwright, The Dynamics of Process-Product Life Cycles, *Harvard Business Review*, March–April 1979.
65 Robert E. Cannon, *A Tutorial on Product Life Cycle*, MROtoday, Pfingsten Publishing LLC, Fort Atkinson, WI, 2003.
66 APICS Online Dictionary, http://www.apics.org/resources/dictionary/.
67 Ibid.
68 Ibid.
69 Brian Willcox CFPIM, CIRM, Action MRPII, Study notes for Detailed Scheduling and Planning, 2001.
70 APICS Online Dictionary, http://www.apics.org/resources/dictionary/.
71 John R. Dougherty, Getting Started with Sales and Operations Planning, APICS Master Planning of Resources Reprints, 2000, p. 24.
72 Ibid.
73 Richard Ling, Q&A with the Originator of Sales and Operations Planning, *PeopleTalk, The Journal of the Real-Time Enterprise*, January–March 2004.

Chapter 5

74 Kevin J. Duggan, *Creating Mixed Model Value Streams*, Productivity Press, New York, NY, 2002, p. 18.
75 Richard Schonberger, Lean & Fat Factories, *The Manufacturer*, November 2002.
76 *Logistics Today*, Ten Best Supply Chains, December 1, 2003.
77 Taiichi Ohno, *Toyota Production System*, Productivity Press, Portland, OR, 1988, p. 48–51.
78 John R. Dougherty, Managing MPS Changes Despite Time Fences and Frozen Horizons, *APICS Conference Proceedings*, 1998.
79 James P. Womack and Daniel T. Jones, *Lean Thinking, Banish Waste and Create Wealth in Your Corporation*, Second Edition, Free Press, New York, NY, 2003, p. 55.
80 Kevin J. Duggan, *Creating Mixed Model Value Streams*, Productivity Press, New York, NY, 2002, p. 39.
81 Chris Gray and Tom Wallace, Manage It, *APICS—The Performance Advantage*, October 2003.
82 Sam Tomas, Lean Manufacturing Workshop Series, *Introduction to Lean Manufacturing*, APICS, 2003.
83 Kevin J. Duggan, *Creating Mixed Model Value Streams*, Productivity Press, New York, NY, 2002, p. 34–36.
84 Sami Cassis, Leading Lean, *Factory Logic Newsletter*, 2003, Volume 1, Release 27.
85 Mike Rother and John Shook, *Learning to See*, Version 1.2, The Lean Enterprise Institute, 1999, p. 54.
86 Kevin J. Duggan, *Creating Mixed Model Value Streams*, Productivity Press, New York, NY, 2002, p. 118.
87 John Bicheno, *The New Lean Toolbox, Towards Fast, Flexible Flow*, PICSIE Books, Buckingham, UK, 2004, p. 119.
88 Bill Leedale, IFS North America, Interview, November 17, 2004.
89 John Bicheno, *The New Lean Toolbox, Towards Fast, Flexible Flow*, PICSIE Books, Buckingham, UK, 2004, p. 109.

90 Richard Schonberger, Kanban at the Nexus, *Production and Inventory Management Journal*, Third/Fourth Quarter 2002.

91 Jeffrey K. Liker, *The Toyota Way*, McGraw-Hill, New York, NY, 2004, p. 110.

92 Uday Karmarkar, Getting Control of Just-In-Time, *Harvard Business Review*, Sept–Oct 1989.

93 Jeffrey K. Liker, *The Toyota Way*, McGraw-Hill, New York, NY, 2004, p. 122.

94 Ibid., p. 110.

95 Stephen Moncrief, CPIM, Push and Pull, *APICS—The Performance Advantage*, June 2003.

96 Doug Bartholomew, Scheduling for Complexity, Automotive manufacturers take different approaches to sequencing vehicles on the assembly line, *Industry Week*, April 2002.

97 John Bicheno, *The New Lean Toolbox, Towards Fast, Flexible Flow*, PICSIE Books, Buckingham, UK, 2004, p. 109.

98 Doug Bartholomew, Lean Efforts Get Software Assist, *Industry Week*, November 1, 2003.

99 John Bicheno, Lean Manufacturing Workshop Series, Lean Scheduling, APICS, 2003.

100 Vincent Bozzone, *Speed to Market: Lean Manufacturing for Job Shops*, Second Edition, AMACOM, New York, NY, 2002, p. xxii.

101 Donald W. Fogarty, John H. Blackstone, Jr., Thomas R. Hoffman, *Production and Inventory Management*, South-Western Publishing, Cincinnati, OH, 1991, p. 662.

102 Professor Leon McGinnis, Georgia Institute of Technology, interview, August 10, 2004.

103 Dave Turbide, interview, July 9, 2004.

104 Rick Whiting, Databases – Boeing, *Information Week*, March 22, 2004.

105 James P. Kelleher, The Zero Inventories Concepts as Applied to Job Shops, APICS Zero Inventory Philosophy and Practices Seminar Proceedings, 1984.

106 Terence T. Burton and Steven M. Boeder, *The Lean Extended Enterprise*, J. Ross Publishing, Boca Raton, FL, 2003, p. 139.

107 Vincent Bozzone, *Speed to Market: Lean Manufacturing for Job Shops*, Second Edition, AMACOM, New York, NY, 2002, p. 6.

108 John Bicheno, Lean Manufacturing Workshop Series, *Lean Scheduling*, APICS, 2003.

109 Dave Turbide, *Simplified Market Pull Scheduling*, White Paper, December 2003.

110 O. Kermit Hobbs, Jr., Application of JIT Techniques In a Discrete Batch Job Shop, *Production and Inventory Management Journal*, First Quarter 1994.

111 Mitchell Millstein, Putting an Eye on the Scheduling Function, *APICS—The Performance Advantage*, September 2001.

112 The Free On-Line Dictionary of Computing, February 9, 2002.

113 Tom Knight, Invistics, Interview, August 3, 2004.

114 Eli Goldratt, *The Haystack Syndrome*, North River Press, New Haven, CT, 1990, p. 58.

115 Donald W. Fogarty, John H. Blackstone, Jr., Thomas R. Hoffman, *Production and Inventory Management*, South-Western Publishing, 1991, p. 656.

116 Eli Goldratt, *The Race*, North River Press, New Haven, CT, 1986, p. 120.

117 Eli Goldratt, *The Haystack Syndrome*, North River Press, New Haven, CT, 1990, p. 98.

118 Eli Goldratt, Eli Schragenheim, and Carol Ptak, *Necessary But Not Sufficient*, The North River Press, New Haven, CT, 2000, p. 197.
119 Steven W. Thompson, Lean, TOC or Six Sigma—Which tune should a company dance to?, SME Online publication, www.sme.org.
120 Woodruff, D.L., M.L. Spearman, and W.J. Hopp, CONWIP: A Pull Alternative to Kanban, *International Journal of Production Research* 28, 1990, no. 5, p. 879–894.
121 Tom Knight, Invistics, Interview, November 16, 2004.
122 Rajan Suri, *Quick Response Manufacturing: A Companywide Approach to Reducing Lead Times*, Productivity Press, Portland, OR, 1998, p. 243–257.
123 John Bicheno, Lean Scheduling, APICS Lean Manufacturing Workshop Series, 2003.
124 Rajan Suri, *Quick Response Manufacturing: a companywide approach to reducing lead times*, Productivity Press, Portland, OR, 1998, p. 235.
125 Nancy Bartels, Advanced Planning's Ongoing Evolution, *Manufacturing Business Technology*, December 2004, p. 34.
126 Roberto Michel, Can Lean Live Online, *Manufacturing Business Technology*, June 2004.
127 Jeffrey K. Liker and Karl Burr, *Advanced Planning Systems as an Enabler of Lean Manufacturing*, i2 Technologies White Paper, March 24, 1999.
128 Jeffrey K. Liker, *The Toyota Way*, McGraw-Hill, New York, NY, 2004, p. 162.
129 Jeffrey K. Liker and Karl Burr, *Advanced Planning Systems as an Enabler of Lean Manufacturing*, i2 Technologies White Paper, March 24, 1999.
130 Mike Rother and John Shook, *Learning* To See, Lean Enterprise Institute, Brookline, MA, 2003, p. 34.

Chapter 6

131 Frederick Newell, *Why CRM Doesn't Work*, Bloomberg Press, Princeton, NJ, 2003, p. 7.
132 Dave Caruso, Throw Ill-Advised Caution to the Wind, *Manufacturing Business Technology*, May 2004, Vol. 22, No. 5, p. 17.
133 *Product Lifecycle Management*, A CIMdata Report, 2002
134 Jim Fulcher, Only the Nodes Know, *Manufacturing Business Technology*, March 2004, Vol. 22, No. 3, p. 22.
135 Tim Stevens, Factories of the Future—Integrated Product Development, *Industry Week*, June 1, 2002.
136 *Virtual Enterprise Integration: Creating a Sustainable Manufacturing Life Cycle*, CASA/SME Blue Book, Computer and Automated Systems Association of the Society of Manufacturing Engineers, 2001, p. 20.
137 Preston G. Smith and Donald G. Reinertsen, *Developing Products in Half The Time*, John Wiley and Sons, New York, NY, 1998, p. 267.
138 Ibid., p. 50.
139 John Schneiter, Taming the 5,000 Pound Gorilla, *Industry Week*, March 19, 2002.
140 How Lean Manufacturing Principles Speed Product Design, Machine Design via ProQuest Information and Learning Company, originally published April 1, 2004.
141 Gene Thomas, From Make-Sell to Sell-Make, *APICS—The Performance Advantage*, August 2001, p. 31.
142 Ibid., p. 32–34.

143 Richard W. Bourke, Product Configurators: Key Enablers for Mass Customization, *MidRange ERP*, August 2000.

144 G. Berton Latamore, The Burden of Choice, *APICS—The Performance Advantage*, January 2001, p. 42.

145 *Scope of Today's Mission-Critical Enterprise Applications*, UGS PLM White-Paper, March 2003.

146 Dave Caruso, ERP as infrastructure?, *Manufacturing Business Technology*, January 2003.

147 Michael Treacy and Daniel Wiersma, *The Discipline of Market Leaders*, Addison Wesley, Reading, MA, 1997, p. 31.

Chapter 7

148 Michael Hammer and James Champy, *Reengineering the Corporation*, Revised Edition, HarperBusiness, New York, NY, 2001, p. 95.

149 Tony Baer, Demand Strategies, *www.onstrategies.com*, interview, September 15, 2004.

150 Ibid.

151 Joshua Greenbaum, SAP to the Rescue?, *Managing Automation*, January 17, 2005.

152 Tony Baer, Grand Unification Theory Redux, OnStrategies electronic newsletter, September 10, 2003.

153 Roberto Michel, No Simple Revival, *Manufacturing Business Technology*, July 2004, p. 24.

Chapter 8

154 John Sviokla, Knowledge Pays, *CIO Magazine*, Feb. 15, 2001.

155 *Architecture That Delivers A Business Advantage*, BroadVision White Paper, 2002.

156 Daniel W. Rasmus, *Collaboration, Content and Communities: An Update*, Giga Information Group, May 31, 2002.

157 Terri King, www.saltmine.com, Interview, Nov. 17, 2004.

158 Megan Santosus, The Secret to KM Success, CIO Magazine Knowledge Management Research Center, http://www.cio.com/research/knowledge/edit/k070804_secret.html, July 8, 2004.

159 Copyright QlikTech International, used by express permission of QlikTech, *www.qliktech.com*.

160 Copyright IFS North America, Inc., used by express permission of IFS, *www.ifsna.com*.

161 Copyright IFS North America, Inc., used by express permission of IFS, *www.ifsna.com*.

Chapter 9

162 Taiichi Ohno, *Toyota Production System*, Productivity Press, Portland, OR, 1988, p. 48–51.

163 Mike Rother and John Shook, *Learning To See*, Lean Enterprise Institute, Brookline, MA, 2003, p. 5.

164 Kevin Rosell, Epicor Software Corporation, Interview, March 31, 2004.

165 Brian Maskell & Bruce Baggaley, *Practical Lean Accounting*, Productivity Press, New York, NY, 2004, p. 25, 77.

166 Paul Swamidass, Thomas Walter Center for Technology Management, *Bar Code Users and Their Performance*, White Paper, July 1998.

167 Steve Geary and Kate Vitasek, Cause and Effect, *APICS Magazine*, May 2003, Vol. 13, No. 5, p. 72.

168 Vivek Ranadive, *The Power of Now*, McGraw Hill, New York, NY, 1999, p. 3–20.

169 Adam Bartkowski, CEO, Apriso Software, Interview, July 23, 2004.

170 Tony Baer, The Reality of Real-Time Intelligence, *Manufacturing Business Technology*, July 2004, p. 62.

171 Adam Bartkowski, The Execution Economy and Bottom-Out Enterprise Software, Apriso Corporation, Press Background Information, 2003.

172 Dave Caruso, Emerging SCM Directive: Think Global, *Manufacturing Business Technology*, July 2004, p. 38.

173 ABI Research Sees RFID Consultants Starting to Attack Vertical Markets, *Manufacturing Business Technology*, electronic publication, www.manufacturingsystems.com, accessed Aug. 24, 2004.

174 Sami Cassis, *Lean Manufacturing Systems—the Next Generation*, Factory Logic White Paper, Sept. 2003.

175 Thomas Wailgum, Tag, You're Late, *CIO Magazine*, Nov. 15, 2004, p. 50.

176 http://www.csc.calpoly.edu/~dstearns/SchlohProject/intro.html.

177 Marty Weil, RFID Confusion and Possibility, *APICS—The Performance Advantage*, February 2004, p. 51.

178 Ibid.

179 Thomas Wailgum, Tag, You're Late, *CIO Magazine*, Nov. 15, 2004, p. 50.

180 Marty Weil, Life After the Deadline, *APICS Magazine*, March 2005, p. 51.

181 The Data Avalanche, *Logistics Europe*, October 2003, p. 18.

182 Jeff Woods, *Prepare for Disillusionment with RFID*, Gartner Website, June 29, 2004.

183 William Christopher, Finding the Controlling Simplicities to Simplify and Improve Cost Management, *Journal of Cost Management*, March/April 2003, p. 46.

184 MobileTech Solutions, *Route Automation Guide*, January 2001.

Chapter 10

185 Eli Goldratt, *The Haystack Syndrome*, North River Press, Great Barrington, MA, 1990, p. 55.

186 Mohanbir Sawhney, Damn the ROI, Full Speed Ahead, *CIO Magazine*, July 15, 2002.

187 Tracy Mayor, Value Made Visible, CIO Magazine, May 1, 2000.

188 *Quantifying IT Intangible Benefits*, A Glomark Whitepaper, 2004, www.glomark.com

189 Brian Maskell and Bruce Baggaley, *Practical Lean Accounting*, Productivity Press, New York, NY, 2004, p. 2.

190 Richard Schonberger, Kanban at the Nexus, *Production and Inventory Management Journal*, APICS, Third/Fourth Quarter, 2002.

191 Guide to a Balanced Scorecard Performance Management Methodology, Procurement Executives Association, no publisher identified, 1998, p. 1.

192 Michael Cowley and Ellen Domb, *Beyond Strategic Vision, Effective Corporate Action with Hoshin Planning*, Butterworth-Heinemann, Burlington, MA, 1997.

193 Robert S. Kaplan and David P. Norton, Using the Balanced Scorecard as a Strategic Management System, *Harvard Business Review*, January/February 1996.

194 Warren Bennis, *On Becoming a Leader*, Addison Wesley, Reading, MA, p. 45, 180.

195 William Christopher, Interview, October 18, 2004.

196 Mike Rother and John Shook, *Learning To See*, Lean Enterprise Institute, Brookline, MA, 2003, p. 8.

197 Christopher Meyer, How the Right Measures Help Teams Excel, *Harvard Business Review*, May/June 1994.

198 Merriam-Webster Online Dictionary, *www.m-w.com*.

199 Bob Kerr, High Performance Solutions, Interview, July 15, 2004.

200 Robert S. Kaplan and David P. Norton, Using the Balanced Scorecard as a Strategic Management System, *Harvard Business Review*, January/February 1996.

201 Michele L. Bechtell, *The Management Compass, Steering the Corporation Using Hoshin Planning*, American Management Association, New York, NY, 1995, p. 5.

202 Ibid. p. 21.

203 Pascal Dennis, *Lean Production Simplified*, Productivity Press, New York, NY, 2002, p. 123.

204 Ibid. p. 127.

205 James P. Womack and Daniel T. Jones, *Lean Thinking, Banish Waste and Create Wealth in your Corporation*, Second Edition, Free Press, New York, NY, 2003, p. 27.

206 Michael Cowley and Ellen Domb, *Beyond Strategic Vision: Effective Corporate Action with Hoshin Planning*, Butterworth-Heinemann, Newton, MA, 1997.

207 Richard J. Schonberger, *Let's Fix It, Overcoming the Crisis in Manufacturing*, The Free Press, New York, NY, 2001, p. 112.

208 *Waging War on Complexity, How to Master the Matrix Organizational Structure*, A T Kearney White Paper, February 2002.

209 Richard J. Schonberger, *Let's Fix It, Overcoming the Crisis in Manufacturing*, The Free Press, New York, NY, 2001, p. 107.

210 William Christopher, Interview, October 18, 2004.

211 Brian H. Maskell and Gay Gooderham, *Software Systems that Support Performance Measurement*, Brian Maskell and Associates White Paper, June 12, 2003.

212 *Waging War On Complexity, How to Master the Matrix Organizational Structure*, A T Kearney White Paper, February 2002.

Chapter 11

213 Jennifer Salopek, Give Change a Chance, *APICS—the Performance Advantage*, November/December 2001, p. 28.

214 Jim Brown and Olin Thompson, *Project Failure—The Numbers, Why and What it Means*, Technology Evaluation Research Center, Sept. 20, 2004, www.technology-evaluation.com.

215 Managing Complexity, *The Economist*, Nov. 25, 2004.

216 Mark Engleman, SoftSelect Systems www.softselect.com, Enterprise Application Software Improvement Management, Using SoftSelect Tools and Methodologies, Aug. 18, 2002.
217 A Field Day, *Red Herring*, Sept. 13, 2002.
218 Joris Evers, IDG News Service, Microsoft Puts Brakes on Next Business Apps, *InfoWorld*, June 25, 2004.
219 Jim Shepherd, Microsoft Paints Project Green a New Hue, *AMR Research Alert Highlight*, March 9, 2005.
220 Peoplesoft products to stay until 2013, *USA Today*, reported in *Manufacturing Business Technology* midday report, Jan. 19, 2005.
221 Jim Womack, Lean Information Management electronic newsletter, Lean Enterprise Institute, Nov. 5, 2004.
222 Michael H. Hugos, Agility Is a Frame of Mind, *ComputerWorld*, Nov. 29, 2004.
223 Boris Groendahl, Gartner Sees Shift to Bite Size Business Software, *ComputerWorld*, Nov. 1, 2004.
224 Ephraim Schwartz, Progress Means Better Progress, Infoworld.com, March 15, 2004, p. 10.
225 Michael H. Hugos, Agility Is a Frame of Mind, *ComputerWorld*, Nov. 29, 2004.
226 J. Chalifour, Demand at the Fount of Open Source, TechnologyEvaluation.com, December 10, 2004.
227 Ibid.
228 Erik Keller, Open Source is the next big disruptive technology, *Manufacturing Business Technology*, October 2004.
229 Managing Complexity, The *Economist*, Nov. 25, 2004.
230 Eric Steven Raymond, *The Cathedral and the Bazaar*, Thyrsus Enterprises, 2000, http://www.catb.org/~esr/writings/cathedral-bazaar/cathedral-bazaar
231 Eli Goldratt, Carol Ptak, and Eli Schragenheim, *Necessary But Not Sufficient*, The North River Press, Great Barrington, MA, 2000, p. 125.
232 Roger Brooks, Oliver Wight Americas, Run the Business, Grow the Business, and Improve the Capabilities, presentation to APICS Portland Chapter, March 21, 2003.
233 Dominique Deckmyn, *ComputerWorld*, Most enterprise computing efforts fail, March 21, 2000, *http://www.computerworld.com/news/2000/story/0,11280,41973,00.html*.
234 *A Guide to the Project Management Body of Knowledge*, Project Management Institute, Newton Square, PN, 2000 Edition.
235 Ibid.
236 Samuel Greengard, Project Portfolio Management Goes Mainstream, *Business Finance*, March 2004.
237 Samuel Greengard, Made For Midsize, *Industry Week*, July 1, 2002.
238 Ron Axam and Darren Jerome, A Guide to ERP Success, *EAI Journal*, February 2003, p. 36.

Postscript

239 Robert M. Pirsig, *Zen and the Art of Motorcycle Maintenance, An Inquiry into Values*, Perennial, 1974.
240 Terence T. Burton and Steven M. Boeder, *The Lean Extended Enterprise*, J. Ross Publishing, Boca Raton, FL 2003.

241 Shunryu Suzuki, *Zen Mind, Beginner's Mind, Introduction to The Buddha and His Teachings*, Barnes and Noble, 1993, p. 228.

242 Rafael Aguayo, *Dr. Deming, The American Who Taught the Japanese About Quality*, Fireside, New York, NY, 1991, p. xi.

243 Jeffrey K. Liker, *The Toyota Way*, McGraw-Hill, New York, NY, 2004, p. 223.

244 Ibid.

245 Norman Bodek, Interview, Nov. 22, 2004.

246 Stephen R. Covey, *The Seven Habits of Highly Effective People*, Fireside, New York, NY, 1989.

Index